Color
Appearance
Models

MARK D. FAIRCHILD

ADDISON-WESLEY

An imprint of Addison Wesley Longman, Inc.

Reading, Massachusetts • Harlow, England
Menlo Park, California • Berkeley, California
Don Mills, Ontario • Sydney • Bonn • Amsterdam
Tokyo • Mexico City

The publisher offers discounts on this book when ordered in quantity for special sales. For more information, please contact:

>   Corporate & Professional Publishing Group
>   Addison Wesley Longman, Inc.
>   One Jacob Way
>   Reading, Massachusetts 01867

Library of Congress Cataloging-in-Publication Data

Fairchild, Mark D.
    Color appearance models / Mark D. Fairchild.
        p.   cm.
    Includes bibliographical references and index.
    ISBN 0-201-63464-3
    1. Color vision.   I. Title.
    QP483.F35   1997
    612.8'4—dc21   97-26404
                                                            CIP

ISBN 0-201-63464-3
Text printed on recycled and acid-free paper.

1 2 3 4 5 6 7 8 9 10—DOC—01 00 99 98 97
First Printing, November 1997

To those who most make my life special—
Lisa, Acadia, Sierra, and Cirrus.

How much of beauty—of color as well as form—on which our eyes daily rest goes unperceived by us?

—*Henry David Thoreau*

Contents

Preface   xi

Acknowledgments   xiv

Introduction   xv

CHAPTER 1
Human Color Vision   1

    1.1   Optics of the Eye   3
    1.2   The Retina Revisited   8
    1.3   Visual Signal Processing   16
    1.4   Mechanisms of Color Vision   21
    1.5   Spatial and Temporal Properties of Color Vision   30
    1.6   Color-vision Deficiencies   34
    1.7   Key Features for Color Appearance Modeling   38

CHAPTER 2
Psychophysics   41

    2.1   Psychophysics Defined   43
    2.2   Historical Context   45
    2.3   Hierarchy of Scales   48
    2.4   Threshold Techniques   50
    2.5   Matching Techniques   53
    2.6   One-dimensional Scaling   54
    2.7   Multidimensional Scaling   56
    2.8   Design of Psychophysical Experiments   58
    2.9   Importance of Psychophysics in Color Appearance Modeling   60

CHAPTER 3
Colorimetry   61

    3.1   Basic and Advanced Colorimetry   63
    3.2   Why Is Color?   64
    3.3   Light Sources and Illuminants   65
    3.4   Colored Materials   73

3.5   The Human Visual Response   77
3.6   Tristimulus Values and Color-matching Functions   81
3.7   Chromaticity Diagrams   90
3.8   CIE Color Spaces   91
3.9   Color-difference Specification   94
3.10   The Next Step   95

CHAPTER 4
Color-appearance Terminology   97
4.1   Importance of Definitions   99
4.2   Color   100
4.3   Hue   101
4.4   Brightness and Lightness   102
4.5   Colorfulness and Chroma   103
4.6   Saturation   104
4.7   Unrelated and Related Colors   105
4.8   Definitions in Equations   106
4.9   Brightness-Colorfulness Versus Lightness-Chroma   107

CHAPTER 5
Color-order Systems   111
5.1   Overview and Requirements   113
5.2   The Munsell Book of Color   115
5.3   The Swedish Natural Color System   119
5.4   The Colorcurve System   122
5.5   Other Color-order Systems   123
5.6   Uses of Color-order Systems   126
5.7   Color-naming Systems   129

CHAPTER 6
Color-appearance Phenomena   131
6.1   What Are Color-appearance Phenomena?   133
6.2   Simultaneous Contrast, Crispening, and Spreading   135
6.3   Bezold-Brücke Hue Shift (Hue Changes with Luminance)   139
6.4   Abney Effect (Hue Changes with Colorimetric Purity)   140
6.5   Helmholtz-Kohlrausch Effect (Brightness Depends on Luminance and Chromaticity)   141
6.6   Hunt Effect (Colorfulness Increases with Luminance)   144
6.7   Stevens Effect (Contrast Increases with Luminance)   145
6.8   Helson-Judd Effect (Hue of Nonselective Samples)   146
6.9   Bartleson-Breneman Equations (Image Contrast Changes with Surround)   149

6.10   Discounting-the-Illuminant   150
6.11   Other Context and Structural Effects   151
6.12   Color Constancy?   156

CHAPTER 7
Viewing Conditions   159
7.1   Configuration of the Viewing Field   161
7.2   Colorimetric Specification of the Viewing Field   165
7.3   Modes of Viewing   168
7.4   Unrelated and Related Colors Revisited   172

CHAPTER 8
Chromatic Adaptation   173
8.1   Light, Dark, and Chromatic Adaptation   176
8.2   Physiology   180
8.3   Sensory and Cognitive Mechanisms   186
8.4   Corresponding-colors Data   189
8.5   Models   193
8.6   Computational Color Constancy   195

CHAPTER 9
Chromatic-adaptation Models   197
9.1   von Kries Model   201
9.2   Retinex Theory   204
9.3   Nayatani et al. Model   205
9.4   Guth's Model   208
9.5   Fairchild's Model   211

CHAPTER 10
Color Appearance Models   215
10.1   Definition of Color Appearance Model   217
10.2   Construction of Color Appearance Models   218
10.3   CIELAB   219
10.4   Why Not Use Just CIELAB?   228
10.5   What About CIELUV?   229

CHAPTER 11
The Nayatani et al. Model   231
11.1   Objectives and Approach   233
11.2   Input Data   234
11.3   Adaptation Model   235
11.4   Opponent-color Dimensions   237
11.5   Brightness   238
11.6   Lightness   239
11.7   Hue   240

11.8   Saturation   240
11.9   Chroma   241
11.10 Colorfulness   241
11.11 Inverse Model   242
11.12 Phenomena Predicted   242
11.13 Why Not Use Just the Nayatani Model?   244

CHAPTER 12
The Hunt Model   245
12.1   Objectives and Approach   247
12.2   Input Data   248
12.3   Adaptation Model   249
12.4   Opponent-color Dimensions   255
12.5   Hue   255
12.6   Saturation   257
12.7   Brightness   258
12.8   Lightness   260
12.9   Chroma   260
12.10 Colorfulness   260
12.11 Inverse Model   261
12.12 Phenomena Predicted   263
12.13 Why Not Use Just the Hunt Model?   264

CHAPTER 13
The RLAB Model   267
13.1   Objectives and Approach   269
13.2   Input Data   271
13.3   Adaptation Model   272
13.4   Opponent-color Dimensions   274
13.5   Lightness   276
13.6   Hue   276
13.7   Chroma   277
13.8   Saturation   277
13.9   Inverse Model   279
13.10 Phenomena Predicted   280
13.11 Why Not Use Just the RLAB Model?   281

CHAPTER 14
Other Models   283
14.1   Overview   285
14.2   ATD Model   286
14.3   LLAB Model   292

CHAPTER 15
Testing Color Appearance Models   301
    15.1 Overview   303
    15.2 Qualitative Tests   304
    15.3 Corresponding-colors Data   308
    15.4 Magnitude-estimation Experiments   310
    15.5 Direct Model Tests   312
    15.6 CIE Activities   318
    15.7 A Pictorial Review of Color Appearance Models   320

CHAPTER 16
Traditional Colorimetric Applications   325
    16.1 Color Rendering   327
    16.2 Color Differences   329
    16.3 Indices of Metamerism   332
    16.4 A General System of Colorimetry?   334

CHAPTER 17
Device-independent Color Imaging   337
    17.1 The Problem   340
    17.2 Levels of Color Reproduction   341
    17.3 General Solution   346
    17.4 Device Calibration and Characterization   347
    17.5 The Need for Color Appearance Models   352
    17.6 Definition of Viewing Conditions   353
    17.7 Viewing-conditions-independent Color Space   355
    17.8 Gamut Mapping   355
    17.9 Color Preferences   359
    17.10 Inverse Process   360
    17.11 Example System   360
    17.12 ICC Implementation   362

CHAPTER 18
The Future   367
    18.1 Will There Be One Color Appearance Model?   369
    18.2 Other Color Appearance Models   370
    18.3 Ongoing Research to Test Models   370
    18.4 Ongoing Model Development   371
    18.5 What to Do Now   371

APPENDIX A
The CIE Color Appearance Model (1997)   373
    A.1   Historical Development, Objectives, and Approach   375
    A.2   Bradford-Hunt 96S (Simple) Model   378

A.3   Bradford-Hunt 96C (Comprehensive) Model   382
A.4   The CIE TC1-34 Model, CIECAM97s   384
A.5   The ZLAB Color Appearance Model   389
A.6   Outlook   392

References   395

Index   409

Preface

*The law of proportion according to which the
several colors are formed, even if a man knew, he
would be foolish in telling, for he could not give
any necessary reason, nor indeed any tolerable or
probable explanation of them.*

—Plato

Despite Plato's warning, this book is about one of the major unresolved issues
in the field of color science, the efforts that have been made toward its resolution, and the techniques that can be used to address current technological
problems. That issue is the prediction of the color appearance experienced by
an observer when viewing stimuli in natural, complex settings. Useful solutions
to this problem have impacts in a number of industries, such as lighting, materials, and imaging.

In lighting, color appearance models can be used to predict the color-
rendering properties of various light sources, thereby allowing specification of
quality rather than just efficiency. In materials fields (coatings, plastics, textiles,
and so on), color appearance models can be used to specify tolerances across a
wider variety of viewing conditions than is currently possible. The imaging industries have produced the biggest demand for accurate and practical color
appearance models. The rapid growth in color-imaging technology, particularly the desktop publishing market, has led to the emergence of color-
management systems. It is widely acknowledged that such systems require
color appearance models to allow images originating in one medium and
viewed in a particular environment to be acceptably reproduced in a second
medium and viewed under different conditions.

While the need for color appearance models is recognized, their development has been at the forefront of color science and confined to the discourse of
academic journals and conferences. This book brings the fundamental issues
and current solutions in the area of color appearance modeling together in a
single place for those needing to solve practical problems.

Everyone knows what color is, but the accurate description and specification of colors is quite another story. In 1931, the Commission Internationale
de l'Éclairage (CIE) recommended a system for color measurement that estab-

lished the basis for modern colorimetry. That system allowed the specification of color matches through CIE XYZ tristimulus values. However, it was immediately recognized that more advanced techniques were required. The CIE recommended the CIELAB and CIELUV color spaces in 1976 to enable uniform international practice for the measurement of color differences and the establishment of color tolerances.

While the CIE system of colorimetry has been applied successfully for nearly 70 years, it is limited to the comparison of stimuli that are identical in every spatial and temporal respect and viewed under matched viewing conditions. CIE XYZ values describe whether two stimuli match. CIELAB values can be used to describe the perceived differences between stimuli in a single set of viewing conditions. Color appearance models extend the current CIE systems to allow the description of what color stimuli look like under a variety of viewing conditions. The application of such models opens up a world of possibilities for the accurate specification, control, and reproduction of color.

Understanding color-appearance phenomena and developing models to predict them have been the topics of a great deal of research—particularly in the last 10 to 15 years. Color appearance remains a topic of much active research that is often being driven by technological requirements. Despite the fact that the CIE is not yet able to recommend a single color appearance model as the best that is available for all applications, many specialists need to implement some form of a model to solve their research, development, and engineering needs.

One application area is the development of color-management systems based on the ICC profile format that is being developed by the International Color Consortium and incorporated into essentially all modern computer operating systems. However, implementation of color management using ICC profiles requires the application of color appearance models with no specific instructions on how to do so. And unfortunately, the fundamental concepts, phenomena, and models of color appearance are not recorded in a single source. Currently, anyone interested in the field must search out the primary references across a century of scientific journals and conference proceedings. This is due to the large amount of active research in the area.

While searching for and keeping track of primary references is fine for those doing research on color appearance models, it should not be necessary for every scientist, engineer, and software developer interested in the field. So the goal of this book is to provide, in a single source, an overview of color appearance and details of many of the most widely used models. The general approach has been to provide an overview of the fundamentals of color measurement and the phenomena that necessitate the development of color appearance models. This eases the transition into the formulation of the various models and their applications that appear later in the book. This approach has proven quite useful in various university courses, short courses, and seminars in which the full range of material must be presented in a limited time.

Following is a preview of each chapter:

- Chapters 1 through 3 review the fundamental concepts of human color vision, psychophysics, and the CIE system of colorimetry that are prerequisites to an understanding of the development and implementation of color appearance models.
- Chapters 4 through 7 present the fundamental definitions, descriptions, and phenomena of color appearance. They review the historical literature that has led to modern research and development of color appearance models.
- Chapters 8 and 9 concentrate on one of the most important component mechanisms of color appearance: chromatic adaptation. The models of chromatic adaptation described in Chapter 9 are the foundation of the color appearance models described in later chapters.
- Chapter 10 presents the definition of color appearance models and outlines their construction using the CIELAB color space as an example.
- Chapters 11 through 13 provide detailed descriptions of the Nayatani et al., Hunt, and RLAB color appearance models, along with the advantages and disadvantages of each.
- Chapter 14 reviews the ATD and LLAB appearance models that are of increasing interest for some applications.
- Chapters 15 and 16 describe tests of the various models through a variety of visual experiments and colorimetric applications of the models.
- Chapter 17 presents an overview of device-independent color imaging, the application that has provided the greatest technological push for the development of color appearance models.
- Chapter 18 includes thoughts on the future directions for color appearance modeling research and application.
- The appendix provides a snapshot of the status of the CIE color appearance models being developed for further testing by CIE TC1-34 as this book goes to press.

The field of color appearance modeling is still in its infancy and likely to continue developing in the near future. However, Chapters 1 through 10 of this book provide overviews of fundamental concepts, phenomena, and techniques that will change little, if at all, in the coming years. Thus these chapters should serve as a steady reference. The later chapters describe models, tests, and applications that will be subject to evolutionary changes as research progresses. However, they do provide a useful snapshot of the current state of affairs and provide a basis from which it should be much easier to keep track of future developments. To assist readers in this task, a World Wide Web page has been set up that lists important developments and publications since the publication of this book (see *http://www.awl.com/cseng/* and/or *http://www.cis.rit.edu/people/faculty/fairchild*). A spreadsheet with example calculations can also be found there.

> 'Yes,' I answered her last night;
> 'No,' this morning sir, I say,
> Colours seen by candle-light
> Will not look the same by day.

> —Elizabeth Barrett Browning

Acknowledgments

A project like this book is never really completed by a single person. I particularly thank my family for the undying support that encouraged its completion. The research and learning that led to it are directly attributable to my students. Much of the research would not have been completed without their tireless work, and I would not have learned about color appearance models were it not for their keen desire to learn more about them from me. I am deeply indebted to all of my students and friends—those who did research with me, those who worked at various times in the Munsell Color Science Laboratory, and those who participated in my university and short courses at all levels. There is no way to list all of them without making an omission, so I will take the easy way out and thank them as a group.

I also am indebted to those who reviewed various chapters while this book was being prepared and who provided useful insights, suggestions, and criticisms. These reviewers include Paula J. Alessi, Edwin Breneman, Ken Davidson, Ron Gentile, Robert W. G. Hunt, Lindsay MacDonald, Mike Pointer, Michael Stokes, Jeffrey Wang, Eric Zeise, and Valerie Zelenty.

Thanks also to all of the industrial and government sponsors of our research and education in the Munsell Color Science Laboratory at R.I.T. In particular, thanks to Thor Olson of Management Graphics for the donation of the Opal image recorder and the loan of the 120-camera back used to output the color images included in this book. Last, but not least, I thank Colleen Desimone for her excellent work as the Munsell Color Science Laboratory secretary, which makes life at the office much easier.

M.D.F.
Honeoye Falls, N.Y.

Introduction

Like beauty, color is in the eye of the beholder. For as long as human scientific inquiry has been recorded, the nature of color perception has been a topic of great interest. Despite tremendous evolution of technology, fundamental issues of color perception remain unanswered. Many scientific attempts to explain color rely purely on the physical nature of light and objects. However, without the human observer there is no color. It is often asked whether a tree falling in the forest makes a sound if no one is there to observe it. Perhaps equal philosophical energy should be spent wondering what color its leaves are if no one is there to see them.

What Is a Color Appearance Model?

It is common to say that certain wavelengths of light, or certain objects, are a given color. This is an attempt to relegate color to the purely physical domain. It is more correct to state that those stimuli are perceived to be of a certain color when viewed under specified conditions. Attempts to specify color as a purely physical phenomenon fall within the domain of spectrophotometry and spectroradiometry. When the lowest-level sensory responses of an average human observer are factored in, one enters the domain of colorimetry. When the many other variables that influence color perception are considered, in order to better describe our perceptions of stimuli, one is within the domain of color appearance modeling—the subject of this book.

Consider the following observations:

The headlights of an oncoming automobile are nearly blinding at night but barely noticeable during the day.

As light grows dim, colors fade from view while objects remain readily apparent.

Stars disappear from sight during the daytime.

The walls of a freshly painted room appear significantly different from the color of the sample that was used to select the paint in a hardware store.

Artwork surrounded by different color mat boards takes on significantly different appearances.

Printouts of images do not match the originals displayed on a
computer monitor.

Scenes appear more colorful and of higher contrast on a sunny day.

Blue and green objects (for example, board game pieces) become
indistinguishable under dim incandescent illumination.

It is nearly impossible to select appropriate-colored socks (e.g.,
black, brown, or blue) in the early morning light.

There is no such thing as a gray or brown lightbulb.

There are no colors described as reddish-green or yellowish-blue.

None of the above observations can be explained by physical measurements of materials and/or illumination alone. Rather, such physical measurements must be combined with measurements of the prevailing viewing
conditions and models of human visual perception in order to make reasonable
predictions of these effects. This aggregate of measurements is precisely what
color appearance models are designed to embrace. Each of these observations,
and many more like them, can be explained by various color appearance phenomena and models; they cannot be explained by the established techniques of
color measurement, sometimes referred to as *basic colorimetry.*

This book details the differences between basic colorimetry and color appearance models and provides fundamental background on human visual perception and color appearance phenomena. It also describes the application of
color appearance models to current technological problems such as digital
color reproduction. Upon completion of this book, one should be able to fairly
easily explain each of the previous observations.

Basic colorimetry provides the fundamental color-measurement techniques that are used to specify stimuli in terms of their sensory potential for an
average human observer. These techniques are absolutely necessary as the foundation for color appearance models. However, on their own, the techniques of
basic colorimetry can be used only to specify whether two stimuli, viewed under identical conditions, match in color for an average observer. Advanced colorimetry aims to extend the techniques of basic colorimetry to enable the
specification of color-difference perceptions and, ultimately, color appearance.

Several established techniques for color-difference specification are available that have been formulated and refined over the past few decades. These
have also reached the point that a few, agreed-upon standards are used world-
wide. Color appearance models aim to go the final step by allowing the mathematical description of the appearance of stimuli in a wide variety of viewing
conditions. Such models have been the subject of much research over the past
two decades and more recently become required for practical applications.
Various models have been proposed. These are beginning to find their way into
color-imaging systems through the refinement of color-management techniques. This requires that an ever-broadening array of scientists, engineers,

programmers, imaging specialists, and others understand the fundamental philosophy, construction, and capabilities of color appearance models as described in the ensuing chapters.

So as not to make the learning process too difficult, here are some clues to the explanation of the color-appearance observations listed near the beginning of this introduction.

- The change of appearance of oncoming headlights can be largely explained by the processes of light adaptation and described by Weber's law.
- The fading of color in dim light while objects remain clearly visible is explained by the transition from trichromatic cone vision to monochromatic rod vision.
- The incremental illumination of a star on the daytime sky is not large enough to be detected. The same physical increment on the darker nighttime sky is easily perceived because the visual threshold to luminance increments has changed between the two viewing conditions.
- The paint chip doesn't match the wall due to changes in the size, surround, and illumination of the stimulus.
- Changes in the color of a surround or background profoundly influence the appearance of stimuli. This can be particularly striking for photographs and other artwork.
- Assuming the computer monitor and printer are accurately calibrated and characterized, differences in media, white point, luminance level, and surround can still force the printed image to look significantly different from the original.
- The Hunt effect and Stevens effect describe the apparent increase in colorfulness and contrast of scenes with increases in illumination level.
- Low levels of incandescent illumination do not provide the energy required by the short-wavelength sensitive mechanisms of the human visual system (the least sensitive of the color mechanisms) to distinguish green objects from blue objects.
- In the early morning light, the ability to distinguish dark colors is diminished.
- The perceptions of gray and brown occur only as related colors; thus they cannot be observed as light sources that are the brightest element of a scene.
- The hue perceptions red and green (or yellow and blue) are encoded in a bipolar fashion by our visual system and thus cannot exist together.

Given those clues, it is time to read on and further unlock the mysteries of color appearance.

CHAPTER 1

Human
Color
Vision

1.1   Optics of the Eye
1.2   The Retina Revisited
1.3   Visual Signal Processing
1.4   Mechanisms of Color Vision
1.5   Spatial and Temporal Properties of Color Vision
1.6   Color-vision Deficiencies
1.7   Key Features for Color Appearance Modeling

COLOR APPEARANCE MODELS aim to extend basic colorimetry to the level of specifying the perceived color of stimuli in a wide variety of viewing conditions. To fully appreciate the formulation, implementation, and application of color appearance models, several fundamental topics in color science must first be understood. These are covered in the first few chapters of this book. Since color appearance represents several of the dimensions of our visual experience, any system designed to predict correlates to these experiences must be based, to some degree, on the form and function of the human visual system. All of the color appearance models described in this book are derived with human visual function in mind. It becomes much simpler to understand the formulations of the various models if the basic anatomy, physiology, and performance of the visual system is understood. Thus this book begins with a treatment of the human visual system.

Because the scope of a single chapter is limited, this treatment of the human visual system is an overview of the topics most important for an appreciation of color appearance modeling. The field of vision science is immense and fascinating. Readers are encouraged to explore the literature and the many useful texts on human vision in order to gain further insight and details. Of particular note are the review paper on the mechanisms of color vision by Lennie and D'Zmura (1988), the text on human color vision by Kaiser and Boynton (1996), and the more general text on the foundations of vision by Wandell (1995). Much of the material covered in this chapter is treated in more detail in those references.

1.1   Optics of the Eye

Our visual perceptions are initiated and strongly influenced by the anatomical structure of the eye. Figure 1-1 shows a schematic representation of the optical structure of the human eye, with some key features labeled. The human eye acts like a camera. The cornea and lens act together like a camera lens to focus an image of the visual world on the retina, located at the back of the eye, which acts like the film or other image sensor of a camera. These and other structures have a significant impact on our perception of color.

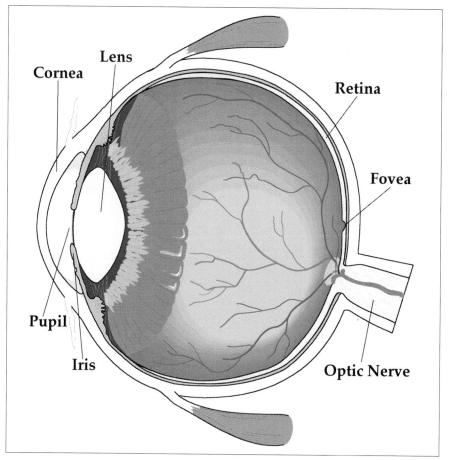

Figure 1-1. Schematic diagram of the human eye with some key structures labeled.

The Cornea

The *cornea* is the transparent outer surface of the front of the eye through which light passes. It serves as the most significant image-forming element of the eye. This is because its curved surface at the interface with air represents the largest change in index of refraction in the eye's optical system.

The cornea is avascular, receiving its nutrients from marginal blood vessels and the fluids surrounding it. Refractive errors, such as nearsightedness (myopia), farsightedness (hyperopia), or astigmatism, can be attributed to variations in the shape of the cornea and are sometimes corrected with laser surgery that reshapes the cornea.

The Lens

The *lens* is a layered, flexible structure that varies in index of refraction. It is a naturally occurring gradient-index optical element with the index of refraction higher in the center of the lens than at the edges. This feature serves to reduce some of the aberrations that might normally be present in a simple optical system.

The shape of the lens is controlled by the ciliary muscles. When we gaze at a nearby object, the lens becomes "fatter" and thus has increased optical power to allow us to focus on the near object. When we gaze at a distant object, the lens becomes "flatter," thereby resulting in the decreased optical power required to bring far away objects into sharp focus.

As we age, the internal structure of the lens changes, thus resulting in a loss of flexibility. Generally, at about 50 years of age, the lens has completely lost its flexibility and observers can no longer focus on near objects (this is called presbyopia, or "old eye"). It is at this point that most people need reading glasses or bifocals.

Concurrent with the hardening of the lens is an increase in its optical density. The lens absorbs and scatters short wavelength (blue and violet) energy. As it hardens, the level of this absorption and scattering increases. In other words, the lens becomes more and more yellow with age. Various mechanisms of chromatic adaptation generally make us unaware of these gradual changes. However, we are all looking at the world through a yellow filter that not only changes with age, but that significantly differs from observer to observer. The effect is most noticeable when performing critical color matching or comparing color matches with other observers. It is particularly apparent with purple objects. An older lens absorbs most of the blue energy reflected from a purple object but does not affect the reflected red energy, so older observers tend to report that the object is significantly redder than reported by younger observers.

Important issues regarding the characteristics of lens aging and its influence on visual performance are discussed by Pokorny et al. (1987), Werner and Schefrin (1993), and Schefrin and Werner (1993).

The Humors

The volume between the cornea and lens is filled with the *aqueous humor,* which is essentially water. The region between the lens and retina is filled with *vitreous humor,* which is also a fluid, but with a higher viscosity similar to that of gelatin. Both humors exist in a state of slightly elevated pressure (relative to air pressure). This ensures that the flexible eyeball retains its shape and dimensions in order to avoid the deleterious effects of wavering retinal images. The flexibility of the entire eyeball serves to increase its resistance to injury. It is much more difficult to break a structure that gives way under impact than one of equal "strength" that attempts to remain rigid.

Since the indices of refraction of the humors are roughly equal to that of water, and those of the cornea and lens are only slightly higher, the rear surface of the cornea and the entire lens have relatively little optical power.

The Iris

The *iris* is the sphincter muscle that controls pupil size. It is pigmented, thus giving the eye its specific color. Eye color is determined by the concentration and distribution of melanin within the iris. The pupil, which is the hole in the middle of the iris through which light passes, defines the level of illumination on the retina.

Pupil size is largely determined by the overall level of illumination. However, it is important to note that it can also vary with nonvisual phenomena, such as arousal. (This effect can be observed by enticingly shaking a toy in front of a cat and paying attention to its pupils.) Thus it is difficult to accurately predict pupil size from the prevailing illumination. In practical situations, pupil diameter varies from about 3 mm to about 7 mm. This change in pupil diameter results in approximately a five-fold change in pupil area and therefore in retinal illuminance.

The visual sensitivity change with pupil area is further limited by the fact that marginal rays are less effective at stimulating visual response in the cones than are central rays (the Stiles-Crawford effect). The change in pupil diameter alone is not sufficient to explain excellent human visual function over prevailing illuminance levels that can vary over 10 orders of magnitude.

The Retina

The optical image formed by the eye is projected onto the retina. The *retina* is a thin layer of cells, approximately the thickness of tissue paper, located at the back of the eye. It incorporates the visual system's photosensitive cells and initial signal processing and transmission "circuitry." These cells are neurons—part of the central nervous system—and can appropriately be considered a part of the brain. The photoreceptors—*rods* and *cones*—serve to transduce the information present in the optical image into chemical and electrical signals that can be transmitted to the later stages of the visual system. These signals are then processed by a network of cells and transmitted to the brain through the optic nerve. More detail on the retina is presented in the next section.

Behind the retina is a layer called the *pigmented epithelium*. This dark pigment layer serves to absorb any light that happens to pass through the retina without being absorbed by the photoreceptors. The function of the pigmented epithelium is to prevent light from being scattered back through the retina and thus reducing the sharpness and contrast of the perceived image. Nocturnal animals give up this improved image quality in exchange for a highly reflective *tapetum* that reflects the light back in order to provide a second chance for the photoreceptors to absorb the energy. This property is why the eyes of a deer, or other nocturnal animal, caught in the headlights of an oncoming automobile appear to glow.

THE FOVEA

Perhaps the most important structural area on the retina is the fovea. The *fovea* is the area on the retina where we have the best spatial and color vision. When we look at, or fixate, an object in our visual field, we move our head and eyes such that the image of the object falls on the fovea. As you are reading this text, you are moving your eyes to make the various words fall on your fovea as you read them.

To illustrate how drastically spatial acuity falls off as the stimulus moves away from the fovea, try to read the preceding text in this paragraph while fixating on the period at the end of this sentence. It is probably difficult, if not impossible, to read text that is only a few lines away from the point of fixation.

The fovea covers an area that subtends about two degrees of visual angle in the central field of vision. To help visualize two degrees of visual angle, a general rule is that the width of your thumbnail, held at arm's length, is approximately one degree of visual angle.

THE MACULA

The fovea is protected by a yellow filter known as the *macula*. The macula serves to protect this critical area of the retina from intense exposures to short wavelength energy. It might also serve to reduce the effects of chromatic aberration that cause a short wavelength image to be rather severely out of focus most of the time.

Unlike the lens, the macula does not become more yellow with age. However, there are significant differences in the optical density of the macular pigment from observer to observer and, in some cases, between a single observer's left and right eyes. The yellow filters of the lens and macula, through which we all view the world, are the major source of variability in color vision between observers with normal color vision.

The Optic Nerve

A last key structure of the eye is the *optic nerve*. The optic nerve is a nerve made up of the *axons* (outputs) of the ganglion cells, the last level of neural processing in the retina. It is interesting to note that the optic nerve consists of approximately one million fibers that carry information generated by approximately 130 million photoreceptors. Thus there is a clear compression of the visual signal prior to its transmission to higher levels of the visual system. A one-to-one "pixel map" of the visual stimulus is never available for processing by the brain's higher visual mechanisms. This processing is explored in greater detail later in the chapter. Since the optic nerve takes up all of the space that would normally be populated by photoreceptors, there is a small area in each eye in which no visual stimulation can occur. This area is called the *blind spot.*

The structures described here have a clear impact in shaping and defining the information available to the visual system that ultimately results in the perception of color appearance. The action of the pupil serves to define retinal illuminance levels that, in turn, have a dramatic impact on color appearance. The yellow filtering effects of the lens and macula modulate the spectral responsivity of our visual system and introduce significant inter-observer variability. The spatial structure of the retina helps to define the extent and nature of various visual fields that are critical for defining color appearance. The neural networks in the retina reiterate that visual perception in general, and color appearance specifically, cannot be treated as simple, pointwise image-processing problems. Several of these important features are discussed in more detail in the following sections on the retina, visual physiology, and visual performance.

1.2   The Retina Revisited

Figure 1-2 illustrates a cross-sectional representation of the retina. The retina includes several layers of neural cells beginning with the photoreceptors—the *rods* and *cones.* A vertical signal-processing chain through the retina can be constructed by examining the connections of photoreceptors to bipolar cells, which are in turn connected to ganglion cells, which form the optic nerve. Even this simple pathway results in the signals from multiple photoreceptors being compared and combined. This is because multiple photoreceptors provide input to many of the bipolar cells and multiple bipolar cells provide input to many of the ganglion cells. More important, this simple concept of retinal signal processing ignores two other significant types of cells. These are the *horizontal cells,* which connect photoreceptors and bipolar cells laterally to one another, and the *amacrine cells,* which connect bipolar cells and ganglion cells laterally to one another. Figure 1-2 provides only a slight indication of the extent of these various interconnections.

The specific processing that occurs in each type of cell is not completely understood and is beyond the scope of this chapter. However, it is important to realize that the signals transmitted from the retina to the higher levels of the brain via the ganglion cells are not simple pointwise representations of the receptor signals. Rather, they consist of sophisticated combinations of the recep-

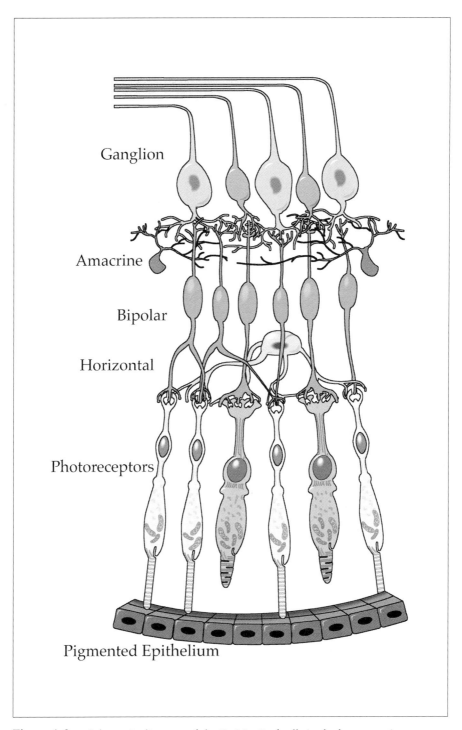

Figure 1-2. Schematic diagram of the "wiring" of cells in the human retina.

tor signals. To envision the complexity of the retinal processing, keep in mind that each synapse between neural cells can effectively perform a mathematical operation (add, subtract, multiply, divide) in addition to the amplification, gain control, and nonlinearities that can occur within the neural cells. Thus the network of cells within the retina can serve as a sophisticated image computer. This is how the information from 130 million photoreceptors can be reduced to signals in approximately one million ganglion cells without loss of visually meaningful data.

It is interesting to note that light passes through all of the neural machinery of the retina prior to reaching the photoreceptors. This has little impact on visual performance, since these cells are transparent and in a fixed position and thus are not perceived. It also allows the significant amounts of nutrients required and waste produced by the photoreceptors to be processed through the back of the eye.

Rods and Cones

Figure 1-3 provides a representation of the two classes of retinal photoreceptors: the rods and cones. Rods and cones derive their respective names from their prototypical shape. Rods tend to be long and slender, while peripheral cones are conical. This distinction is misleading, however, because foveal cones, which are tightly packed due to their high density in the fovea, are long and slender, resembling peripheral rods.

The more important distinction between rods and cones is in visual function. Rods serve vision at low luminance levels (e.g., less than 1 cd/m^2), while cones serve vision at higher luminance levels. Thus the transition from rod to cone vision is one mechanism that allows our visual system to function over a

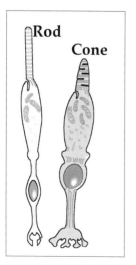

Figure 1-3.   Illustrations of prototypical rod and cone photoreceptors.

large range of luminance levels. At high luminance levels (e.g., greater than 100 cd/m^2), the rods are effectively saturated and only the cones function. In the intermediate luminance levels, both rods and cones function and contribute to vision. Vision when only rods are active is called *scotopic vision*. Vision served only by cones is called *photopic vision*. Vision in which both rods and cones are active is called *mesopic vision*.

Rods and cones also differ substantially in their spectral sensitivities, as illustrated in Figure 1-4(a). There is only one type of rod receptor with a peak spectral responsivity at approximately 510 nm. There are three types of cone receptors with peak spectral responsivities spaced throughout the visual spectrum.

THREE TYPES OF CONES

The three types of cones are most properly referred to as *L, M,* and *S cones.* These names refer to the long-wavelength, middle-wavelength, and short-wavelength sensitive cones, respectively. Sometimes the cones are denoted with other symbols such as RGB or ργβ, suggestive of red, green, and blue sensitivities. As can be seen in Figure 1-4(a), this concept is erroneous and the L, M, and S names are more appropriately descriptive.

Note from the figure that the spectral responsivities of the three cone types broadly overlap; this design is significantly different from the "color separation" responsivities that are often built into physical imaging systems. Such sensitivities, typically incorporated in imaging systems for practical reasons, are the fundamental reason that accurate color reproduction is often difficult, if not impossible, to achieve.

The three types of cones clearly serve color vision. Since there is only one type of rod, the rod system is incapable of color vision. This can easily be observed by viewing a normally colorful scene at very low luminance levels. Figure 1-4(b) illustrates the two CIE spectral luminous efficiency functions: the $V'(\lambda)$ function for scotopic (rod) vision and the $V(\lambda)$ function for photopic (cone) vision. These functions represent the overall sensitivity of the two systems with respect to the perceived brightness of the various wavelengths. Since there is only one type of rod, the $V'(\lambda)$ function is identical to the spectral responsivity of the rods and depends on the spectral absorption of *rhodopsin,* the photosensitive pigment in rods. The $V(\lambda)$ function, however, represents a combination of the three types of cone signals rather than the responsivity of any single cone type.

Note the difference in peak spectral sensitivity between scotopic and photopic vision. With scotopic vision, we are more sensitive to shorter wavelengths. This effect, known as the *Purkinje shift,* can be observed by finding two objects, one blue and the other red, that appear the same lightness when viewed in daylight. When the same two objects are viewed under very low luminance levels, the blue object will appear quite light while the red object will appear nearly black because of the scotopic spectral sensitivity function.

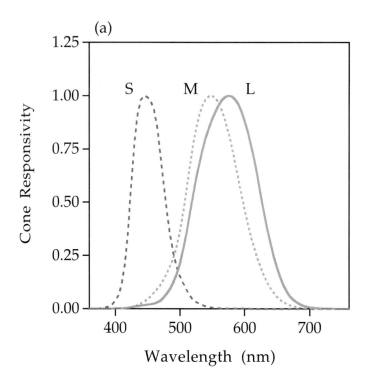

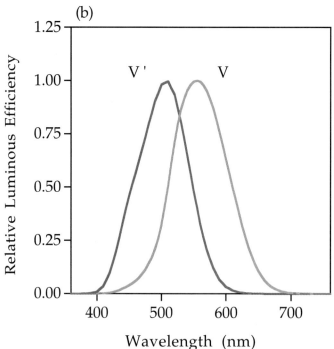

Figure 1-4. (a) Spectral responsivities of the L, M, and S cones and (b) the CIE spectral luminous efficiency functions for scotopic, V′(λ), and photopic, V(λ), vision.

Another important feature about the three cone types is their relative distribution in the retina. It turns out that the S cones are relatively sparsely populated throughout the retina and completely absent in the most central area of the fovea. There are far more L and M cones than S cones, and there are approximately twice as many L cones as M cones. The relative populations of the L:M:S cones are approximately 40:20:1. These relative populations must be considered when combining the cone responses [plotted with individual normalizations in Figure 1-4(a)] to predict higher-level visual responses.

Figure 1-5 provides a schematic representation of the foveal photoreceptor mosaic with false coloring to represent a hypothetical distribution. The L cones are in red, M cones in green, and S cones in blue. Figure 1-5 is presented simply as a convenient visual representation of the cone populations and should not be taken literally.

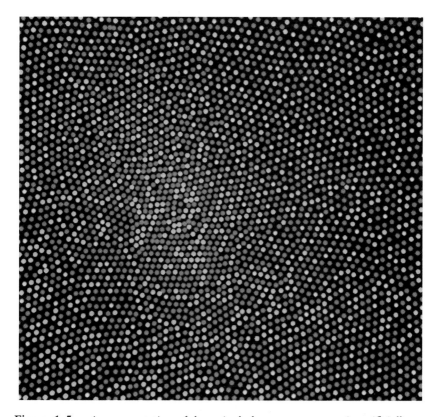

Figure 1-5.   A representation of the retinal photoreceptor mosaic artificially colored to represent the relative proportions of L (red), M (green), and S (blue) cones in the human retina. Modeled after Williams et al. (1991).

ROD FUNCTION

As illustrated in Figure 1-5, there are no rods in the fovea. This feature of the visual system can be observed when trying to look directly at a small, dimly illuminated object, such as a faint star at night. The object seems to disappear because its image falls on the foveal area where there are no rods to detect the dim stimulus.

Figure 1-6 shows the distribution of rods and cones across the retina. Several important features of the retina can be observed in this figure. First, notice the extremely large numbers of photoreceptors. In some retinal regions, there are about 150,000 photoreceptors per square millimeter of retina! Also notice that there are far more rods (around 120 million per retina) than cones (around 7 million per retina). This might seem somewhat counterintuitive since cones function at high luminance levels and produce high visual acuity, while rods function at low luminance levels and produce significantly reduced visual acuity. (This is analogous to low-speed fine-grain photographic film versus high-speed coarse-grain film.) The solution to this apparent mystery lies in the fact that single cones feed into ganglion cell signals. In contrast, rods pool their responses over hundreds of receptors (feeding into a single ganglion cell) in order to produce increased sensitivity at the expense of acuity. This differ-

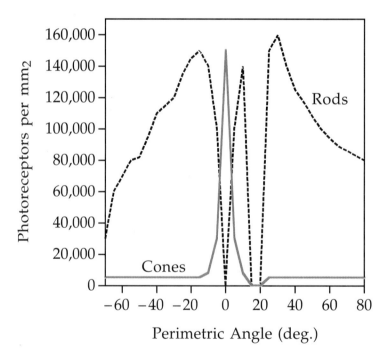

Figure 1-6.   Density (receptors per square millimeter) of rod and cone photoreceptors as a function of location on the human retina.

ence also partially explains how the information from so many receptors can be transmitted through one million ganglion cells.

Figure 1-6 also illustrates that cone receptors are highly concentrated in the fovea and more sparsely populated throughout the peripheral retina, while there are no rods in the central fovea. The lack of rods in the central fovea allows that valuable space to be used to produce the highest possible spatial acuity with the cone system.

THE BLIND SPOT

A final feature to be noted in the figure is the *blind spot*. This is the area, 12°–15° from the fovea, where the optic nerve is formed and there is no room for photoreceptors. Figure 1-7 provides some stimuli that can be used to demonstrate the existence of the blind spot.

One reason the blind spot generally goes unnoticed is that it is located on opposite sides of the visual field in each of the two eyes. However, even when one eye is closed, the blind spot is not generally noticed.

To observe your blind spot, close your left eye and fixate the cross in Figure 1-7(a) with your right eye. Then adjust the viewing distance of the book until the spot to the right of the cross disappears when it falls on the blind spot. Note that what you see when the spot disappears is not a black region; rather it appears to be an area of blank paper. This is an example of a phenomenon known as *filling-in*. Since your brain no longer has any signal indicating a change in the visual stimulus at that location, it simply fills in the most probable stimulus, in this case a uniform white piece of paper. The strength of this filling-in can be illustrated by using Figure 1-7(b) to probe your blind spot. In this case, with your left eye closed, fixate the cross with your right eye and adjust the viewing distance until the gap in the line disappears when it falls on your blind spot. Amazingly, the perception is that of a continuous line, since that is now the most probable visual stimulus.

Figure 1.7   Stimuli used to illustrate presence of the blind spot and "filling-in" phenomena. Close your left eye. Fixate the cross with your right eye and adjust the viewing distance until (a) the spot falls on your blind spot or (b) the gap in the line falls on your blind spot. Notice the perception in that area in each case.

If you prefer to perform these exercises using your left eye, simply turn the book upside down to find the blind spot on the other side of your visual field.

The filling-in phenomenon goes a long way to explain the function of the visual system. The signals present in the ganglion cells represent only local changes in the visual stimulus. Effectively, only information about spatial or temporal transitions (i.e., edges) is transmitted to the brain. Perceptually, this code is sorted out by examining the nature of the changes and filling in the appropriate uniform perception until a new transition is signaled. This coding provides tremendous savings in bandwidth to transmit the signal and can be thought of as somewhat similar to run-length encoding that is sometimes used in digital imaging.

1.3   Visual Signal Processing

The neural processing of visual information is quite complex within the retina and becomes significantly, if not infinitely, more complex at later stages. This section provides a brief overview of the paths that some of this information takes.

It is helpful to begin with a general map of the steps along the way. The optical image on the retina is first transduced into chemical and electrical signals in the photoreceptors. These signals are then processed through the network of retinal neurons (horizontal, bipolar, amacrine, and ganglion cells) described previously. The ganglion-cell axons gather to form the optic nerve, which projects to the lateral geniculate nucleus (LGN) in the thalamus. The LGN cells, after gathering input from the ganglion cells, project to visual area one (V1) in the occipital lobe of the cortex. At this point, the information processing begins to become amazingly complex. Approximately 30 visual areas have been defined in the cortex with names such as V2, V3, V4, and MT. Signals from these areas project to several other areas and vice versa. The cortical processing includes many instances of feed-forward, feed-back, and lateral processing. Somewhere in this network of information, our ultimate perceptions are formed. A few more details of these processes are described in the following paragraphs.

Light incident on the retina is absorbed by photopigments in the various photoreceptors. In rods, the photopigment is **rhodopsin.** Upon absorbing a photon, rhodopsin changes in structure. This sets off a chemical chain reaction that ultimately results in the closing of ion channels in the rod's cell walls, which produce an electrical signal based on the relative concentrations of various ions (e.g., sodium and potassium) inside and outside the cell wall.

A similar process takes place in cones. Rhodopsin is made up of *opsin* and *retinal.* Cones have similar photopigment structures. However, in cones the "cone-opsins" have slightly different molecular structures that result in the

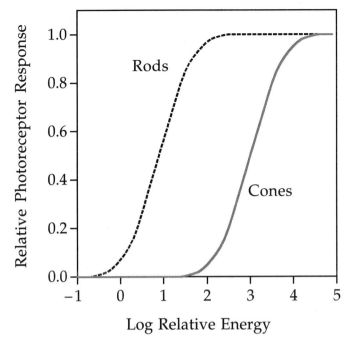

Figure 1-8.   Relative energy responses for the rod and cone photoreceptors.

various spectral responsivities observed in the cones. Each type of cone (L, M, or S) contains a different form of "cone-opsin." Figure 1-8 illustrates the relative responses of the photoreceptors as a function of retinal exposure.

It is interesting to note that these functions show characteristics similar to those found in all imaging systems. At the low end of the receptor responses is a threshold below which the receptors do not respond. There is then a fairly linear portion of the curves followed by response saturation at the high end. Such curves are representations of the photocurrent at the receptors and represent the very first stage of visual processing. These signals are then processed through the retinal neurons and synapses until a transformed representation is generated in the ganglion cells for transmission through the optic nerve.

Receptive Fields

For various reasons, including noise suppression and transmission speed, the amplitude-modulated signals in the photoreceptors are converted into frequency-modulated representations at the ganglion-cell level and higher. In these, and indeed most, neural cells, the magnitude of the signal is represented in terms of the number of spikes of voltage per second fired by the cell rather than by the voltage difference across the cell wall. To represent the physiological properties of these cells, the concept of receptive fields becomes useful.

A *receptive field* is a graphical representation of the area in the visual field to which a given cell responds. In addition, the nature of the response (e.g., positive, negative, or spectral bias) is typically indicated for various regions in the receptive field. As a simple example, the receptive field of a photoreceptor is a small circular area representing the size and location of that particular receptor's sensitivity in the visual field. Figure 1-9 represents some prototypical receptive fields for ganglion cells. They illustrate center-surround antagonism, which is characteristic at this level of visual processing. The receptive field in Figure 1-9(a) illustrates a positive central response, which is typically generated by a positive input from a single cone. This response is surrounded by a negative surround response, which is typically driven by negative inputs from several neighboring cones. Thus the response of this ganglion cell is made up of inputs from a number of cones with both positive and negative signs. The result is that the ganglion cell does not simply respond to points of light. Rather, it serves as an edge detector (actually a "spot" detector). Those familiar with digital image processing can think of the ganglion cell responses as similar to the output of a convolution kernel designed for edge detection.

Figure 1-9(b) illustrates that a ganglion cell response of opposite polarity is equally likely. The response in Figure 1-9(a) is called an *on-center ganglion cell,* while that in Figure 1-9(b) is called an *off-center ganglion cell.* Often on-center and off-center cells will occur at the same spatial location, fed by the same photoreceptors, thereby resulting in an enhancement of the system's dynamic range.

Note that the ganglion cells represented in Figure 1-9 will have no response to uniform fields (given that the positive and negative areas are balanced). This illustrates one aspect of the image compression carried out in the retina. The brain is not bothered with redundant visual information; only information about changes in the visual world is transmitted. This

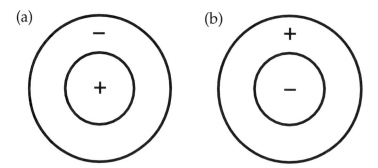

Figure 1-9.   Typical center-surround antagonistic receptive fields: (a) on-center; (b) off-center.

spatial information processing in the visual system is the fundamental basis of the important impact of the background on color appearance. Figure 1-9 illustrates spatial opponency in ganglion cell responses. Figure 1-10 shows, however, that in addition to spatial opponency, there is often spectral opponency in ganglion cell responses. Figure 1-10(a) shows a red-green opponent response with the center fed by positive input from an L cone and the surround fed by negative input from M cones. Figure 1-10(b) illustrates the off-center version of this cell. Thus, before the visual information has even left the retina, processing has occurred with a profound effect on color appearance.

Figures 1-9 and 1-10 illustrate typical ganglion cell receptive fields. There are other types and varieties of ganglion cell responses, but they all share these basic concepts. Visual signals, on their way to the primary visual cortex, pass through the LGN. While the ganglion cells do terminate at the LGN, making synapses with LGN cells, there appears to be a one-to-one correspondence between ganglion cells and LGN cells. Thus the receptive fields of LGN cells are identical to those of ganglion cells. The LGN appear to act as a relay station for the signals. However, it probably serves some visual function, since there are neural projections from the cortex back to the LGN that could serve as some type of switching or adaptation feedback mechanism. The axons of LGN cells project to V1 in the visual cortex.

Processing in Visual Area One

In V1 of the cortex, the encoding of visual information becomes significantly more complex. Much as the outputs of various photoreceptors are combined and compared to produce ganglion cell responses, the outputs of various LGN cells are compared and combined to produce cortical responses. As the signals move further up in the cortical processing chain, this process repeats itself,

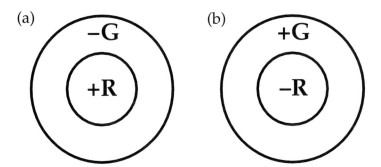

Figure 1-10. Examples of (a) red-green and (b) green-red spectrally and spatially antagonistic receptive fields.

with the level of complexity increasing very rapidly to the point that receptive fields begin to lose meaning. In V1, cells can be found that selectively respond to various things, including

- Oriented edges or bars
- Input from one eye or the other, or both
- Various spatial frequencies
- Various temporal frequencies
- Particular spatial locations
- Various combinations of these features

In addition, cells can be found that seem to linearly combine inputs from LGN cells and others with nonlinear summation. All of these various responses are necessary to support visual capabilities, such as the perceptions of size, shape, location, motion, depth, and color. Given the complexity of cortical responses in V1 cells, it is not difficult to imagine how complex visual responses can become in an interwoven network of approximately 30 visual areas.

Figure 1-11 schematically illustrates a small portion of the connectivity of the various cortical areas that have been identified. Bear in mind that this

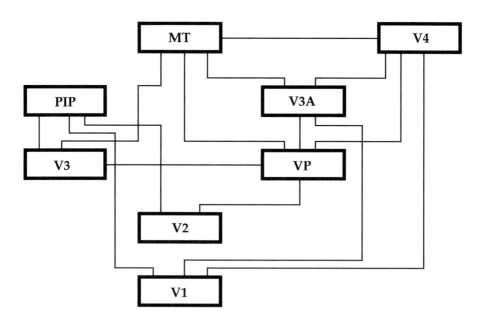

Figure 1-11.   Partial flow diagram to illustrate the many streams of visual information processing in the visual cortex. Information can flow in both directions along each connection.

figure is showing connections of areas, not cells. There are on the order of 10^9 cortical neurons serving visual functions. At these stages, it becomes exceedingly difficult to explain the function of single cortical cells in simple terms. In fact, the function of a single cell might not have meaning, since the representation of various perceptions must be distributed across collections of cells throughout the cortex. Rather than attempting to explore the physiology further, the following sections describe some of the overall perceptual and psychophysical properties of the visual system that help to specify its performance.

1.4   Mechanisms of Color Vision

Historically, there have been many theories that attempt to explain the function of color vision. A brief look at some of the more modern ones provides useful insight into current concepts.

Trichromatic Theory

In the later half of the nineteenth century, the *trichromatic theory of color vision* was developed based on the work of Maxwell, Young, and Helmholtz. They recognized that there must be three types of receptors, approximately sensitive to the red, green, and blue regions of the spectrum, respectively. The trichromatic theory simply assumed that three images of the world were formed by these three sets of receptors and then transmitted to the brain, where the ratios of the signals in each of the images was compared in order to sort out color appearances.

The trichromatic (three-receptor) nature of color vision was not in doubt, but the process of three images being transmitted to the brain is inefficient and the theory fails to explain several visually observed phenomena.

Hering's Opponent-Colors Theory

At around the same time the trichromatic theory was proposed, Hering proposed the opponent-colors theory of color vision based on many subjective observations of color appearance. These observations included appearance of hues, simultaneous contrast, afterimages, and color vision deficiencies.

Hering noted that certain hues were never perceived to occur together. For example, a color perception is never described as reddish-green or yellowish blue, while combinations of red and yellow, red and blue, green and yellow, and green and blue are readily perceived. This suggested to Hering that there was something fundamental about the red-green and yellow-blue pairs that causes them to oppose one another.

Similar observations were made of simultaneous contrast in which an object placed on a red background appears greener, an object on a green background appears redder, on a yellow background appears bluer, and on a blue background appears more yellow. Figure 1-12 demonstrates the opponent nature of visual afterimages. The afterimage of red is green, of green is red, of yellow is blue, and of blue is yellow.

Hering also observed that people with color vision deficiencies lose the ability to distinguish hues in red-green or yellow-blue pairs.

All of these observations provide clues regarding the processing of color information in the visual system. Hering proposed that there were three types of receptors, but his receptors had bipolar responses to light-dark, red-green, and yellow-blue. At the time, this was thought to be physiologically implausible and Hering's opponent-colors theory did not receive appropriate acceptance.

The Modern Opponent-Colors Theory

In the middle of the twentieth century, Hering's opponent-colors theory enjoyed a revival of sorts when quantitative data supporting it began to appear. For example, Svaetichin (1956) found opponent signals in electrophysiological measurements of responses in the retinas of goldfish (which happen to be trichromatic!). DeValois et al. (1958) found similar opponent physiological re-

Figure 1-12. Stimulus for the demonstration of opponent afterimages. Fixate upon the black spot in the center of the four colored squares for about 30 sec. Then move your gaze to fixate the black spot in the uniform white area. Note the colors of the afterimages relative to the colors of the original stimuli.

sponses in the LGN cells of the macaque monkey. Jameson and Hurvich (1955) also added quantitative psychophysical data through their hue-cancellation experiments with human observers that allowed measurement of the relative spectral sensitivities of opponent pathways. These data, combined with the overwhelming support of much additional research since that time, have led to the development of the modern opponent-colors theory of color vision (sometimes called a *stage theory*) as illustrated in Figure 1-13.

Figure 1-13 illustrates that the first stage of color vision, the receptors, is indeed trichromatic, as hypothesized by Maxwell, Young, and Helmholtz. However, contrary to simple trichromatic theory, the three "color-separation" images are not transmitted directly to the brain. Instead, the neurons of the retina (and perhaps higher levels) encode the color into opponent signals. The outputs of all three cone types are summed (L + M + S) to produce an achromatic response that matches the CIE V(λ) curve, as long as the summation is taken in proportion to the relative populations of the three cone types. Differencing of the cone signals allows construction of red-green (L − M + S) and yellow-blue (L + M − S) opponent signals. The transformation from LMS signals to the opponent signals serves to decorrelate the color information carried in the three channels, thus allowing more efficient signal transmission and reducing difficulties with noise. The three opponent pathways also have distinct spatial and temporal characteristics that are important for predicting color appearance. They are discussed further in Section 1.5.

The importance of the transformation from trichromatic to opponent signals for color appearance is reflected in the prominent place it finds within the formulation of all color appearance models. Figure 1-13 includes not only a schematic diagram of the neural "wiring" that produces opponent responses, but also the relative spectral responsivities of these mechanisms both before and after opponent encoding.

Adaptation Mechanisms

However, it is not enough to consider the processing of color signals in the human visual system as a static "wiring diagram." The dynamic mechanisms of adaptation that serve to optimize the visual response to the particular viewing environment at hand must also be considered. Thus an overview of the various types of adaptation is in order. Of particular relevance to the study of color appearance are the mechanisms of dark, light, and chromatic adaptation.

DARK ADAPTATION
Dark adaptation is the change in visual sensitivity that occurs when the prevailing level of illumination is decreased, such as when one walks into a darkened theater on a sunny afternoon. At first, the entire theater appears completely dark. But after a few minutes, one is able to clearly see objects in the theater such as the aisles, seats, and other people. This happens because the

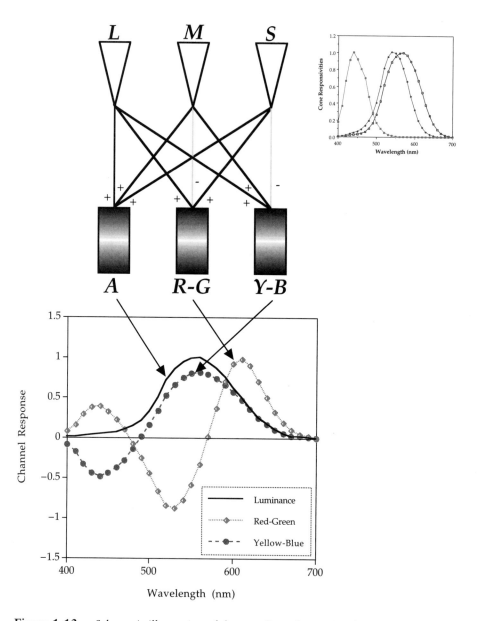

Figure 1-13. Schematic illustration of the encoding of cone signals into opponent-colors signals in the human visual system.

24

visual system is responding to the lack of illumination by becoming more sensitive and therefore capable of producing a meaningful visual response at the lower illumination level.

Figure 1-14 shows the recovery of visual sensitivity (decrease in threshold) after transition from an extremely high illumination level to complete darkness. At first, the cones gradually become more sensitive until, after a couple of minutes, the curve levels off. Then, for about 10 minutes, visual sensitivity is roughly constant. By then, the rods, which have a longer recovery time, have recovered enough sensitivity to outperform the cones and thus begin controlling overall sensitivity. Rod sensitivity continues to improve until, after about 30 minutes, it becomes asymptotic.

Recall that the five-fold change in pupil diameter is not sufficient to serve vision over the large range of illumination levels typically encountered. Therefore neural mechanisms must produce some adaptation. Mechanisms thought to be responsible for various types of adaptation include the following:

- Depletion and regeneration of photopigment
- The rod-cone transition

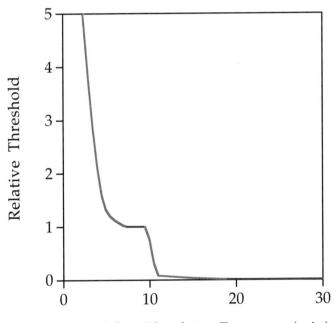

Figure 1-14.   Dark-adaptation curve showing the recovery of threshold after a bleaching exposure. The break in the curve illustrates the point at which the rods become more sensitive than the cones.

- Gain control in the receptors and other retinal cells
- Variation of pooling regions across photoreceptors
- Spatial and temporal opponency
- Gain control in opponent and other higher-level mechanisms
- Neural feedback
- Response compression
- Cognitive interpretation

LIGHT ADAPTATION

Light adaptation is essentially the inverse process of dark adaptation. However, considering it separately is important because its visual properties differ. Light adaptation occurs, for example, when one leaves a darkened theater and returns outdoors on a sunny afternoon. In this case, the visual system must become less sensitive in order to produce useful perceptions, since there is significantly more visible energy available.

The same physiological mechanisms serve light adaptation. However, there is an asymmetry in the forward and reverse kinetics that result in the time course of light adaptation being on the order of 5 minutes rather than 30 minutes. Figure 1-15 illustrates the utility of light adaptation. The visual system has a limited output dynamic range, say 100:1, available for the signals that produce visual perceptions. The world in which we function, however, includes illumination levels covering at least 10 orders of magnitude ranging from, for example, a starlit night to a sunny afternoon. Fortunately, it is almost never important to view the entire range of illumination levels at the same time. If a single response function were used to map the large range of stimulus intensities into the visual system's output, then only a small range of the available output would be used for any given scene. Such a response is shown by the dashed line in Figure 1-15. Clearly, such a response function would limit the perceived contrast of any given scene and visual sensitivity to changes would be severely degraded due to signal-to-noise issues.

On the other hand, light adaptation serves to produce a family of visual response curves, represented by the solid lines in Figure 1-15. These curves map the useful illumination range in any given scene into the full dynamic range of the visual output, thus resulting in the best possible visual perception for each situation. Light adaptation can be thought of as the process of sliding the visual response curve along the illumination level axis in Figure 1-15 until the optimum level for the given viewing conditions is reached. Light and dark adaptation can be thought of as analogous to an automatic exposure control in a photographic system.

CHROMATIC ADAPTATION

Chromatic adaptation is closely related to light and dark adaptation. Again, similar physiological mechanisms are thought to produce chromatic adaptation. *Chromatic adaptation* is the largely independent sensitivity control of the three

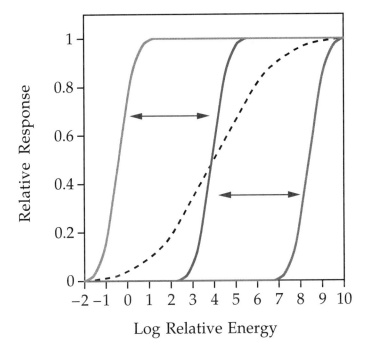

Figure 1-15. Illustration of the process of light adaptation whereby a very large range of stimulus intensity levels can be mapped into a relatively limited-response dynamic range. Solid curves show a family of adapted responses. Dashed curve shows a hypothetical response with no adaptation.

mechanisms of color vision. This is illustrated schematically in Figure 1-16, which shows that the overall height of the three-cone spectral responsivity curves can vary independently. While chromatic adaptation is often discussed and modeled as independent sensitivity control in the cones, there is no reason to believe that it does not occur in opponent and other color mechanisms as well.

Chromatic adaptation can be observed by examining a white object, such as a piece of paper, under various types of illumination (e.g., daylight, fluorescent, and incandescent). Daylight contains relatively far more short-wavelength energy than does fluorescent light, and incandescent illumination contains relatively far more long-wavelength energy than does fluorescent light. However, the paper approximately retains its white appearance under all three light sources. This is because the S-cone system becomes relatively less sensitive under daylight to compensate for the additional short-wavelength energy, while the L-cone system becomes relatively less sensitive under incandescent illumination to compensate for the additional long-wavelength energy.

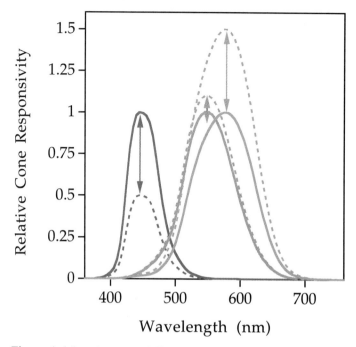

Figure 1-16. Conceptual illustration of the process of chromatic adaptation as the independent sensitivity regulation of the three cone responsivities.

Chromatic adaptation can be thought of as analogous to an automatic white-balance in video cameras. Figure 1-17 provides a visual demonstration of chromatic adaptation in which the two halves of the visual field are conditioned to produce disparate levels of chromatic adaptation. Given its fundamental importance in color appearance modeling, chromatic adaptation is covered in more detail in Chapter 8.

Visual Mechanisms Impacting Color Appearance

Many important cognitive visual mechanisms impact color appearance. These are described in further detail in Chapters 6–8. They include memory color, color constancy, discounting the illuminant, and object recognition.

- *Memory color* refers to the phenomenon that recognizable objects often have a prototypical color that is associated with them. For example, most people have a memory for the typical color of green grass and can produce a stimulus of this color if requested to do so in an experiment. Interestingly, the memory color often is not found in the actual objects. For example, green grass and

Figure 1-17.   A demonstration of retinally localized chromatic adaptation. Fixate the black spot in between the uniform blue and yellow areas for about 30 seconds. Then shift your gaze to the white spot in the center of the barn image. Note that the barn image appears approximately uniform after this adaptation. Original barn image from Kodak Photo Sampler PhotoCD.

blue sky are typically remembered as being more saturated than the actual stimuli.

- *Color constancy* refers to the everyday perception that the colors of objects remain unchanged across significant changes in illumination color and luminance level. Color constancy is served by the mechanisms of chromatic adaptation and memory color and can easily be shown to be very poor when careful observations are made.

- *Discounting the illuminant* refers to an observer's ability to automatically interpret the illumination conditions and perceive the colors of objects after discounting the influences of illumination color.

- *Object recognition* is generally driven by the spatial, temporal, and light-dark properties of the objects rather than by chromatic properties (Davidoff, 1991).

Thus once the objects are recognized, the mechanisms of memory color and discounting the illuminant can fill in the appropriate color. Such mechanisms have fascinating impacts on color appearance and become of critical importance when performing color comparisons across different media.

Clearly, visual information processing is extremely complex and not yet fully understood (perhaps it never will be). It is of interest to consider the increasing complexity of cortical visual responses as the signal moves through the visual system. Single-cell electrophysiological studies have found cortical cells with extremely complex driving stimuli. For example, cells in monkeys that respond only to images of monkey paws or faces have been occasionally found in physiological experiments. The existence of such cells begs the question of how complex a single-cell response can become. It is not possible for every perception to have its own cortical cell. Thus, at some point in the visual system, the representation of perceptions must be distributed, with combinations of various signals producing various perceptions. Such distributed representations open up the possibilities for numerous permutations on a given perception, such as color appearance. It is clear from the large number of stimulus variables that affect color appearance that our visual system often experiments with these permutations.

1.5   Spatial and Temporal Properties of Color Vision

No dimension of visual experience can be considered in isolation. The color appearance of a stimulus is not independent of its spatial and temporal characteristics. For example, a black-and-white stimulus flickering at an appropriate temporal frequency can be perceived as quite colorful. The spatial and temporal characteristics of the human visual system are typically explored through measurement of *contrast sensitivity functions* (CSFs). CSFs in vision science are analogous to modulation transfer functions (MTFs) in imaging science. However, CSFs cannot legitimately be considered MTFs, since the human visual system is highly nonlinear. A CSF is defined by the threshold response to contrast (sensitivity is the inverse of threshold) as a function of spatial or temporal frequency. *Contrast* is typically defined as the difference between maximum and minimum luminance in a stimulus divided by the sum of the maximum and minimum luminances, and CSFs are typically measured with stimuli that vary sinusoidally across space or time. Thus a uniform pattern has a contrast of zero and sinusoidal patterns with troughs that reach a luminance of zero have a contrast of 1.0 no matter what their mean luminance is.

Figure 1-18 illustrates typical spatial CSFs for luminance (black-white) contrast and chromatic (red-green and yellow-blue at constant luminance) contrast. The luminance CSF is band-pass in nature, with a peak sensitivity around 5 cycles per degree. This function approaches zero at zero cycles per degree, thus illustrating the tendency for the visual system to be insensitive to uniform fields. It also approaches zero at about 60 cycles per degree, the point at which detail can no longer be resolved by the optics of the eye or the photoreceptor mosaic. The band-pass CSF correlates with the concept of center-surround antagonistic receptive fields that would be most sensitive to an intermediate range of spatial frequency. The chromatic mechanisms are of a low-pass nature and have significantly lower cutoff frequencies. This indicates the reduced availability of chromatic information for fine details (high spatial frequencies) that is often taken advantage of in image coding and compression schemes (e.g., NTSC or JPEG).

The low-pass characteristics of the chromatic mechanisms also illustrate that edge detection/enhancement does not occur along these dimensions. The blue-yellow chromatic CSF has a lower cutoff frequency than does the red-green chromatic CSF due to the scarcity of S cones in the retina. Note also that the luminance CSF is significantly higher than the chromatic CSFs. This indicates that the visual system is more sensitive to small changes in luminance

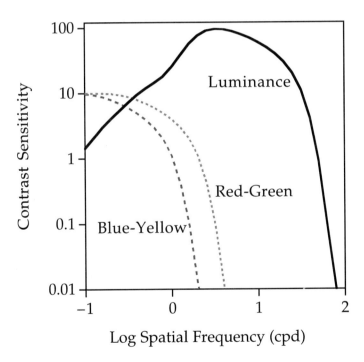

Figure 1-18. Spatial contrast sensitivity functions for luminance and chromatic contrast.

contrast compared to chromatic contrast. The spatial CSFs for luminance and chromatic contrast are generally not directly incorporated in color appearance models, although there is significant interest in doing so. Zhang and Wandell (1996) presented an interesting technique for incorporating these types of responses into the CIELAB color space calculations.

Figure 1-19 illustrates the spatial properties of color vision with a spatial analysis of a typical image. Figure 1-19(a) shows the original image. The luminance information is presented alone in Figure 1-19(b), and the residual chromatic information is presented alone in Figure 1-19(c). Note that far more spatial detail can be visually obtained from the luminance image than from the chromatic residual image. This is further illustrated in Figure 1-19(d), in which the image has been reconstructed using the full-resolution luminance image combined with the chromatic image after subsampling by a factor of 4. This form of image compression produces no noticeable degradation in perceived resolution or color.

Figure 1-20 illustrates typical temporal CSFs for luminance and chromatic contrast. They share many characteristics with the spatial CSFs shown in Figure 1-18. Again, the luminance temporal CSF is higher in both sensitivity and cutoff frequency than are the chromatic temporal CSFs. Also, it shows

Figure 1-19.   Illustration of the spatial properties of color vision: (a) original image; (b) luminance information only; (c) chromatic information only; (d) reconstruction with full-resolution luminance information combined with chromatic information subsampled by a factor of four. Original motorcycles image from Kodak Photo Sampler PhotoCD.

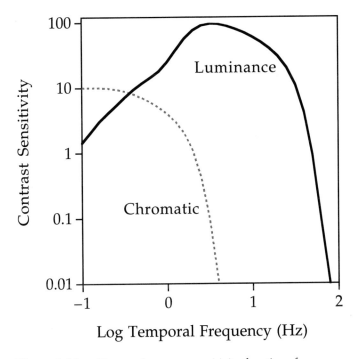

Figure 1-20. Temporal contrast sensitivity functions for
luminance and chromatic contrast.

band-pass characteristics that suggest the enhancement of temporal transients
in the human visual system. While temporal CSFs are not directly incorporated
in color appearance models, they might be important to consider when viewing time-varying images such as digital video clips that might be rendered at
differing frame rates.

It is important to realize that the functions in Figures 1-18 and 1-20 are
typical and not universal. As stated at the beginning of this section, the dimensions of human visual perception cannot be examined independently. The spatial and temporal CSFs interact with one another. A spatial CSF measured at
different temporal frequencies will vary tremendously, as will a temporal CSF
measured at various spatial frequencies. These functions also depend on other
variables such as luminance level, stimulus size, retinal locus, and so on.

The Oblique Effect

An interesting spatial vision phenomenon is the oblique effect. The *oblique effect* is the phenomenon whereby visual acuity is better for gratings oriented at
0° or 90° (relative to the line connecting the two eyes) than for gratings oriented at 45°. This phenomenon is considered in the design of rotated halftone

screens that are set up such that the most visible pattern is oriented at 45°. The effect can be observed by taking a black-and-white halftone newspaper image and adjusting the viewing distance until the halftone dots are just barely imperceptible. If the image is kept at this viewing distance and then rotated 45°, the halftone dots will become clearly perceptible (since they will then be oriented at 0° or 90°).

CSFs and Eye Movements

The spatial and temporal CSFs are also closely linked to the study of eye movements. A static spatial pattern becomes a temporally varying pattern when observers move their eyes across the stimulus. Both the spatial and temporal luminance CSFs approach zero as either form of frequency variation approaches zero, so it follows that a completely static stimulus is invisible. This is indeed the case. If the retinal image can be fixed using a video feedback system attached to an eye tracker, the stimulus does disappear after a few seconds. (Sometimes this can be observed by carefully fixating an object and noting that the perceptions of objects in the periphery begin to fade away after a few seconds. The centrally fixated object does not fade away, since the ability to hold the eye completely still has variance greater than the high spatial resolution in the fovea.)

To avoid this rather unpleasant phenomenon in typical viewing, our eyes are constantly moving. Large eye movements take place to allow the viewing of different areas of the visual field with the high-resolution fovea. Also, there are small constant eye movements that serve to keep the visual world nicely visible. This also explains why the shadows of retinal cells and blood vessels are generally not visible, since they do not move *on* the retina, but rather move *with* the retina. The history of eye movements has significant impact on adaptation and appearance through integrated exposure of various retinal areas and the need for movements to preserve apparent contrast.

1.6   Color-vision Deficiencies

There are various types of inherited and acquired color-vision deficiencies. Kaiser and Boynton (1996) provide a current and comprehensive overview of the topic. This section concentrates on the most common inherited deficiencies.

Protanopia, Deuteranopia, and Tritanopia

Some color-vision deficiencies are caused by the lack of a particular type of cone photopigment. Since there are three types of cone photopigments, there are three general classes of these color-vision deficiencies: protanopia, deuteranopia, and tritanopia.

An observer with protanopia, called a protanope, is missing the L-cone photopigment and therefore is unable to discriminate reddish and greenish hues, since the red-green opponent mechanism cannot be constructed.

A deuteranope is missing the M-cone photopigment and therefore also cannot distinguish reddish and greenish hues, due to the lack of a viable red-green opponent mechanism. Protanopes and deuteranopes can be distinguished by their relative luminous sensitivity, since that is constructed from the summation of different cone types. The protanopic luminous sensitivity function is shifted toward shorter wavelengths.

A tritanope is missing the S-cone photopigment and therefore cannot discriminate yellowish and bluish hues due to the lack of a yellow-blue opponent mechanism.

Anomalous Trichromacy

There are also anomalous trichromats who have abnormal trichromatic vision. In this case, the ability to discriminate particular hues is reduced either due to shifts in the spectral sensitivities of the photopigments or the contamination of photopigments (e.g., some L-cone photopigment in the M-cones, and so on). Among the anomalous trichromats are those with any of the following:

- Protanomaly, that is, either they are weak in L-cone photopigment or the L-cone absorption is shifted toward shorter wavelengths
- Deuteranomaly, that is, either they are weak in M-cone photopigment or the M-cone absorption is shifted toward longer wavelengths
- Tritanomaly, that is, either they are weak in S-cone photopigment or the S-cone absorption is shifted toward longer wavelengths

There are also some cases of cone monochromatism (effectively, only one cone type) and rod monochromatism (no cone responses).

While it is impossible for a person with normal color vision to experience what the visual world looks like to a person with a color-vision deficiency, it is possible to illustrate the hues that become indistinguishable. Figure 1-21 provides such a demonstration. The figure was produced by transforming the color-normal image [Figure 1-21(a)] to cone signals. Then one cone signal at a time was deleted prior to transforming the image back to device signals used to generate the images. This allowed an illustration of the various colors that would be confused by protanopes, deuteranopes, and tritanopes. The study of

Figure 1-21.   Images illustrating the color-discrimination capabilities that are missing from observers with various color-vision deficiencies: (a) original image; (b) protanope; (c) deuteranope; (d) tritanope. Original birds image from Kodak Photo Sampler PhotoCD.

color-vision deficiencies is of more than academic interest in the field of color appearance modeling and color reproduction. This is illustrated in Table 1-1, which shows the approximate percentages of the population with various types of color-vision deficiencies.

As can be seen from the table, color-vision deficiencies are not rare, particularly in the male population (about 8%). Hence, in many applications it might be important to account for the possibility of color-deficient observers.

Color-vision Deficiencies and Gender

Why the disparity between the occurrence of color-vision deficiencies in males and females? This can be traced to the genetic basis of such deficiencies. It turns out that the most common forms of color-vision deficiencies are sex-linked genetic traits.

The genes for photopigments are present on the X chromosome. Females inherit one X chromosome from their mother and one from their father. Only one of these need have the genes for the normal photopigments in order to produce normal color vision. On the other hand, males inherit an X chromosome from their mother and a Y chromosome from their father. If the single X chro-

Table 1-1.   Approximate percentage occurrences of various color-vision deficiencies. Based on data in Hunt (1991a).

Type	Male %	Female %
Protanopia	1.0	0.02
Deuteranopia	1.1	0.01
Tritanopia	0.002	0.001
Cone Monochromatism	~0	~0
Rod Monochromatism	0.003	0.002
Protanomaly	1.0	0.02
Deuteranomaly	4.9	0.38
Tritanomaly	~0	~0
TOTAL	8.0	0.4

mosome does not include the genes for the photopigments, the son will have a color-vision deficiency. If a female is color deficient, it means she has two deficient X chromosomes and all male children are destined to have a color-vision deficiency. So the genetic "deck of cards" is stacked against males when it comes to inheriting deficient color vision.

Knowledge regarding the genetic basis of color vision has grown tremendously in recent years. John Dalton was an early investigator of deficient color vision. He studied his own vision, which was historically thought to be protanopic based on his observations, and came up with a theory as to the cause of his deficiencies. Dalton hypothesized that his color-vision deficiency was caused by a coloration of his vitreous humor that caused it to act like a filter. He arranged to donate his eyes upon his death to have them dissected so as to experimentally confirm his theory. Unfortunately, Dalton's theory was incorrect. However, his eyes are preserved in a museum in Manchester, UK. D. M. Hunt et al. (1995) performed DNA tests on Dalton's preserved eyes and were able to show that Dalton was a deuteranope rather than a protanope, but with an L-cone photopigment having a spectral responsivity shifted toward the shorter wavelengths. They were also able to complete a colorimetric analysis to show that their genetic results were consistent with the observations originally recorded by Dalton.

Screening Observers Who Make Color Judgments

Given the fairly high rate of occurrence of color-vision deficiencies, it is necessary to screen observers prior to allowing them to make critical color-appearance or color-matching judgments. Various tests are available, but two types are of practical importance: pseudoisochromatic plates and the Farnsworth-Munsell 100-Hue Test.

PSEUDOISOCHROMATIC PLATES

Pseudoisochromatic plates (e.g., *Ishihara's Tests for Colour-Blindness*) are color plates made up of dots of random lightness that include a pattern or number in the dots formed out of an appropriately contrasting hue. The random lightness of the dots is a design feature to avoid discrimination of the patterns based on lightness difference only.

The plates are presented under properly controlled illumination to observers who are asked to respond by either tracing the pattern or reporting the number observed. Various plates are designed with color combinations that would be difficult to discriminate for observers with the different types of color-vision deficiencies. These tests are commonly administered as part of a normal ophthalmological examination and can be obtained from optical suppliers and general scientific suppliers. Screening with a set of pseudoisochromatic plates should be considered as a minimum evaluation for anyone carrying out critical color judgments.

FARNSWORTH-MUNSELL 100-HUE TEST

The Farnsworth-Munsell 100-Hue Test, available through the Munsell Color Company, consists of four sets of chips that must be arranged in an orderly progression of hue. Observers with various types of color-vision deficiencies will make errors in the arrangement of the chips at various locations around the hue circle. The test can be used to distinguish between the different types of deficiencies and also to evaluate the severity of color discrimination problems. It also can be used to identify observers who have normal color vision but poor color discrimination for all colors.

1.7 Key Features for Color Appearance Modeling

This chapter provides a necessarily brief overview of the form and function of the human visual system, concentrating on the features that are important in the study, modeling, and prediction of color-appearance phenomena. This section gives a short review of the key features.

Important elements in the optics of the eye include the lens, macula, and cone photoreceptors. The lens and macula impact color matching through their action as yellow filters. They affect inter-observer variability, since their optical density varies significantly by person. The cones serve as the first stage of color vision. They transform the spectral power distribution on the retina into a three-dimensional signal that defines what is available for processing at higher levels in the visual system. This is the basis of metamerism, the fundamental governing principle of colorimetry.

The numerical distribution of the cones (L:M:S of about 40:20:1) is important in constructing the opponent signals present in the visual system.

Proper modeling of these steps requires the ratios to be accounted for appropriately. The spatial distribution of rods and cones and their lateral interactions are critical in the specification of stimulus size and retinal locus. A color appearance model for stimuli viewed in the fovea would differ from one for peripheral stimuli. The spatial interaction in the retina, represented by horizontal and amacrine cells, is critical for mechanisms that produce color-appearance effects due to changes in background, surround, and level of adaptation.

The encoding of color information through the opponent channels along with the adaptation mechanisms before, during, and after this stage are perhaps the most important feature of the human visual system that must be incorporated into color appearance models. Each such model incorporates a chromatic adaptation stage, an opponent processing stage, and nonlinear response functions. Some also incorporate light and dark adaptation effects and interactions between the rod and cone systems.

The cognitive mechanisms of vision, such as memory color and discounting the illuminant, also have a profound impact on color appearance. These and other color-appearance phenomena are described in greater detail in Chapters 6–8.

CHAPTER 2

Psychophysics

2.1 Psychophysics Defined
2.2 Historical Context
2.3 Hierarchy of Scales
2.4 Threshold Techniques
2.5 Matching Techniques
2.6 One-dimensional Scaling
2.7 Multidimensional Scaling
2.8 Design of Psychophysical Experiments
2.9 Importance of Psychophysics in Color Appearance
     Modeling

AN UNDERSTANDING OF THE basic function of the human visual system is necessary for appreciation of the formulation, implementation, and application of color appearance models. However, the need for a basic understanding of the principles of psychophysics might not seem so clear. Psychophysical techniques have produced most of our knowledge of human color vision and color appearance phenomena. These are the underpinnings of colorimetry and its extension through color appearance models. Also, psychophysical techniques are used to test, compare, and generate data for improving color appearance models. Thus to fully understand the use and evaluation of such models, a basic appreciation of the field of psychophysics is essential. Psychophysical techniques such as those described in this chapter can help to prove that the implementation of a color appearance model truly improves a system.

This chapter provides an overview of experimental design and data analysis techniques for visual experiments. Carefully conducted visual experiments allow accurate quantitative evaluation of perceptual phenomena that are often thought of as being completely subjective. Such results can be of immense value in a wide variety of fields, including color measurement and the evaluation of perceived image quality. Also described are issues regarding the choice and design of viewing environments; an overview of various classes of visual experiments; and a review of experimental techniques for threshold, matching, and scaling experiments, as are data reduction and analysis procedures.

The treatment of psychophysics presented in this chapter is based on the ASTM *Standard Guide for Designing and Conducting Visual Experiments* (ASTM, 1996), which was based on materials from the RIT course work of this book's author. There are several excellent texts on psychophysics that provide additional details on the topics covered in this chapter. Of particular note are those of Gescheider (1985), Bartleson and Grum (1984), Torgerson (1958), and Thurstone (1959). Unfortunately, the last three references are out of print and can be found only in libraries. An interesting overview of the application of psychophysics to image quality has been presented by Engeldrum (1995).

2.1   Psychophysics Defined

Psychophysics is the scientific study of the relationships between the physical measurements of stimuli and the sensations and perceptions that those stimuli evoke. It can be considered a discipline of science similar to the more traditional disciplines such as physics, chemistry, and biology.

The tools of psychophysics are used to derive quantitative measures of perceptual phenomena that are often considered subjective. It is important to note that the results of properly designed psychophysical experiments are just as objective and quantitative as the measurement of length with a ruler (or any other physical measurement). One important difference is that the uncertainties associated with psychophysical measurements tend to be significantly larger than those of physical measurements. However, the results are equally useful and meaningful as long as those uncertainties are considered (as they always should be for physical measurements as well). Psychophysics is used to study all dimensions of human perception. Since the topic of this book is color appearance, visual psychophysics is specifically discussed.

Two Classes of Visual Experiments

Visual experiments tend to fall into two broad classes:

1. Threshold and matching experiments, which are designed to measure visual sensitivity to small changes in stimuli (or perceptual equality)
2. Scaling experiments, which are intended to generate a relationship between the physical and perceptual magnitudes of a stimulus

It is critical to first determine which class of experiment is appropriate for a given application. Threshold experiments are appropriate for measuring sensitivity to changes and the detectability of stimuli. For example, they could be used to determine whether an image compression algorithm was truly visually lossless or if the performances of two color appearance models differ perceptibly in some practical application.

Scaling experiments are appropriate when it is necessary to specify the relationships between stimuli. For example, they could be used to develop a quantitative relationship between the perceived quality of a printed image and the spatial addressability of the printer. In color appearance modeling, the results of scaling experiments are used to derive relationships between physically measurable colorimetric quantities (e.g., CIE XYZ tristimulus values) and perceptual attributes such as lightness, chroma, and hue.

2.2   Historical Context

As with any scientific discipline, one can better appreciate psychophysics by studying its historical development. While scientists have been making and recording careful observations of their perceptions for centuries, the formal discipline of psychophysics is fewer than 150 years old. Important milestones in the history of psychophysics can be represented in the work of Weber, Fechner, and Stevens.

Weber's Work

In the early part of the nineteenth century, E. H. Weber investigated the perception of the heaviness of lifted weights. He asked observers to lift a given weight, and then he added to the weight (with all of the necessary experimental controls) until the observers could just distinguish the new weight from the original. This is a measurement of the threshold for change in weight. Weber noted that as the initial weight increased, the change in weight required to reach a threshold increased proportionally. If the initial magnitude of the stimulus (weight in this case) is denoted I and the change required to achieve a threshold is denoted ΔI, Weber's results can be expressed by stating that the ratio $\Delta I/I$ is constant. In fact, this general relationship holds approximately true for many perceptual stimuli and has come to be known as Weber's law.

Weber's result is quite intuitive. For example, imagine carrying a few sheets of paper and then adding a 20-page document to the load. Clearly the difference between the two weights would be perceived. Now imagine carrying a briefcase full of books and papers and then adding another 20-page document to the case. Most likely the added weight of 20 more pages would go unnoticed. That is because a greater change is required to reach a perceptual threshold when the initial stimulus intensity is higher. Weber's law can also be used to explain why stars cannot be seen during the daytime. At night, the stars represent a certain increment in intensity, ΔI, over the background illumination of the sky, I, that exceeds the visual threshold; therefore they can be seen. During the day, the stars still produce the same increment in intensity, ΔI, over the background illumination. However, the background intensity of the daytime sky, I, is much larger than at night. Therefore the ratio $\Delta I/I$ is far lower during the day than at night. So low, in fact, that the stars cannot be perceived during the day.

Fechner's Work

The next milestone is the work of Fechner. Fechner built on the work of Weber to derive a mathematical relationship between stimulus intensity and perceived magnitude. While his motivation was to solve the mind-body problem by proving that functions of the mind could be physically measured, he inadvertently became known as the father of psychophysics through his publication of *Elements of Psychophysics* in 1860 (Fechner, 1966).

Fechner started with two basic assumptions:

1. Weber's law was indeed valid.
2. A just-noticeable difference (JND) can be considered a unit of perception.

Weber's results showed that the JNDs measured on the physical scale were not equal as the stimulus intensity increased; rather, they increased in proportion to the stimulus intensity. Fechner strived to derive a transformation of the physical stimulus intensity scale to a perceptual magnitude scale on which the JNDs were of equal size for all perceptual magnitudes. This problem essentially is to solve the differential equation posed by Weber's law. Since JNDs followed a geometric series on the stimulus intensity scale, the solution is straightforward that a logarithmic transformation will produce JNDs of equal incremental size on a perceptual scale. Thus the JNDs represented by equal ratios on the physical scale become transformed into equal increments on a perceptual scale according to what has come to be known as Fechner's law.

Simply put, Fechner's law states that the perceived magnitude of a stimulus is proportional to the logarithm of the physical stimulus intensity. This relationship is illustrated in Figure 2-1. Fechner's law results in a compressive nonlinear relationship between the physical measurement and perceived magnitude that illustrates a decreasing sensitivity (i.e., slope of the curve in Figure 2-1) with increasing stimulus intensity. If Fechner's law were strictly valid, then the relationship would follow the same nonlinear form for all perceptions. While the general trend of a compressive nonlinearity is valid for most perceptions, various perceptions take on relationships with differently shaped functions. This means that Fechner's law is not completely accurate. However, there are numerous examples in the vision science literature of instances in which Fechner's law (or at least Weber's law) is obeyed.

Stevens's Work

Addressing the lack of generality in Fechner's result, Stevens (1961) published an intriguing paper entitled "To honor Fechner and repeal his law." Stevens studied the relationship between physical stimulus intensity and perceptual magnitude for over 30 different types of perceptions using a magnitude estimation technique. He found that his results produced straight lines when the logarithm of perceived magnitude was plotted as a function of the logarithm of stimulus intensity. However, the straight lines for various perceptions had different slopes. Straight lines on log-log coordinates are equivalent to power functions on linear coordinates with the slopes of the lines on the log-log co-

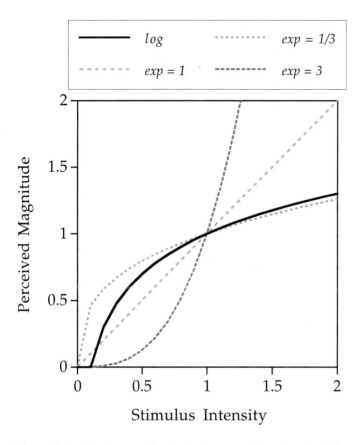

Figure 2-1   Various psychophysical response functions, including the logarithmic function suggested by Fechner and power law relationships with various exponents as suggested by Stevens.

ordinates equivalent to the exponents of the power functions on linear axes. Thus Stevens hypothesized that the relationship between perceptual magnitude and stimulus intensity followed a power law with various exponents for different perceptions rather than Fechner's logarithmic function. This result is often referred to as the Stevens power law.

Figure 2-1 illustrates three power law relationships with differing exponents. When the exponent is less than unity, a power law follows a compressive nonlinearity typical of most perceptions. When the exponent is equal to unity, the power law becomes a linear relationship. While there are few examples of perceptions that are linearly related to physical stimulus intensity, one important one is the relationship between perceived length and physical length over short distances. A power law with an exponent greater than unity results in an expansive nonlinearity. Such perceptual relationships do exist when the stimulus might be harmful and thus result in the perception of pain. A compressive function for the pain perception could be quite dangerous, since the observer would become less and less sensitive to the stimulus as it became more and more dangerous.

The Stevens power law can be used to model many perceptual phenomena and can be found in fundamental aspects of color measurement, such as the relationship between CIE *XYZ* tristimulus values and the predictors of lightness and chroma in the CIELAB color space that are based on a cube-root, compressive power law nonlinearity.

2.3   Hierarchy of Scales

When deriving perceived magnitude scales, one must understand the properties of the resulting scales. Often a psychophysical technique will produce a scale with only limited mathematical utility. In such cases, inappropriate mathematical manipulations must *not* be applied to the scale. Four key types of scales have been defined, given in order of increasing mathematical power and complexity: nominal, ordinal, interval, and ratio scales.

Nominal Scales

A *nominal scale* is a relatively trivial scale in that it scales items simply by name. For color, a nominal scale could consist of reds, yellows, greens, blues, and neutrals. Scaling in this case would simply require deciding which color belonged in which category. Only naming can be performed with nominal data.

Ordinal Scales

An *ordinal scale* is a scale in which elements are sorted in ascending or descending order based on greater or lesser amount of a particular attribute. A set of color swatches could be sorted by hue. Then, in each hue range, the colors could be sorted from the lightest to the darkest. The swatch colors are likely not to be evenly spaced, so there might be, for example, three dark, one medium, and two light green swatches. If these are then numbered from one to six in order of increasing lightness, an ordinal scale results. Such a scale contains no information as to how much lighter one of the, say, green swatches is than another green swatch, and it is clear that they are not evenly spaced. For an ordinal scale, all that matters is that the samples be arranged in increasing or decreasing amounts of an attribute. The spacing between samples can be large or small and can change up and down the scale. Logical operations such as greater-than, less-than, or equal-to can be performed with ordinal scales.

Interval Scales

An *interval scale* is a scale that has equal intervals. On an interval scale, for example, if a pair of samples is separated by two units and a second pair at some other point on the scale is also separated by two units, the differences between the pairs will be perceptually equal. However, there is no meaningful zero point on an interval scale.

In addition to the mathematical operations listed for the previous scales, addition and subtraction can be performed with interval data. The Celsius and Fahrenheit temperature scales are interval scales.

Ratio Scales

A *ratio scale* is a scale that has all the properties of the first three, plus a meaningfully defined zero point. Thus it is possible to properly equate ratios of numbers on a ratio scale.

Ratio scales in visual work are often difficult and sometimes impossible to obtain. Sometimes this is because a meaningful zero point does not exist. For example, an interval scale of image quality is relatively easy to derive, but try to imagine an image with zero quality. Similarly, it is relatively straightforward to derive interval scales of hue, but there is no physically meaningful zero hue.

All of the mathematical operations that can be performed on an interval scale can also be performed on a ratio scale. In addition, multiplication and division can be performed.

Example of the Use of Scales

It is helpful to reinforce the concepts of the hierarchy of scales by an example. Suppose you need to measure the heights of all the people in a room. This could be done using any of the four scales. If only a nominal scale were available, you could choose a first subject and assign a name to his or her height, say Joe. Then you could examine the height of each additional subject relative to the height of Joe. If another person had the same height as Joe (assuming some reasonable tolerance), that person's height would also be assigned the name Joe. Heights that differed from Joe's would be given a different name. This process could be completed by comparing the heights of everyone in the room until each unique height was assigned a unique name. Note that there is no information regarding who is taller or shorter than anyone else. The only information available is whether subjects share the same height (and therefore name).

If, instead, one were to use an ordinal scale to measure height, Joe could be arbitrarily assigned a height of zero. If the next subject was taller than Joe, he or she would be assigned any number larger than Joe's, say 10. If a third subject was found to be taller than Joe, but shorter than the second subject, that person would be assigned a number between zero and 10. This would continue until everyone in the room was assigned a number to represent their heights. Since the magnitude of the numbers was assigned arbitrarily, nothing can be said about how much shorter or taller one subject is than another. However, the subjects could be put in order from shortest to tallest.

If one were to use an interval scale, Joe could again be arbitrarily assigned a height of zero. However, other subjects could then be assigned heights relative to Joe's in terms of meaningful increments, such as +3 cm (taller than Joe) or −2 cm (shorter than Joe). If subjects A and B had heights of +3 cm and −2 cm, respectively, on this interval scale, it could be determined that subject A is 5 cm taller than subject B. Note, however, that there is still no information to indicate how tall either subject is. The only information available with the interval scale is differences between subjects.

Finally, if one were to use a ratio scale (the normal method), Joe might be measured and found to be 182 cm tall. Then subjects A and B would have heights of 185 cm and 180 cm, respectively. If another subject came along who was 91 cm tall, it could be concluded that Joe was twice as tall as this subject. Since zero cm tall is a physically meaningful zero point, a true ratio scale is available for the measurement of height and thus multiplications and divisions of scale values can be performed.

2.4   Threshold Techniques

Threshold experiments are designed to determine the just-perceptible change in a stimulus, sometimes referred to as a just-noticeable difference (JND). Threshold techniques are used to measure an observer's sensitivity to changes in a given stimulus. Absolute thresholds are defined as the JND for a change from no stimulus, while difference thresholds represent the JND from a particular stimulus level greater than zero.

Thresholds are reported in terms of the physical units used to measure the stimulus. For example, a brightness threshold might be measured in luminance units of cd/m^2. Sensitivity is defined as the inverse of the threshold, since a low threshold implies high sensitivity. Threshold techniques are useful for defining visual tolerances such as those for perceived color differences.

Types of Threshold Techniques

Several basic types of threshold experiments are presented in this section in order of increasing complexity of experimental design and utility of the data generated. Many modifications of these techniques have been developed for particular applications. Experimenters strive to design experiments that remove as much control of the results from the observers as possible, thus minimizing the influence of variable observer judgment criteria. Generally this comes at the cost of implementing more complicated experimental procedures. Threshold techniques include the following:

- Method of adjustment
- Method of limits
- Method of constant stimuli

METHOD OF ADJUSTMENT

The *method of adjustment* is the simplest and most straightforward technique for deriving threshold data. With it, the observer controls the stimulus magnitude and adjusts it to a point that is just perceptible (absolute threshold) or just perceptibly different (difference threshold) from a starting level. The threshold is taken to be the average setting across a number of trials by one or more observers.

This technique has the advantage that it is quick and easy to implement. However, a major disadvantage is that the observer is in control of the stimulus. This can bias the results due to variability in observers' criteria and adaptation effects. If an observer approaches the threshold from above, adaptation might result in a higher threshold than if it were approached from below.

The method of adjustment is often used to obtain a first estimate of the threshold to be used in the design of more-sophisticated experiments. It is also commonly used in matching experiments, including asymmetric matching experiments used in color-appearance studies.

METHOD OF LIMITS

The *method of limits* is only slightly more complex than the method of adjustment. With this method, the experimenter presents the stimuli at predefined discrete intensity levels in either ascending or descending series. For an ascending series, the experimenter presents a stimulus, beginning with one that is certain to be imperceptible, and asks the observers to respond yes if they perceive it and no if they do not. If they respond no, the experimenter increases the stimulus intensity and presents another trial. This continues until the observer responds yes.

A descending series begins with a stimulus intensity that is clearly perceptible and continues until the observers respond no—that is, they cannot perceive the stimulus. The threshold is taken to be the average stimulus intensity at which the transition from no to yes (or yes to no) responses occurs for a number of ascending and descending series. Averaging over both types of series minimizes adaptation effects. However, the observers are still in control of their criteria, since they can respond yes or no at their own discretion.

METHOD OF CONSTANT STIMULI

With the *method of constant stimuli,* the experimenter chooses several stimulus intensity levels (typically about 5 or 7) around the level of the threshold. Each of these stimuli is then presented multiple times in random order. Over the trials, the frequency with which each stimulus level is perceived is determined. From such data, a *frequency-of-seeing curve,* or *psychometric function,* can be derived that allows the determination of the threshold and its uncertainty. The threshold is generally taken to be the intensity at which the stimulus is perceived on 50% of the trials. Psychometric functions can be derived either for a single observer (through multiple trials) or for a population of observers (one or more trials per observer). Two types of response can be obtained:

- Yes-no (or pass-fail)
- Forced choice

Yes-no Method   In the *yes-no method* of constant stimuli procedure, the observers are asked to respond yes if they detect the stimulus (or stimulus change) and no if they do not. The psychometric function is then simply the percentage of yes responses as a function of stimulus intensity. Fifty percent yes responses would be taken as the threshold level.

Alternatively, this procedure can be used to measure visual tolerances above threshold by providing a reference stimulus intensity (e.g., a color-difference anchor pair) and asking observers to pass stimuli that fall below the intensity of the reference (e.g., a smaller color difference) and fail those that fall above it (e.g., a larger color difference). The psychometric function is then taken to be the percentage of fail responses as a function of stimulus intensity, and the 50% fail level is deemed to be the point of visual equality.

Forced-choice Method   The *forced-choice method* eliminates the influence of varying observer criteria on the results. This is accomplished by presenting the stimulus in one of two intervals defined by either a spatial or temporal separation. The observers are then asked to indicate in which of the two intervals the stimulus was presented. The observers are not allowed to respond that the stimulus was not present; they must guess one of the two intervals if they are unsure (hence the name, forced choice). The psychometric function is then plotted as the percentage of correct responses as a function of stimulus intensity. The function ranges from 50% correct when the observers are simply guessing to 100% correct for stimulus intensities at which they can always detect the stimulus. Thus the threshold is defined as the stimulus intensity at which the observers are correct 75% of the time and therefore detecting the stimulus 50% of the time. As long as the observers respond honestly, their criteria, whether liberal or conservative, cannot influence the results.

Staircase Procedures   Staircase procedures are a modification of the forced-choice procedure. They are designed to measure only the threshold point on the psychometric function. They are particularly applicable to situations in which the stimulus presentations can be fully automated. A stimulus is presented, and the observer is asked to respond. If the response is correct, the same stimulus intensity is presented again. If the response is incorrect, the stimulus intensity is increased for the next trial. Generally, if the observer responds correctly on three consecutive trials, the stimulus intensity is decreased. The stimulus intensity steps are decreased until some desired precision in the threshold is reached. The sequence of three correct or one incorrect response prior to changing the stimulus intensity will result in a convergence to a stimulus intensity that is correctly identified on 79% of the trials ($0.79^3 = 0.5$), very close to the nominal threshold level of 75%. Often several independent staircase procedures are run simultaneously to further randomize the experiment. A staircase procedure could also be run with yes-no responses.

Probit Analysis of Threshold Data

Threshold data that generate a psychometric function can be most usefully analyzed using *Probit analysis*. Probit analysis is used to fit a cumulative normal distribution to the data (psychometric function). The threshold point and its uncertainty can then be easily determined from the fitted distribution. There are also several significance tests that can be performed to verify the suitability of the analyses. Finney (1971) provides details on the theory and application of Probit analysis. Also, several commercially available statistical software packages can be used to perform such analyses.

2.5   Matching Techniques

Matching techniques are similar to threshold techniques except that the goal is to determine when two stimuli are not perceptibly different. Measures of the variability in matching are sometimes used to estimate thresholds. Matching experiments provided the basis for CIE colorimetry through the metameric matches used to derive color-matching functions. For example, suppose a given color is perceptually matched by an additive mixture of red, green, and blue primary lights that do not mix to produce a spectral energy distribution that is identical to the test color. Then the match is considered *metameric*. The physical properties of such matches can be used to derive the fundamental responsivities of the human visual system and ultimately be used to derive a system of tristimulus colorimetry as outlined in Chapter 3.

Asymmetric Matching

Matching experiments are often used in the study of chromatic adaptation and color appearance as well. In such cases, asymmetric matches are made.

An *asymmetric match* is a color match made across some change in viewing conditions. For example, a stimulus viewed in daylight illumination might be matched to another stimulus viewed under incandescent illumination to derive a pair of corresponding colors for this change in viewing conditions. Such data can then be used to formulate and test color appearance models designed to account for such changes in viewing conditions.

One special case of an asymmetric matching experiment is the *haploscopic experiment*. In this experiment, one eye views a test stimulus in one set of viewing conditions and the other eye simultaneously views a matching stimulus in a different set of viewing conditions. The observer simultaneously views both stimuli and produces a match.

Memory Matching

Another type of matching experiment that is sometimes used in the study of color appearance is *memory matching*. In such experiments, observers produce a match to a previously memorized color. Typically such matches are asymmetric so as to study viewing-conditions dependencies. Occasionally, memory matches are made to mental stimuli such as an ideal achromatic (gray) color or a unique hue (e.g., a unique red with no blue or yellow content).

2.6 One-dimensional Scaling

Scaling experiments are intended to derive relationships between perceptual magnitudes and physical measures of stimulus intensity. Depending on the type and dimensionality of the scale required, several approaches are possible. Normally the type of scale required and the scaling method to be used are decided upon before any visual data are collected. One-dimensional scaling requires that both the attribute to be scaled and the physical variation of the stimulus be one-dimensional. Observers are asked to make their judgments on a single perceptual attribute (e.g., how light is one sample compared to another, what is the quality of the difference between a pair of images, and so on).

A variety of scaling techniques have been devised for the measurement of one-dimensional psychophysical scales, including the following, which are described in the following paragraphs:

- Rank order experiment
- Graphical rating
- Category scaling
- Paired comparison experiment
- Partition scaling
- Magnitude estimation or production
- Ratio estimation or production

In a *rank order experiment,* the observer is asked to arrange a given set of samples according to increasing or decreasing magnitudes of a particular perceptual attribute. When there are a large number of observers, the data may be averaged and reranked to obtain an ordinal scale. To obtain an interval scale, one must make certain assumptions about the data and perform additional analyses. In general, it is somewhat unwise to attempt to derive interval scales from rank order data.

Graphical rating allows direct determination of an interval scale. Observers are presented stimuli and asked to indicate the magnitude of their perceptions on a one-dimensional scale that has defined endpoints. For example, in a lightness scaling experiment a line might be drawn with one end labeled white and the other end labeled black. When the observers are presented with a medium gray that is perceptually halfway between white and black, they would make a mark on the line at the midpoint. If the sample was closer to white than black, they would make a mark at the appropriate physical location along the line, closer to the end labeled white. The interval scale is taken to be the mean location on the graphical scale for each stimulus. This technique relies on the well-established fact that the perception of length over short distances is linear with respect to physically measured length.

Category scaling is a popular technique for deriving ordinal or interval scales for large numbers of stimuli. An observer is asked to separate a large number of samples into various categories. When there are several observers, the number of times each particular sample is placed in a category is recorded. For this to be an effective scaling method, the samples need to be similar enough so that they are not always placed in the distinct categories by different observers or by the same observer on different occasions. Interval scales may be obtained by this method by assuming that the perceptual magnitudes are normally distributed and by making use of the standard normal distribution according to the law of categorical judgments (Torgerson, 1954).

A *paired comparison experiment* can be performed when the number of different stimuli is smaller. With this method, all samples are presented to the observer in all the possible pairwise combinations, usually one pair at a time (sometimes with a third stimulus as a reference). The proportion of times a particular sample is judged greater in some attribute than each other sample is calculated and recorded. Interval scales can be obtained from such data by applying the law of comparative judgments (Thurstone, 1927). Thurstone's law of comparative judgments and its extensions can be usefully applied to ordinal data (such as paired comparisons and category scaling) to derive meaningful interval scales. The perceptual magnitudes of the stimuli are normally distributed on the resulting scales. Thus, if it is safe to assume that the perceptual magnitudes are normally distributed on the true perceptual scale, these analyses derive the desired scale. They also allow useful evaluation of the statistical significance of differences between stimuli, since the power of the normal distribution can be utilized. Torgerson (1958) and Bartleson and Grum (1984) describe these and other related analyses in detail.

Partition scaling provides a rather direct method for deriving interval scales. Commonly, intervals are equated through bisection. The observer is given two different samples (A and B) and asked to select a third such that the difference between it and A appears equal to the difference between it and B. A full interval scale may be obtained by successive bisections.

Magnitude estimation or production can derive a ratio scale directly. In such an experiment, the observer is asked to assign numbers to the stimuli according to the magnitude of the perception. Alternatively, observers are given a number and asked to produce a stimulus that has that perceptual magnitude. This is one of the few techniques that can be used to generate a ratio scale. It can also be used to generate data for multidimensional scaling by asking observers to scale the differences between pairs of stimuli.

Ratio estimation or production is a slightly more complicated technique than magnitude estimation or production. Here, the observer is asked for judgments in one of two ways: (1) select or produce a sample that bears some prescribed ratio to a standard or (2) given two or more samples, state the apparent ratios among them. A typical experiment is to give the observers a sample and ask them to find, select, or produce a test sample that is one-half or twice the standard in some attribute. For most practical visual work, this method is too difficult to use either because of the sample preparation or the judgment by the observers. However, it can be also used to generate a ratio scale.

2.7   Multidimensional Scaling

Multidimensional scaling (MDS) is similar to one-dimensional scaling, except it does not require the assumption that the attribute to be scaled be one-dimensional. The dimensionality is found as part of the analysis. In MDS, the data are placed in an interval or ordinal scale showing the similarities or dissimilarities between each stimulus. The resulting output is a multidimensional geometric configuration of the perceptual relationships between the stimuli, as on a map.

The dissimilarity data required for MDS can be obtained conveniently by using paired comparison and triadic combination experiments. In a paired comparison experiment, all samples in all possible pairs are presented and the observer is asked to make a magnitude estimation of the perceived difference between each pair. The resulting estimates for each pairwise combination can then be subjected to MDS analyses. In a triadic combination experiment, observers are presented with each possible combination of the stimuli, taken three at a time. They are next asked to judge which two of the stimuli in each triad are most similar to one another and which two are most different. The data can then be converted into the frequencies of times that each pair is judged most similar or most different. These frequency data can be combined into either a similarity or dissimilarity matrix for use in MDS analyses.

MDS analysis techniques take such similarity or dissimilarity data as input and produce a multidimensional configuration of points representing the relationships and dimensionality of the data. Such techniques must be used when either the perception in question is multidimensional (such as color—hue, lightness, and chroma) or the physical variation in the stimuli is multidimensional. Kruskal and Wish (1978) provide details of these techniques. There are several issues with respect to MDS analyses. There are two classes of MDS techniques: metric, which requires interval data, and nonmetric, which requires only ordinal data. Both classes result in interval-scale output. Various MDS software packages process input data according to specific assumptions regarding the input data, treatment of individual cases, goodness-of-fit metrics (stress), distance metrics (e.g. Euclidean or city-block), and so on. Several commercial statistical software packages provide MDS capabilities.

A classic example of MDS analysis is the construction of a map from data representing the distances between cities (Kruskal and Wish, 1978). In this example, a map of the United States is constructed from the dissimilarity matrix of distances between eight cities gathered from a road atlas, as illustrated in Table 2-1.

Table 2-1.   Dissimilarity matrix consisting of distances between U.S. cities.

	ATL	*BOS*	*CHI*	*DAL*	*DEN*	*LA*	*SEA*	*NYC*
ATL	—	—	—	—	—	—	—	—
BOS	1,037	—	—	—	—	—	—	—
CHI	674	963	—	—	—	—	—	—
DAL	795	1,748	917	—	—	—	—	—
DEN	1,398	1,949	996	781	—	—	—	—
LA	2,182	2,979	2,054	1,387	1,059	—	—	—
SEA	2,618	2,976	2,013	2,078	1,307	1,131	—	—
NYC	841	206	802	1,552	1,771	2,786	2,815	—

Table 2-2.   Output two-dimensional coordinates for each city in the United
States in the MDS example.

City	Dimension 1	Dimension 2
Atlanta	−0.63	0.40
Boston	−1.19	−0.31
Chicago	−0.36	−0.15
Dallas	0.07	0.55
Denver	0.48	0.00
Los Angeles	1.30	0.36
Seattle	1.37	−0.66
New York City	−1.04	−0.21

The dissimilarity data are then analyzed via MDS. Stress (RMS error) is
used as a measure of goodness-of-fit in order to determine the dimensionality
of the data. In this example, the stress of a one-dimensional fit is about 0.12,
while the stress in two or more dimensions is essentially zero. This indicates
that a two-dimensional fit, as expected, is appropriate. The results output include the coordinates in each of the two dimensions for each of the cities, as
listed in Table 2-2.

Plotting the coordinates of each city in the two output dimensions results
in a familiar map of the United States, as shown in Figure 2-2. However, note
that dimension 1 goes from east to west and dimension 2 goes from north to
south, thus resulting in a map that has the axes reversed from a traditional
map. This illustrates a feature of MDS, namely, that the definition of the
output dimensions requires post hoc analysis by the experimenter. MDS experiments can be used to explore the dimensionality and structure of color-
appearance spaces (e.g., Indow, 1988).

2.8   Design of Psychophysical Experiments

The previous sections provide an overview of some of the techniques used to
derive psychophysical thresholds and scales. However, there are many more issues that arise in the design of psychophysical experiments that have a significant impact on the experimental results, particularly where color appearance is
concerned. Many of these experimental factors are the key variables that have
illustrated the need to extend basic colorimetry with the development of color
appearance models. A complete description of all of the variables involved in
visual experiments could easily fill several books, and many of the critical issues for color-appearance phenomena are described in more detail in later
chapters. At this point, a simple listing of some of the factors that require con-

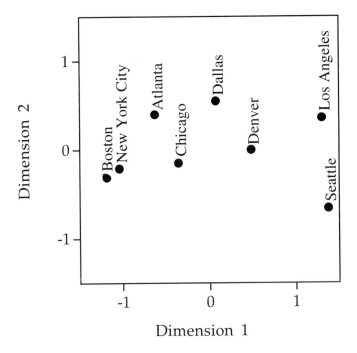

Figure 2-2. The output of a multidimensional scaling (MDS) program used to generate a map of the United States from input data on the proximities of cities.

sideration should be sufficient to bring these issues to light. Important factors in visual experiments include the following:

Adaptation state
Background conditions
Cognitive factors
Complexity of observer task
Context
Control and history of eye
  movements
Controls
Duration of observation
  sessions
Feedback
Illumination color
Illumination geometry
Illumination level
Image content
Instructions

Number of images
Number of observation sessions
Number of observers
Observer acuity
Observer age
Observer experience
Observer motivation
Range effects
Regression effects
Repetition rate
Rewards
Screening for color-vision
  deficiencies
Statistical significance of results
Surround conditions

All of these factors, and probably many more, can profoundly affect psychophysical results and should be carefully specified and/or controlled. Such issues need to be addressed by both those performing experiments and those trying to interpret and utilize the results for various applications.

2.9   Importance of Psychophysics in Color Appearance Modeling

A fundamental understanding of the processes involved in psychophysical experiments provides useful insight for understanding the need for and development and evaluation of color appearance models. Psychophysical experiments provided much of the information reviewed in Chapter 1 on the human visual system. Psychophysics is the basis of colorimetry presented in Chapter 3. The results of psychophysical experiments are also presented in Chapters 6, 8, and 15, which deal with, respectively, color-appearance phenomena, chromatic adaptation, and the testing of color appearance models. Simply put, without extensive psychophysical experimentation, none of the information required to create and use color appearance models would exist.

CHAPTER 3

Colorimetry

3.1   Basic and Advanced Colorimetry
3.2   Why Is Color?
3.3   Light Sources and Illuminants
3.4   Colored Materials
3.5   The Human Visual Response
3.6   Tristimulus Values and Color-matching Functions
3.7   Chromaticity Diagrams
3.8   CIE Color Spaces
3.9   Color-difference Specification
3.10 The Next Step

COLORIMETRY IS THE fundamental underpinning of color-appearance specification. This chapter reviews the well-established practice of colorimetry according to the CIE system established in 1931. This system allows the specification of color matches for an average observer. It has, amazingly, withstood an onslaught of technological pressures and remained a useful international standard for over 65 years (Wright, 1981b; Fairchild, 1993b).

CIE colorimetry, however, provides only the starting point. Color appearance models enhance this system in an effort to predict the actual appearances of stimuli in various viewing conditions rather than simply predicting whether two stimuli will match. This chapter provides a general review of the concepts of colorimetry so as to set the stage for development of various color appearance models. It is not intended to be a complete reference on colorimetry, since there are a number of excellent texts on the subject available. For introductions to colorimetry, see Hunt (1991a), Berger-Schunn (1994), and Hunter and Harold (1987). The precise definition of colorimetry is in the primary reference, CIE Publication 15.2 (CIE, 1986). For complete details in an encyclopedic reference volume, see the classic book by Wyszecki and Stiles (1982), *Color Science*.

3.1   Basic and Advanced Colorimetry

Colorimetry is the measurement of color. Wyszecki (1973) described an important distinction between basic colorimetry and advanced colorimetry (see also Wyszecki, 1986). This distinction is the basis of this book and warrants attention. It is perhaps most enlightening to quote Wyszecki's exact words describing basic colorimetry:

> Colorimetry, in its strict sense, is a tool used to making a prediction on whether two lights (visual stimuli) of different spectral power distributions will match in colour for certain given conditions of observation. The prediction is made by determining the tristimulus values of the two visual stimuli. If the tristimulus values of a stimulus are identical to those of the other stimulus, a colour match will be observed by an average observer with normal color vision.

Wyszecki (1973) went on to describe advanced colorimetry:

Colorimetry in its broader sense includes methods of assessing the appearance of colour stimuli presented to the observer in complicated surroundings as they may occur in everyday life. This is considered the ultimate goal of colorimetry, but because of its enormous complexity, this goal is far from being reached. On the other hand, certain more restricted aspects of the overall problem of predicting colour appearance of stimuli seem somewhat less elusive. The outstanding examples are the measurement of colour differences, whiteness, and chromatic adaptation. Though these problems are still essentially unresolved, the developments in these areas are of considerable interest and practical importance.

This chapter describes the well-established techniques of basic colorimetry that form the foundation for color appearance modeling. It also describes some of the widely used methods for color-difference measurement, one of the first objectives of advanced colorimetry. Wyszecki's distinction between basic and advanced colorimetry serves to highlight the purpose of this book—an account of the research and modeling aimed at extending basic colorimetry toward the ultimate goals of advanced colorimetry.

3.2   Why Is Color?

To begin to understand the measurement of color, one must first consider the nature of color. Figure 3-1 illustrates the answer to the question, "Why is color?" "Why" is a more appropriate question than the more typical "what." This is because color is not a simple thing that can be easily described to someone who has never experienced it. Color cannot even be defined without resort to examples (see Chapter 4). It is an attribute of visual sensation. The color appearance of objects depends on three components, which are shown in the triangle of color in Figure 3-1.

The first component is a source of visible electromagnetic energy necessary to initiate the sensory process of vision. The second component is an object, whose chemical properties modulate the electromagnetic energy. The third component is the human visual system. The modulated energy is imaged by the eye, detected by photoreceptors, and processed by the neural mechanisms of the human visual system to produce the perception of color. Note that the light source and visual system are also linked in Figure 3-1. This is done to indicate the influence that the light source itself has on color appearance through chromatic adaptation, and so on.

Since all three components of the triangle of color in Figure 3-1 are required to produce color, they must also be quantified in order to produce a reliable system of physical colorimetry. Light sources are quantified through their spectral power distribution and standardized as illuminants. Material objects are specified by the geometric and spectral distribution of the energy they reflect or transmit. The human visual system is quantified through its color-matching properties that represent the first stage response (cone absorption) in the system. Thus colorimetry, as a combination of all of these areas, draws upon techniques and results from the fields of physics, chemistry, psychophysics, physiology, and psychology. The following three sections discuss in detail the three components of the triangle of color.

3.3   Light Sources and Illuminants

The first component of the triangle of color in Figure 3-1 is the light source. Light sources provide the electromagnetic energy required to initiate visual responses. The specification of the color properties of light sources is performed by two methods for basic colorimetry: measurement and standardization. The distinction between these two methods is clarified in the definition of light sources and illuminants.

A *light source* is an actual physical emitter of visible energy. Examples of light sources are incandescent lightbulbs, the sky at any given moment, and fluorescent tubes. An *illuminant* is simply a standardized table of values that rep-

Figure 3-1.   The triangle of color. Color exists due to the interaction of three components: light sources, objects, and the human visual system.

resent a spectral power distribution typical of some particular light source. CIE illuminants A, D65, and F2 are standardized representations of typical incandescent, daylight, and fluorescent sources, respectively. Some illuminants have corresponding sources that are physical embodiments of the standardized spectral power distributions. For example, CIE source A is a particular type of tungsten source that produces the relative spectral power distribution of CIE illuminant A. Other illuminants do not have corresponding sources. For example, CIE illuminant D65 is a statistical representation of an average daylight with a correlated color temperature of approximately 6,500 K. Thus there is no CIE source D65 capable of producing the illuminant D65 spectral power distribution. The importance of distinguishing between light sources and illuminants in color-appearance specification is discussed in Table 7-1. There are likely to be significant differences between the spectral power distributions of a CIE illuminant and a light source designed to simulate it. So the actual spectral power distribution of the light source must be used in colorimetric calculations of stimuli used in color-appearance specification.

Role of Spectroradiometry

The measurement of the spectral power distributions of light sources is the realm of spectroradiometry. *Spectroradiometry* is the measurement of radiometric quantities as a function of wavelength. In color measurement, the wavelength region of interest encompasses electromagnetic energy of wavelengths from approximately 400 nm (violet) to 700 nm (red). A variety of radiometric quantities can be used to specify the properties of a light source. Of particular interest in color-appearance measurement are irradiance and radiance. Both are measurements of the power of light sources with basic units of watts.

Irradiance is the radiant power per unit area incident onto a surface and has units of watts per square meter (W/m^2). Spectral irradiance adds the wavelength dependency and has units of W/m^{2}nm, sometimes expressed as W/m^3. *Radiance* differs from irradiance in that it is a measure of the power emitted from a source (or surface), rather than incident upon a surface, per unit area per unit solid angle with units of watts per square meter per steradian (W/m^{2}sr). Spectral radiance includes the wavelength dependency having units of W/m^{2}sr nm or W/m^{3}sr.

Radiance has interesting properties in that it is preserved through optical systems (neglecting absorption) and is independent of distance. Thus the human visual system responds commensurably to radiance. This makes radiance a key measurement in color-appearance specification. The retina itself responds commensurably to the irradiance incident upon it, but in combination with the optics of the eyeball, retinal irradiance is proportional to the radiance of a surface. This can be demonstrated by viewing an illuminated surface from various distances and observing that the perceived brightness does not change (consistent with the radiance of the surface). The irradiance at the eye from a given surface falls off with the square of distance from the surface. Radiance does not fall off in this fashion. This is because the decrease in power incident on the pupil is directly canceled by a proportional decrease in the solid angle subtended by the pupil with respect to the surface in question. The spectral radiance, $L(\lambda)$, of a surface with a spectral reflectance factor of $R(\lambda)$ can be calculated from the spectral irradiance, $E(\lambda)$, falling upon the surface. This is done by using Equation 3-1, with the assumption that the surface is a Lambertian diffuser (i.e., equal radiance in all directions):

$$L(\lambda) = \frac{R(\lambda)E(\lambda)}{\pi}$$

(3-1)

A spectral power distribution (e.g., see Figure 3-2) is simply a plot, or table, of a radiometric quantity as a function of wavelength. Since the overall power levels of light sources can vary over many orders of magnitude, spectral power distributions are often normalized to facilitate comparisons of color properties. The traditional approach is to normalize a spectral power distribution such that it has a value of 100 (or sometimes 1.0) at a wavelength of 560 nm (arbitrarily chosen as near the center of the visible spectrum). Such normalized spectral power distributions are called *relative spectral power distributions* and are unitless.

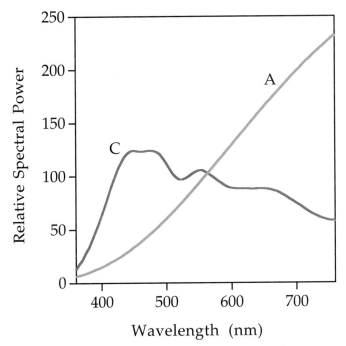

Figure 3-2. Relative spectral power distributions of CIE illuminants A and C.

Black-body Radiators

Another important radiometric quantity is the color temperature of a light source. A special type of theoretical light source, called a *black-body radiator,* or *Planckian radiator,* emits energy due only to thermal excitation and is a perfect emitter of energy.

The energy emitted by a black body increases in quantity and shifts toward shorter wavelengths as the temperature of the black body increases. The spectral power distribution of a black-body radiator can be specified using Planck's equation as a function of a single variable, absolute temperature (in Kelvins). Thus, if the absolute temperature of a black body is known, so is its spectral power distribution. The temperature of a black body is called its *color temperature,* since that uniquely specifies the color of the source. Black-body radiators seldom exist outside specialized laboratories, so color temperature is not a generally useful quantity.

A second quantity, *correlated color temperature* (CCT), is more generally useful. A light source need not be a black-body radiator in order to be assigned a CCT. The CCT of a light source is simply the color temperature of a black-body radiator that has most nearly the same color as the source in question. As examples, an incandescent source might have a CCT of 2,800 K; a typical fluorescent tube, 4,000 K; an average daylight, 6,500 K; and the white-point of a computer graphics display, 9,300 K. As the correlated color temperature of a source increases, it becomes more blue, or less red.

CIE Illuminants

The CIE has established a number of spectral power distributions as CIE illuminants for colorimetry. These include CIE illuminants A, C, D65, D50, F2, F8, and F11:

- CIE illuminant A represents a Planckian radiator with a color temperature of 2,856 K. It is used for colorimetric calculations when incandescent illumination is of interest.
- CIE illuminant C is the spectral power distribution of illuminant A as modified by particular liquid filters defined by the CIE. It represents a daylight simulator with a CCT of 6,774 K.
- CIE illuminants D65 and D50 are part of the CIE D-series of illuminants that have been statistically defined based upon a large number of measurements of real daylight. Illuminant D65 represents an average daylight with a CCT of 6,504 K, and D50 represents an average daylight with a CCT of 5,003 K. D65 is commonly used in colorimetric applications, while D50 is often used in graphic arts applications. CIE D illuminants with other correlated color temperatures can be easily obtained.

- CIE F illuminants (12 in all) represent typical spectral power distributions for various types of fluorescent sources. CIE illuminant F2 represents cool-white fluorescent with a CCT of 4,230 K. Illuminant F8 represents a fluorescent D50 simulator with a CCT of 5,000 K, and illuminant F11 represents a triband fluorescent source with a CCT of 4,000 K. Triband fluorescent sources are popular because of their efficiency, efficacy, and pleasing color-rendering properties.
- The equal-energy illuminant (sometimes called illuminant E) is often of mathematical utility. It is defined with a relative spectral power of 100.0 at all wavelengths.

Table 3-1 includes the spectral power distributions and useful colorimetric data for these CIE illuminants. The relative spectral power distributions of these illuminants are plotted in Figures 3-2 through 3-4.

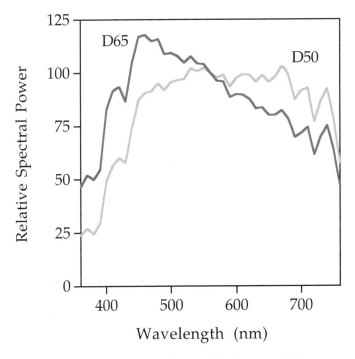

Figure 3-3.   Relative spectral power distributions of CIE illuminants D50 and D65.

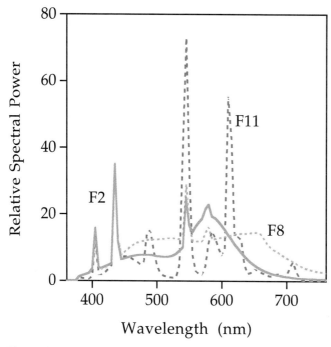

Figure 3-4.   Relative spectral power distributions of CIE illuminants F2, F8, and F11.

Table 3-1.   Relative spectral power distributions and colorimetric data for some example CIE illuminants. Colorimetric data are for the CIE 1931 standard colorimetric observer (2°).

Wavelength (nm)	A	C	D65	D50	F2	F8	F11
360	6.14	12.90	46.64	23.94	0.00	0.00	0.00
365	6.95	17.20	49.36	25.45	0.00	0.00	0.00
370	7.82	21.40	52.09	26.96	0.00	0.00	0.00
375	8.77	27.50	51.03	25.72	0.00	0.00	0.00
380	9.80	33.00	49.98	24.49	1.18	1.21	0.91
385	10.90	39.92	52.31	27.18	1.48	1.50	0.63
390	12.09	47.40	54.65	29.87	1.84	1.81	0.46
395	13.35	55.17	68.70	39.59	2.15	2.13	0.37
400	14.71	63.30	82.75	49.31	3.44	3.17	1.29
405	16.15	71.81	87.12	52.91	15.69	13.08	12.68

Table 3-1 (continued)

Wavelength (nm)	A	C	D65	D50	F2	F8	F11
410	17.68	80.60	91.49	56.51	3.85	3.83	1.59
415	19.29	89.53	92.46	58.27	3.74	3.45	1.79
420	21.00	98.10	93.43	60.03	4.19	3.86	2.46
425	22.79	105.80	90.06	58.93	4.62	4.42	3.33
430	24.67	112.40	86.68	57.82	5.06	5.09	4.49
435	26.64	117.75	95.77	66.32	34.98	34.10	33.94
440	28.70	121.50	104.87	74.82	11.81	12.42	12.13
445	30.85	123.45	110.94	81.04	6.27	7.68	6.95
450	33.09	124.00	117.01	87.25	6.63	8.60	7.19
455	35.41	123.60	117.41	88.93	6.93	9.46	7.12
460	37.81	123.10	117.81	90.61	7.19	10.24	6.72
465	40.30	123.30	116.34	90.99	7.40	10.84	6.13
470	42.87	123.80	114.86	91.37	7.54	11.33	5.46
475	45.52	124.09	115.39	93.24	7.62	11.71	4.79
480	48.24	123.90	115.92	95.11	7.65	11.98	5.66
485	51.04	122.92	112.37	93.54	7.62	12.17	14.29
490	53.91	120.70	108.81	91.96	7.62	12.28	14.96
495	56.85	116.90	109.08	93.84	7.45	12.32	8.97
500	59.86	112.10	109.35	95.72	7.28	12.35	4.72
505	62.93	106.98	108.58	96.17	7.15	12.44	2.33
510	66.06	102.30	107.80	96.61	7.05	12.55	1.47
515	69.25	98.81	106.30	96.87	7.04	12.68	1.10
520	72.50	96.90	104.79	97.13	7.16	12.77	0.89
525	75.79	96.78	106.24	99.61	7.47	12.72	0.83
530	79.13	98.00	107.69	102.10	8.04	12.60	1.18
535	82.52	99.94	106.05	101.43	8.88	12.43	4.90
540	85.95	102.10	104.41	100.75	10.01	12.22	39.59
545	89.41	103.95	104.23	101.54	24.88	28.96	72.84
550	92.91	105.20	104.05	102.32	16.64	16.51	32.61
555	96.44	105.67	102.02	101.16	14.59	11.79	7.52
560	100.00	104.11	100.00	100.00	16.16	11.76	2.83
565	103.58	102.30	98.17	98.87	17.56	11.77	1.96
570	107.18	100.15	96.33	97.74	18.62	11.84	1.67
575	110.80	97.80	96.06	98.33	21.47	14.61	4.43
580	114.44	95.43	95.79	98.92	22.79	16.11	11.28

Table 3-1 (continued)

Wavelength (nm)	A	C	D65	D50	F2	F8	F11
585	118.08	93.20	92.24	96.21	19.29	12.34	14.76
590	121.73	91.22	88.69	93.50	18.66	12.53	12.73
595	125.39	89.70	89.35	95.59	17.73	12.72	9.74
600	129.04	88.83	90.01	97.69	16.54	12.92	7.33
605	132.70	88.40	89.80	98.48	15.21	13.12	9.72
610	136.35	88.19	89.60	99.27	13.80	13.34	55.27
615	139.99	88.10	88.65	99.16	12.36	13.61	42.58
620	143.62	88.06	87.70	99.04	10.95	13.87	13.18
625	147.24	88.00	85.49	97.38	9.65	14.07	13.16
630	150.84	87.86	83.29	95.72	8.40	14.20	12.26
635	154.42	87.80	83.49	97.29	7.32	14.16	5.11
640	157.98	87.99	83.70	98.86	6.31	14.13	2.07
645	161.52	88.20	81.86	97.26	5.43	14.34	2.34
650	165.03	88.20	80.03	95.67	4.68	14.50	3.58
655	168.51	87.90	80.12	96.93	4.02	14.46	3.01
660	171.96	87.22	80.21	98.19	3.45	14.00	2.48
665	175.38	86.30	81.25	100.60	2.96	12.58	2.14
670	178.77	85.30	82.28	103.00	2.55	10.99	1.54
675	182.12	84.00	80.28	101.70	2.19	9.98	1.33
680	185.43	82.21	78.28	99.13	1.89	9.22	1.46
685	188.70	80.20	74.00	93.26	1.64	8.62	1.94
690	191.93	78.24	69.72	87.38	1.53	8.07	2.00
695	195.12	76.30	70.67	89.49	1.27	7.39	1.20
700	198.26	74.36	71.61	91.60	1.10	6.71	1.35
705	201.36	72.40	72.98	92.25	0.99	6.16	4.10
710	204.41	70.40	74.35	92.89	0.88	5.63	5.58
715	207.41	68.30	67.98	84.87	0.76	5.03	2.51
720	210.37	66.30	61.60	76.85	0.68	4.46	0.57
725	213.27	64.40	65.74	81.68	0.61	4.02	0.27
730	216.12	62.80	69.89	86.51	0.56	3.66	0.23
735	218.92	61.50	72.49	89.55	0.54	3.36	0.21
740	221.67	60.20	75.09	92.58	0.51	3.09	0.24
745	224.36	59.20	69.34	85.40	0.47	2.85	0.24
750	227.00	58.50	63.59	78.23	0.47	2.65	0.20
755	229.59	58.10	55.01	67.96	0.43	2.51	0.24
760	232.12	58.00	46.42	57.69	0.46	2.37	0.32

Table 3-1 (continued)

Wavelength (nm)	A	C	D65	D50	F2	F8	F11
X	109.85	98.07	95.05	96.42	99.20	96.43	100.96
Y	100.0	100.0	100.0	100.0	100.0	100.0	100.0
Z	35.58	118.23	108.88	82.49	67.40	82.46	64.37
x	0.4476	0.3101	0.3127	0.3457	0.3721	0.3458	0.3805
y	0.4074	0.3162	0.3290	0.3585	0.3751	0.3586	0.3769
CCT	2,856 K	6,800 K	6,504 K	5,003 K	4,230 K	5,000 K	4,000 K

3.4   Colored Materials

Once the light source or illuminant is specified, the next step in the colorimetry of material objects is the characterization of their interaction with visible radiant energy, as illustrated in the triangle of color in Figure 3-1. The interaction of radiant energy with materials obeys the law of conservation of energy. There are only three fates that can befall radiant energy incident on an object—absorption, reflection, or transmission. The amounts of absorbed, reflected, and transmitted radiant power must sum to the incident radiant energy at each wavelength. This is illustrated in Equation 3-2, where $\Phi(\lambda)$ is used as a generic term for incident radiant flux, $R(\lambda)$ is the reflected flux, $T(\lambda)$ is the transmitted flux, and $A(\lambda)$ is the absorbed flux:

$$\phi(\lambda) = R(\lambda) + T(\lambda) + A(\lambda) \tag{3-2}$$

Reflection, transmission, and absorption are the phenomena that take place when light interacts with matter. Reflectance, transmittance, and absorptance are the quantities measured to describe these phenomena. Since these quantities always must sum to the incident flux, they are typically measured in relative terms as percentages of the incident flux rather than as absolute radiometric quantities. Thus *reflectance* can be defined as the ratio of the reflected energy to the incident energy, *transmittance* as the ratio of transmitted energy to incident energy, and *absorptance* as the ratio of absorbed energy to incident energy.

Note that all of these quantities are ratio measurements. Ratio measurements are the subject of *spectrophotometry*, which is the measurement of ratios of radiometric quantities. Spectrophotometric quantities are expressed as either percentages (0–100%) or as factors (0.0–1.0). Figure 3-5 illustrates the spectral reflectance, transmittance, and absorptance of a red translucent object. Note that since the three quantities sum to 100%, it is typically unnecessary to measure all three. Generally, either reflectance or transmittance is of particular interest in a given application.

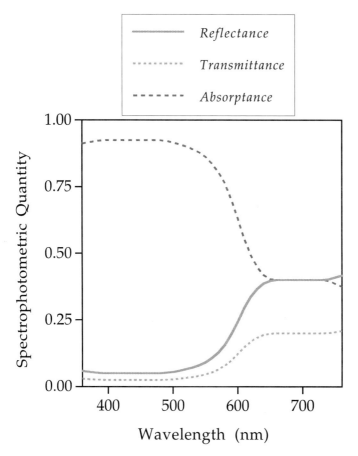

	Reflectance
	Transmittance
	Absorptance

Figure 3-5. Spectral absorptance, reflectance, and transmittance of a red translucent plastic material.

Unfortunately (for colorimetrists), the interaction of radiant energy with objects is not just a simple spectral phenomenon. The reflectance or transmittance of an object is a function not just of wavelength, but also of the illumination and viewing geometry. Such differences can be illustrated by the phenomenon of gloss. Imagine matte, semigloss, and gloss photographic paper or paint. The various gloss characteristics of these materials can be ascribed to the geometric distribution of the specular reflectance from the surface of the object. This is just one geometric appearance effect. Many others exist, such as interesting changes in the color of automotive finishes with illumination and viewing geometry (e.g., metallic, pearlescent, and other "effect" coatings). For such effects to be fully quantified, complete bidirectional reflectance (or transmittance) distribution functions (BRDFs) must be obtained for each possible combination of illumination angle, viewing angle, and wavelength. Measurement of such functions is prohibitively difficult and expensive. It also produces massive quantities of data that are difficult to meaningfully utilize. To avoid this explosion of colorimetric data, the CIE established a small number of standard illumination and viewing geometries for colorimetry.

CIE Standard Illumination and Viewing Geometries

The CIE has defined four standard illumination and viewing geometries for spectrophotometric reflectance measurements. These come as two pairs of optically reversible geometries:

1. Diffuse/normal (d/0) and normal/diffuse (0/d)
2. 45/normal (45/0) and normal/45 (0/45)

The designations indicate the illumination geometry before the slash and the viewing geometry following the slash.

DIFFUSE/NORMAL AND NORMAL/DIFFUSE

In the diffuse/normal geometry, the sample is illuminated from all angles using an integrating sphere and viewed at an angle near the normal to the surface. In the normal/diffuse geometry, the sample is illuminated from an angle near to its normal and the reflected energy is collected from all angles using an integrating sphere. These two geometries are optical reverses of one another and therefore produce the same measurement results (assuming all other instrumental variables are constant).

The measurements made are of total reflectance. In many instruments, an area of the integrating sphere with a black trap can be replaced such that the specular component of reflection is excluded and only diffuse reflectance is measured. The area of the integrating sphere corresponds to the angle of specular (regular) reflection of the illumination in a 0/d geometry or to the angle from which specular reflection would be detected in a d/0 geometry. Such measurements are called "specular component excluded" measurements. This contrasts with the "specular component included" measurements that are made when the entire sphere is intact.

45/NORMAL AND NORMAL/45

The second pair of geometries is the 45/normal (45/0) and normal/45 (0/45) measurement configurations. In a 45/0 geometry, the sample is illuminated with one or more beams of light incident at an angle of 45° from the normal and measurements are made along the normal. In the 0/45 geometry, the sample is illuminated normal to its surface and measurements are made using one or more beams at a 45° angle to the normal.

Like the first pair, these two geometries are optical reverses of one another and produce identical results when all other instrumental variables are equal.

Use of the 45/0 and 0/45 measurement geometries ensures that all components of gloss are excluded from the measurements. Thus these geometries are typically used in applications in which it is necessary to compare the colors of materials having various levels of gloss (e.g., graphic arts and photography). It is critical to note the instrumental geometry used whenever reporting colorimetric data for materials.

The definition of reflectance as the ratio of reflected energy to incident energy is perfectly appropriate for measurements of total reflectance (d/0 or 0/d). However, for bidirectional reflectance measurements (45/0 and 0/45), the ratio of reflected energy to incident energy is exceedingly small, since only a small range of angles of the distribution of reflected energy is detected. Thus more practically useful values for any type of measurement geometry are produced by making reflectance factor measurements relative to a perfect reflecting diffuser.

A *perfect reflecting diffuser* (PRD) is a theoretical material that is both a perfect reflector (100% reflectance) and perfectly Lambertian (radiance equal in all directions). Hence, measurements of reflectance factor are defined as the ratio of the energy reflected by the sample to the energy that would be reflected by a PRD illuminated and viewed in the identical geometry. For the integrating sphere geometries and measuring total reflectance, this definition of reflectance factor is identical to the definition of reflectance. For the bidirectional geometries, measurement of the reflectance factor relative to the PRD results in a zero-to-one scale similar to that obtained for total reflectance measurements. Since PRDs are not physically available, reference standards that are calibrated relative to the theoretical aim are provided by national standardizing laboratories (such as NIST, the National Institute for Standards and Technology, in the United States) and instrument manufacturers.

Fluorescence

One last topic of importance in the colorimetric analysis of materials is fluorescence. Fluorescent materials absorb energy in one region of wavelengths and then emit this energy in a region of other, longer wavelengths. For example, a fluorescent orange material might absorb blue energy and emit it as orange energy.

Fluorescent materials obey the law of conservation of energy as stated in Equation 3-2. However, their behavior differs from other materials, in that some of the absorbed energy is emitted at (normally) longer wavelengths.

A full treatment of the color measurement of fluorescent materials is complex and beyond the scope of this book. In general, however, a fluorescent material is characterized by its total radiance factor, which is the sum of the reflected and emitted energy at each wavelength relative to the energy that would be reflected by a PRD. This definition allows for total radiance factors greater than 1.0, which is often the case. It is important to note that the total radiance factor will depend on the light source used in the measuring instrument, since the amount of emitted energy is directly proportional to the amount of absorbed energy in the excitation wavelengths. Spectrophotometric measurements of reflectance or transmittance of nonfluorescent materials are insensitive to the light source in the instrument, since its characteristics are normalized in the ratio calculations. This important difference highlights the major difficulty in measuring fluorescent materials. Unfortunately, many man-made materials (such as paper and inks) are fluorescent and thus significantly more difficult to measure accurately than their nonfluorescent counterparts.

3.5   The Human Visual Response

The third component in the triangle of color in Figure 3-1 is the human visual response. Measurement or standardization of light sources and materials provides the necessary physical information for colorimetry. What remains is a quantitative technique to predict the response of the human visual system.

In accordance with Wyszecki's (1973) definition of basic colorimetry, quantification of the human visual response focuses on the earliest level of vision—absorption of energy in the cone photoreceptors—through the psychophysics of color matching. The ability to predict when two stimuli match for an average observer—the basis of colorimetry—provides great utility in a variety of applications. While such a system does not specify color appearance, it does provide the basis of color-appearance specification and allows the prediction of matches for various applications and the tools required to set up tolerances on matches necessary for industry.

The properties of human color matching are defined by the spectral responsivities of the three cone types. This is because once the energy is absorbed by the three cone types, the spectral origin of the signals is lost. Further, if the signals from the three cone types are equal for two stimuli, they must match in color when seen in the same conditions, since no additional information is introduced within the visual system to distinguish them.

Thus, if the spectral responsivities of the three cone types are known, two stimuli, denoted by their spectral power distributions $\Phi_1(\lambda)$ and $\Phi_2(\lambda)$, will match in color if the product of their spectral power distributions and each of the three cone responsivities, $L(\lambda)$, $M(\lambda)$, and $S(\lambda)$, integrated over wavelength, are equal. This equality for a visual match is illustrated in Equations 3-3 through 3-5. Two stimuli match if all three of the equalities in these equations hold true.

$$\int_\lambda \phi_1(\lambda)L(\lambda)d\lambda = \int_\lambda \phi_2(\lambda)L(\lambda)d\lambda$$

(3-3)

$$\int_\lambda \phi_1(\lambda)M(\lambda)d\lambda = \int_\lambda \phi_2(\lambda)M(\lambda)d\lambda$$

(3-4)

$$\int_\lambda \phi_1(\lambda)S(\lambda)d\lambda = \int_\lambda \phi_2(\lambda)S(\lambda)d\lambda$$

(3-5)

These equations illustrate the definition of metamerism. Since only the three integrals need be equal for a color match, it is not necessary for the spectral power distributions of the two stimuli to be equal for every wavelength. The cone spectral responsivities, described in Chapter 1, are quite well known today. With such knowledge, one can define a system of basic colorimetry nearly as simply as defining Equations 3-3 through 3-5. However, reasonably accurate knowledge of the cone spectral responsivities is a rather recent scientific development. The need for colorimetry predates this knowledge by several decades. Thus the CIE, in establishing the 1931 system of colorimetry, needed to take a less direct approach.

The System of Photometry

For one to understand the indirect nature of the CIE system of colorimetry, it is useful to first explore the system of photometry. Established in 1924, the system of photometry was intended to develop a spectral weighting function that could be used to describe the perception of brightness matches. (More correctly, the system describes the results of flicker photometry experiments rather than heterochromatic brightness matches, as described further in Chapter 6.) In 1924, the CIE spectral luminous efficiency function, $V(\lambda)$, was established for photopic vision. This function, plotted in Figure 3-6 and set out in Table 3-2, indicates that the visual system is more sensitive (with respect to the perception of brightness) to wavelengths in the middle of the visual spectrum and becomes less and less sensitive to wavelengths near the extremes of the visual spectrum. The $V(\lambda)$ function is used as a spectral weighting function to convert radiometric quantities into photometric quantities via spectral integration, as shown in Equation 3-6:

$$\phi_v = \int_\lambda \phi(\lambda)V(\lambda)d\lambda$$

(3-6)

The term Φ_v in the equation refers to the appropriate photometric quantity defined by the radiometric quantity, $\Phi(\lambda)$, used in the calculation. For example, the radiometric quantities of irradiance, radiance, and reflectance factor are used to derive the photometric quantities of illuminance (lumens/m^2 or lux), lu-

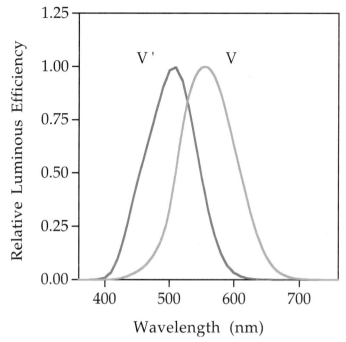

Figure 3-6.   CIE scotopic, V'(λ), and photopic, V(λ), luminous efficiency functions.

minance (cd/m^2), or luminance factor (unitless). All of the optical properties and relationships for irradiance and radiance are preserved for illuminance and luminance. To convert irradiance or radiance to illuminance or luminance, a normalization constant of 683 lumens/watt is required to preserve the appropriate units. For the calculation of the luminance factor, a different type of normalization, described in the next section, is required.

The V(λ) function is clearly not one of the cone responsivities. In fact, according to the opponent-colors theory of color vision, discussed in Chapter 1, such a result wouldn't be expected. Instead, as suggested by this theory, the V(λ) function corresponds to a weighted sum of the three cone responsivity functions. When the cone functions are weighted according to their relative population in the retina and then summed, the overall responsivity matches the CIE 1924 V(λ) function. Thus the photopic luminous response represents a combination of cone signals. This sort of combination is present in the entire system of colorimetry. The use of a spectral weighting function to predict luminance matches is the first step toward a system of colorimetry.

There is also a luminous efficiency function for scotopic vision (rods) known as the V'(λ) function. This function, used for photometry at very low luminance levels, is plotted in Figure 3-6 and presented in Table 3-2 along with

Table 3-2. CIE photopic luminous efficiency function, V(λ), and scotopic luminous efficiency function, V'(λ).

Wavelength (nm)	V(λ)	V'(λ)	Wavelength (nm)	V(λ)	V'(λ)
360	0.0000	0.0000	535	0.9149	0.7305
365	0.0000	0.0000	540	0.9540	0.6500
370	0.0000	0.0000	545	0.9803	0.5655
375	0.0000	0.0000	550	0.9950	0.4810
380	0.0000	0.0006	555	1.0000	0.4049
385	0.0001	0.0014	560	0.9950	0.3288
390	0.0001	0.0022	565	0.9786	0.2682
395	0.0002	0.0058	570	0.9520	0.2076
400	0.0004	0.0093	575	0.9154	0.1644
405	0.0006	0.0221	580	0.8700	0.1212
410	0.0012	0.0348	585	0.8163	0.0934
415	0.0022	0.0657	590	0.7570	0.0655
420	0.0040	0.0966	595	0.6949	0.0494
425	0.0073	0.1482	600	0.6310	0.0332
430	0.0116	0.1998	605	0.5668	0.0246
435	0.0168	0.2640	610	0.5030	0.0159
440	0.0230	0.3281	615	0.4412	0.0117
445	0.0298	0.3916	620	0.3810	0.0074
450	0.0380	0.4550	625	0.3210	0.0054
455	0.0480	0.5110	630	0.2650	0.0033
460	0.0600	0.5670	635	0.2170	0.0024
465	0.0739	0.6215	640	0.1750	0.0015
470	0.0910	0.6760	645	0.1382	0.0011
475	0.1126	0.7345	650	0.1070	0.0007
480	0.1390	0.7930	655	0.0816	0.0005
485	0.1693	0.8485	660	0.0610	0.0003
490	0.2080	0.9040	665	0.0446	0.0002
495	0.2586	0.9430	670	0.0320	0.0001
500	0.3230	0.9820	675	0.0232	0.0001
505	0.4073	0.9895	680	0.0170	0.0001
510	0.5030	0.9970	685	0.0119	0.0000
515	0.6082	0.9660	690	0.0082	0.0000
520	0.7100	0.9350	695	0.0057	0.0000
525	0.7932	0.8730	700	0.0041	0.0000
530	0.8620	0.8110	705	0.0029	0.0000

Table 3-2 (continued)

Wavelength (nm)	V(λ)	V'(λ)	Wavelength (nm)	V(λ)	V'(λ)
710	0.0021	0.0000	740	0.0002	0.0000
715	0.0015	0.0000	745	0.0002	0.0000
720	0.0010	0.0000	750	0.0001	0.0000
725	0.0007	0.0000	755	0.0001	0.0000
730	0.0005	0.0000	760	0.0001	0.0000
735	0.0004	0.0000			

the V(λ) function. Since there is only one type of rod photoreceptor, the V'(λ) function corresponds exactly to the spectral responsivity of the rods after transmission through the ocular media. Figure 3-6 illustrates the shift in peak spectral sensitivity toward shorter wavelengths during the transition from photopic to scotopic vision. This shift, called the *Purkinje shift*, explains why blue objects tend to look lighter than red objects at very low luminance levels. The V'(λ) function is used in a way similar to the V(λ) function.

It has long been recognized in vision research that the V(λ) function might underpredict observed responsivity in the short-wavelength region of the spectrum. To address this issue and standard practice in the vision community, the CIE established an additional function—the CIE 1988 spectral luminous efficiency function, $V_M(\lambda)$ (CIE, 1990).

3.6   Tristimulus Values and Color-matching Functions

Following the establishment of the CIE 1924 luminous efficiency function, V(λ), attention was turned to the development of a system of colorimetry that could be used to specify when two metameric stimuli match in color for an average observer. Since the cone responsivities were unavailable at that time, a system of colorimetry was constructed based on the principles of trichromacy and Grassmann's laws of additive color mixture. The concept of this system is that color matches can be specified in terms of the amounts of three additive primaries required to visually match a stimulus. This is illustrated in the equivalence statement of Equation 3-7:

$$C \equiv R(\mathcal{R}) + G(\mathcal{G}) + B(\mathcal{B}) \tag{3-7}$$

The way Equation 3-7 reads is that a color C is matched by R units of the $\mathcal{R}$ primary, G units of the $\mathcal{G}$ primary, and B units of the $\mathcal{B}$ primary. The script terms $\mathcal{RGB}$ define the particular set of primaries and indicate that for different primary sets, different amounts of the primaries will be required to make a match. The term RGB indicates the amounts of the primaries required to match the color. These amounts are called *tristimulus values*. Since any color can be matched by certain amounts of three primaries, those amounts—or tristimulus values—along with a definition of the primary set, allow the specification of a color. If two stimuli can be matched using the same amounts of the primaries (i.e., they have equal tristimulus values), then they will also match each other when viewed under the same conditions.

Extending the Tristimulus Values

The next step in the derivation of colorimetry is the extension of tristimulus values such that they can be obtained for any given stimulus, defined by a spectral power distribution. Two steps are required to accomplish this. The first is to obtain tristimulus values for matches to spectral colors. The second is to take advantage of Grassmann's laws of additivity and proportionality to sum tristimulus values for each spectral component of a stimulus spectral power distribution in order to obtain the integrated tristimulus values for the stimulus. Conceptually, the tristimulus values of the spectrum (i.e., spectral tristimulus values) are obtained by matching a unit amount of power at each wavelength with an additive mixture of three primaries. Figure 3-7 illustrates a set of spectral tristimulus values for monochromatic primaries at 435.6 nm (*B*), 546.1 nm (*G*), and 700.0 nm (*R*). Spectral tristimulus values for the complete spectrum are also called *color-matching functions,* or sometimes *color-mixture functions.*

Notice that some of the spectral tristimulus values plotted in Figure 3-7 are negative. This implies the addition of a negative amount of power into the match. For example, a negative amount of the *R* primary is required to match a monochromatic 500-nm stimulus. This is because that wavelength is too saturated to be matched by the particular primaries (i.e., it is out of gamut). Clearly, a negative amount of light cannot be added to a match. Negative tristimulus values are obtained by adding the primary to the monochromatic light to desaturate the light and bring it within the gamut of the primaries. Thus a 500-nm stimulus mixed with a given amount of the *R* primary is matched by an additive mixture of appropriate amounts of the *G* and *B* primaries.

The color-matching functions illustrated in Figure 3-7 indicate the amounts of the primaries required to match unit amounts of power at each wavelength. By considering any given stimulus spectral power as an additive mixture of various amounts of monochromatic stimuli, one can obtain the tristimulus values for a stimulus by multiplying the color-matching functions by the amount of energy in the stimulus at each wavelength (Grassmann's proportionality) and then integrating across the spectrum (Grassmann's additivity). Thus the generalized equations for calculating the tristimulus values of a stimulus with spectral power distribution, $\Phi(\lambda)$, are given by Equations 3-8 through 3-10, where $\bar{r}(\lambda)$, $\bar{g}(\lambda)$, and $\bar{b}(\lambda)$ are the color-matching functions:

$$R = \int_\lambda \phi(\lambda)\bar{r}(\lambda)d\lambda \tag{3-8}$$

$$G = \int_\lambda \phi(\lambda)\bar{g}(\lambda)d\lambda \tag{3-9}$$

$$B = \int_\lambda \phi(\lambda)\bar{b}(\lambda)d\lambda \tag{3-10}$$

With the utility of tristimulus values and color-matching functions established, it remains to derive a set of color-matching functions that are representative of the population of observers who have normal color vision. Color-matching functions for individual observers, all with normal color vision, can differ significantly due to variations in lens transmittance; macula transmittance; and cone density, population, and spectral responsivities. Thus, to establish a standardized system of colorimetry, one must obtain a reliable estimate of the average color-matching functions of the population of observers having normal color vision.

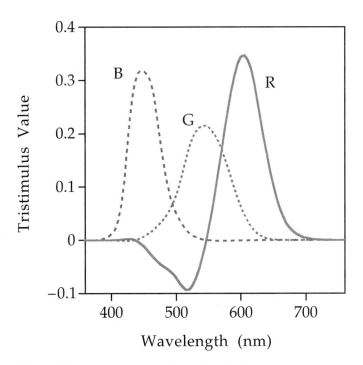

Figure 3-7. Spectral tristimulus values for the CIE RGB system of colorimetry with monochromatic primaries at 435.8 nm, 546.1 nm, and 700.0 nm.

ESTIMATING AVERAGE COLOR-MATCHING FUNCTIONS
In the late 1920s, two sets of experiments were completed to estimate average color-matching functions. These experiments were completed by Wright (1928–29) using monochromatic primaries and Guild (1931) using broadband primaries. Since the primaries from one experiment can be specified in terms of tristimulus values to match them using the other experiment, it is possible to derive a linear transform (3 × 3 matrix transformation) to convert tristimulus values from one set of primaries to another. This transformation also applies to the color-matching functions, since they are themselves tristimulus values. Thus a transformation was derived to place the data from Wright's and Guild's experiments into a common set of primaries. When this was done, the agreement between the two experiments was extremely good, thus verifying the underlying theoretical assumptions in the derivation and use of color-matching functions. Given this agreement, the CIE decided to establish a standard set of color-matching functions based on the mean results of the Wright and Guild experiments. These mean functions were transformed to RGB primaries of 700.0 nm, 546.1 nm, and 435.8 nm, respectively, and are illustrated in Figure 3-7.

In addition, the CIE decided to transform to another set of primaries, the XYZ primaries. This transformation was intended to eliminate the negative values in the color-matching functions and to force one of the color-matching functions to equal the CIE 1924 photopic luminous efficiency function, $V(\lambda)$. The negative values were removed by selecting primaries that could be used to match all physically realizable color stimuli. This can be accomplished only with imaginary primaries that are more saturated than monochromatic lights. This is a straightforward mathematical construct, and it should be noted that although the primaries are imaginary, the color-matching functions derived for those primaries are based on very real color-matching results and the validity of Grassmann's laws. Forcing one of the color-matching functions to equal the $V(\lambda)$ function serves the purpose of incorporating the CIE system of photometry (established in 1924) into the CIE system of colorimetry (established in 1931). This is accomplished by choosing two of the imaginary primaries, X and Z, such that they produce no luminance response, thereby leaving all of the luminance response in the third primary, Y. The color-matching functions for the XYZ primaries are $\bar{x}(\lambda)$, $\bar{y}(\lambda)$, and $\bar{z}(\lambda)$, respectively, called the color-matching functions of the CIE 1931 standard colorimetric observer. They are plotted in Figure 3-8 and listed in Table 3-3 for the practical wavelength range of 360–760 nm in 5-nm increments. (The CIE defines color-matching functions from 360–830 nm in 1-nm increments and with more digits after the decimal point.)

XYZ tristimulus values for colored stimuli are calculated in the same fashion as the RGB tristimulus values described previously. The general equations are given in Equations 3-11 through 3-13, where $\Phi(\lambda)$ is the spectral power distribution of the stimulus; $\bar{x}(\lambda)$, $\bar{y}(\lambda)$, and $\bar{z}(\lambda)$ are the color-matching functions; and k is a normalizing constant.

$$X = k \int_\lambda \phi(\lambda)\overline{x}(\lambda)d\lambda \tag{3-11}$$

$$Y = k \int_\lambda \phi(\lambda)\overline{y}(\lambda)d\lambda \tag{3-12}$$

$$Z = k \int_\lambda \phi\overline{z}(\lambda)d\lambda \tag{3-13}$$

The spectral power distribution of the stimulus is defined in different ways for various types of stimuli. For self-luminous stimuli (e.g., light sources and CRT displays), $\Phi(\lambda)$ is typically spectral radiance or a relative spectral power distribution. For reflective materials, $\Phi(\lambda)$ is defined as the product of the spectral reflectance factor of the material, $R(\lambda)$, and the relative spectral power distribution of the light source or illuminant of interest, $S(\lambda)$, that is, $R(\lambda)S(\lambda)$. For transmitting materials, $\Phi(\lambda)$ is defined as the product of the spectral transmittance of the material, $T(\lambda)$, and the relative spectral power distribution of the light source or illuminant of interest, $S(\lambda)$, that is, $T(\lambda)S(\lambda)$.

The normalization constant, k, in Equations 3-11 through 3-13 is defined differently for relative and absolute colorimetry. In absolute colorimetry, k is

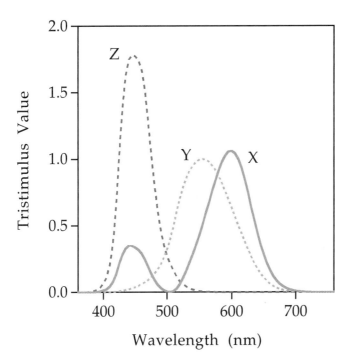

Figure 3-8. Spectral tristimulus values of the CIE 1931 standard colorimetric observer.

Table 3-3.   Color-matching functions for the CIE 1931 standard colorimetric observer (2°).

Wavelength (nm)	$\bar{x}(\lambda)$	$\bar{y}(\lambda)$	$\bar{z}(\lambda)$
360	0.0000	0.0000	0.0000
365	0.0000	0.0000	0.0000
370	0.0000	0.0000	0.0000
375	0.0000	0.0000	0.0000
380	0.0014	0.0000	0.0065
385	0.0022	0.0001	0.0105
390	0.0042	0.0001	0.0201
395	0.0077	0.0002	0.0362
400	0.0143	0.0004	0.0679
405	0.0232	0.0006	0.1102
410	0.0435	0.0012	0.2074
415	0.0776	0.0022	0.3713
420	0.1344	0.0040	0.6456
425	0.2148	0.0073	1.0391
430	0.2839	0.0116	1.3856
435	0.3285	0.0168	1.6230
440	0.3483	0.0230	1.7471
445	0.3481	0.0298	1.7826
450	0.3362	0.0380	1.7721
455	0.3187	0.0480	1.7441
460	0.2908	0.0600	1.6692
465	0.2511	0.0739	1.5281
470	0.1954	0.0910	1.2876
475	0.1421	0.1126	1.0419
480	0.0956	0.1390	0.8130
485	0.0580	0.1693	0.6162
490	0.0320	0.2080	0.4652
495	0.0147	0.2586	0.3533
500	0.0049	0.3230	0.2720
505	0.0024	0.4073	0.2123
510	0.0093	0.5030	0.1582
515	0.0291	0.6082	0.1117
520	0.0633	0.7100	0.0782
525	0.1096	0.7932	0.0573
530	0.1655	0.8620	0.0422

Table 3-3 (continued)

Wavelength (nm)	$\bar{x}(\lambda)$	$\bar{y}(\lambda)$	$\bar{z}(\lambda)$
535	0.2257	0.9149	0.0298
540	0.2904	0.9540	0.0203
545	0.3597	0.9803	0.0134
550	0.4334	0.9950	0.0087
555	0.5121	1.0000	0.0057
560	0.5945	0.9950	0.0039
565	0.6784	0.9786	0.0027
570	0.7621	0.9520	0.0021
575	0.8425	0.9154	0.0018
580	0.9163	0.8700	0.0017
585	0.9786	0.8163	0.0014
590	1.0263	0.7570	0.0011
595	1.0567	0.6949	0.0010
600	1.0622	0.6310	0.0008
605	1.0456	0.5668	0.0006
610	1.0026	0.5030	0.0003
615	0.9384	0.4412	0.0002
620	0.8544	0.3810	0.0002
625	0.7514	0.3210	0.0001
630	0.6424	0.2650	0.0000
635	0.5419	0.2170	0.0000
640	0.4479	0.1750	0.0000
645	0.3608	0.1382	0.0000
650	0.2835	0.1070	0.0000
655	0.2187	0.0816	0.0000
660	0.1649	0.0610	0.0000
665	0.1212	0.0446	0.0000
670	0.0874	0.0320	0.0000
675	0.0636	0.0232	0.0000
680	0.0468	0.0170	0.0000
685	0.0329	0.0119	0.0000
690	0.0227	0.0082	0.0000
695	0.0158	0.0057	0.0000
700	0.0114	0.0041	0.0000
705	0.0081	0.0029	0.0000

Table 3-3 (continued)

Wavelength (nm)	$\bar{x}(\lambda)$	$\bar{y}(\lambda)$	$\bar{z}(\lambda)$
710	0.0058	0.0021	0.0000
715	0.0041	0.0015	0.0000
720	0.0029	0.0010	0.0000
725	0.0020	0.0007	0.0000
730	0.0014	0.0005	0.0000
735	0.0010	0.0004	0.0000
740	0.0007	0.0002	0.0000
745	0.0005	0.0002	0.0000
750	0.0003	0.0001	0.0000
755	0.0002	0.0001	0.0000
760	0.0002	0.0001	0.0000

set equal to 683 lumens/watt, thus making the system of colorimetry compatible with the system of photometry. For relative colorimetry, k is defined by Equation 3-14:

$$k = \frac{100}{\int_{\lambda} S(\lambda)\bar{y}(\lambda)d\lambda}$$

(3-14)

The normalization for relative colorimetry in this equation results in tristimulus values that are scaled from zero to approximately 100 for various materials. It is useful to note that if relative colorimetry is used to calculate the tristimulus values of a light source, the Y tristimulus value is always equal to 100.

There is another, completely inappropriate, use of the term "relative colorimetry" in the graphic arts and other color reproduction industries. In some cases, tristimulus values are normalized to the paper white rather than to a PRD. This results in a Y tristimulus value of 100 for the paper rather than its more typical value of about 85. The advantage of this is that it allows transformations between different paper types, thereby preserving the paper white as the lightest color in an image without having to explicitly keep track of the paper color. Such a practice might be useful, but it is actually a gamut-mapping issue rather than a color-measurement issue. A more appropriate term for this practice might be *normalized colorimetry* so as to avoid confusion with the long-established practice of relative colorimetry.

It is also worth noting that the practice of normalized colorimetry is not always consistent. In some cases, reflectance measurements are made relative to the paper white. This ensures that the Y tristimulus value is normalized between zero for a perfect black and 1.0 (or 100.0) for the paper white. However, the X and Z tristimulus values might still range above 1.0 (or 100.0) depending on the particular illuminant used in the colorimetric calculations. Another approach is to normalize the tristimulus values for each stimulus color, XYZ, by the tristimulus values of the paper, $X_pY_pZ_p$, individually (X/X_p, Y/Y_p, and Z/Z_p). This is analogous to the white-point normalization in CIELAB and is often used to adjust for white-point changes and limited dynamic range in imaging systems. It is important to know which type of normalized colorimetry one is dealing with in various applications.

The relationship between CIE XYZ tristimulus values and cone responses (sometimes called *fundamental tristimulus values*) is of great importance and interest in color appearance modeling. Like the $V(\lambda)$ function, the CIE XYZ color-matching functions all represent a linear combination of cone responsivities. Thus the relationship between the two is defined by a 3×3 linear matrix transformation as described more fully in Chapter 9 (see Figure 9-1). Cone spectral responsivities can be thought of as the color-matching functions for a set of primaries that are constructed such that each primary stimulates only one cone type. It is possible to produce a real primary that stimulates only the S cones. However, no real primaries can be produced that stimulate only the M or L cones, since their spectral responsivities overlap on the visible spectrum. Thus the required primaries are also imaginary and produce all-positive color-matching functions. However, they do not incorporate the $V(\lambda)$ function as a color-matching function (since that requires a primary that stimulates all three cone types). The historical development of and recent progress in the development of colorimetric systems based on physiological cone responsivities has been reviewed by Boynton (1996).

TWO SETS OF COLOR-MATCHING FUNCTIONS
It is important to be aware that there are two sets of color-matching functions established by the CIE. The CIE 1931 standard colorimetric observer was determined from experiments using a visual field that subtended 2 degrees. Thus the matching stimuli were imaged onto the retina completely within the fovea. These color-matching functions are used almost exclusively in color appearance modeling. Often they are called the *2° color-matching functions*, or the *2° observer*. It is of historical interest to note that the 1931 standard colorimetric observer is based on data collected from fewer than 20 observers.

In the 1950s, experiments were completed (Stiles and Burch, 1959) to collect 2° color-matching functions for more observers using more-precise and more-accurate instrumentation. The results showed slight systematic discrepancies, but those were not of sufficient magnitude to warrant a change in the standard colorimetric observer.

At the same time, experiments were completed (Stiles and Burch, 1959) to collect color-matching function data for large visual fields. This was prompted by discrepancies between colorimetric and visual determination of the whiteness of paper. These experiments were completed using a 10° visual field that excluded the central fovea. Thus the color-matching functions include no influence of the macular absorption. The results for large fields were deemed significantly different from the 2° standard, enough to warrant the establishment of the CIE 1964 supplementary standard colorimetric observer, sometimes called the *10° observer.*

The two standard observers differ significantly, so care should be taken to report which observer is used with any colorimetric data. The differences are computationally significant but certainly within the variability of color-matching functions found for either 2° or 10° visual fields. Thus the two standard colorimetric observers can be thought of as representing the color-matching functions of two individuals.

3.7 Chromaticity Diagrams

The color of a stimulus can be specified by a triplet of tristimulus values. To provide a convenient two-dimensional representation of colors, chromaticity diagrams were developed. The transformation from tristimulus values to chromaticity coordinates is accomplished through a normalization that removes luminance information. This transformation is a one-point perspective projection of data points in the three-dimensional tristimulus space onto the unit plane of that space (with a center of projection at the origin), as defined by Equations 3-15 through 3-17:

$$x = \frac{X}{X + Y + Z} \tag{3-15}$$

$$y = \frac{Y}{X + Y + Z} \tag{3-16}$$

$$z = \frac{Z}{X + Y + Z} \tag{3-17}$$

Since there are only two dimensions of information in chromaticity coordinates, the third chromaticity coordinate can always be obtained from the other two because the three always sum to unity. Thus z can be calculated from x and y using Equation 3-18:

$$z = 1.0 - x - y \tag{3-18}$$

Chromaticity coordinates should be used with great care, since they attempt to represent a three-dimensional phenomenon with just two variables. To fully specify a colored stimulus, one must report one of the tristimulus values in addition to two (or three) chromaticity coordinates. Usually the Y tristimulus value is reported, since it represents the luminance information. The equations for obtaining the other two tristimulus values from chromaticity coordinates and the Y tristimulus value are often useful. They are given in Equations 3-19 and 3-20:

$$X = \frac{xY}{y} \tag{3-19}$$

$$Z = \frac{(1.0 - x - y)Y}{y} \tag{3-20}$$

Chromaticity coordinates, alone, provide no information about the color appearance of stimuli, since they include no luminance (or therefore lightness) information and do not account for chromatic adaptation. As an observer's state of adaptation changes, the color corresponding to a given set of chromaticity coordinates can change in appearance dramatically. For example, a change from yellow to blue could occur with a change from daylight to incandescent-light adaptation.

Much effort has been expended in attempts to make chromaticity diagrams more perceptually uniform. While this is an intrinsically doomed effort (i.e., an attempt to convert a nominal scale into an interval scale), it is worth mentioning one of the results. This result is actually the chromaticity diagram currently recommended by the CIE for general use: the CIE 1976 Uniform Chromaticity Scales (UCS) diagram, defined by Equations 3-21 and 3-22:

$$u' = \frac{4X}{X + 15Y + 3Z} \tag{3-21}$$

$$v' = \frac{9Y}{X + 15Y + 3Z} \tag{3-22}$$

The use of chromaticity diagrams should be avoided in most circumstances, particularly when the phenomena being investigated depend highly on the three-dimensional nature of color. For example, the display and comparison of the color gamuts of imaging devices in chromaticity diagrams is misleading to the point of being almost completely erroneous.

3.8   CIE Color Spaces

The general use of chromaticity diagrams has been made largely obsolete by the advent of the CIE color spaces: CIELAB and CIELUV. These spaces extend tristimulus colorimetry to three-dimensional spaces with dimensions that approximately correlate with the perceived lightness, chroma, and hue of a stimulus. This is accomplished by incorporating features to account for chromatic adaptation and nonlinear visual responses. The main aim in the development of these spaces was to provide uniform practices for the measurement of color differences, something that cannot be done reliably in tristimulus or chromaticity spaces.

In 1976, two spaces were recommended for use, since at that time no clear evidence existed to support one over the other. The CIELAB and CIELUV color spaces are described in more detail in Chapter 10. Their equations are briefly summarized in this section. Wyszecki (1986) provides an overview of the development of the CIE color spaces.

CIELAB

The CIE 1976 (L^* a^* b^*) color space, abbreviated CIELAB, is defined by Equations 3-23 through 3-27 for tristimulus values normalized to the white that are greater than 0.008856:

$$L^* = 116(Y/Y_n)^{1/3} - 16 \tag{3-23}$$

$$a^* = 500\left[(X/X_n)^{1/3} - (Y/Y_n)^{1/3}\right] \tag{3-24}$$

$$b^* = 200\left[(Y/Y_n)^{1/3} - (Z/Z_n)^{1/3}\right] \tag{3-25}$$

$$C^*_{ab} = \sqrt{(a^{*2} + b^{*2})} \tag{3-26}$$

$$h_{ab} = \tan^{-1}(b^*/a^*) \tag{3-27}$$

In these equations, X, Y, and Z are the tristimulus values of the stimulus and X_n, Y_n, and Z_n are the tristimulus values of the reference white. L^* represents lightness, a^* approximate redness-greenness, b^* approximate yellowness-blueness, C^*_{ab} chroma, and h_{ab} hue. The L^*, a^*, and b^* coordinates are used to construct a Cartesian color space, as illustrated in Figure 3-9. The L^*, C^*_{ab}, and h_{ab} coordinates are the cylindrical representation of the same space.

The CIELAB space, including the full set of equations for dark colors, is described in greater detail in Chapter 10.

CIELUV

The CIE 1976 (L^* u^* v^*) color space, abbreviated CIELUV, is defined by Equations 3-28 through 3-32. Equation 3-28 is also restricted to tristimulus values normalized to the white that are greater than 0.008856:

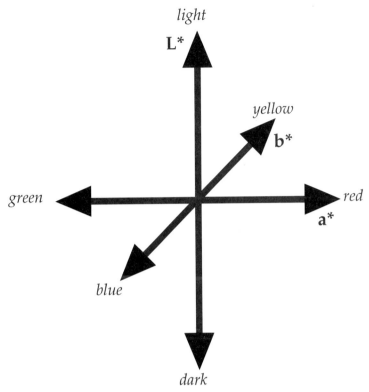

Figure 3-9. Three-dimensional representation of the CIELAB L*, a*, and b* coordinates.

$$L^* = 116(Y/Y_n)^{1/3} - 16 \qquad (3\text{-}28)$$

$$u^* = 13L^*(u' - u'_n) \qquad (3\text{-}29)$$

$$v^* = 13L^*(v' - v'_n) \qquad (3\text{-}30)$$

$$C^*_{uv} = \sqrt{(u^{*2} + v^{*2})} \qquad (3\text{-}31)$$

$$h_{uv} = \tan^{-1}(v^*/u^*) \qquad (3\text{-}32)$$

In these equations, u' and v' are the chromaticity coordinates of the stimulus and u'_n and v'_n are the chromaticity coordinates of the reference white. L^* represents lightness, u^* redness-greenness, v^* yellowness-blueness, C^*_{uv} chroma, and h_{uv} hue. As in CIELAB, the L^*, u^*, and v^* coordinates are used to construct a Cartesian color space and the L^*, C^*_{uv}, and h_{uv} coordinates are the cylindrical representation of the same space.

The CIELAB and CIELUV spaces were both recommended in 1976 as interim solutions to the problem of color-difference specification of reflecting samples. Since then, CIELAB has become almost universally used for color specification and particularly color-difference measurement. At this time there appears to be no reason to use CIELUV over CIELAB.

3.9   Color-difference Specification

Color differences are measured in the CIELAB space as the Euclidean distance between the coordinates for the two stimuli. This is expressed in terms of a CIELAB ΔE^*_{ab}, which can be calculated using Equation 3-33. It can also be expressed in terms of lightness, chroma, and hue differences as illustrated in Equation 3-34 by using the combination of Equations 3-33 and 3-35:

$$\Delta E^*_{ab} = \left[\Delta L^{*2} + \Delta a^{*2} + \Delta b^{*2}\right]^{1/2} \tag{3-33}$$

$$\Delta E^*_{ab} = \left[\Delta L^{*2} + \Delta C^{*2}_{ab} + \Delta H^{*2}_{ab}\right]^{1/2} \tag{3-34}$$

$$\Delta H^*_{ab} = \left[\Delta E^{*2}_{ab} - \Delta L^{*2} - \Delta C^{*2}_{ab}\right]^{1/2} \tag{3-35}$$

The goal of the CIELAB color space design was to have color differences be perceptually uniform throughout the space (i.e., a ΔE^*_{ab} of 1.0 for a pair of red stimuli is perceived to be equal in magnitude to a ΔE^*_{ab} of 1.0 for a pair of gray stimuli). This goal was not strictly achieved.

To improve the uniformity of color-difference measurements, modifications to the CIELAB ΔE^*_{ab} equation have been made based upon various empirical data. One of the most widely used modifications is the CMC color-difference equation (Clarke et al., 1984), which is based on a visual experiment on color-difference perception in textiles. The CIE (1995) has recently evaluated such equations and the available visual data and has recommended a new color-difference equation for industrial use. This system for color-difference measurement is called the CIE 1994 (ΔL^* ΔC^*_{ab} ΔH^*_{ab}) color-difference model, with the symbol ΔE^*_{94} and abbreviation CIE94. Use of the CIE94 equation is preferred over a simple CIELAB ΔE^*_{ab}. CIE94 color differences are calculated using Equations 3-36 through 3-39:

$$\Delta E^*_{94} = \left[\left(\frac{\Delta L^*}{k_L S_L}\right)^2 + \left(\frac{\Delta C^*_{ab}}{k_C S_C}\right)^2 + \left(\frac{\Delta H^*_{ab}}{k_H S_H}\right)^2\right]^{1/2} \tag{3-36}$$

$$S_L = 1 \tag{3-37}$$

$$S_C = 1 + 0.045 C^*_{ab} \qquad\qquad (3\text{-}38)$$

$$S_H = 1 + 0.015 C^*_{ab} \qquad\qquad (3\text{-}39)$$

The parametric factors, k_L, k_C, and k_H, are used to adjust the relative weighting of the lightness, chroma, and hue components, respectively, of color difference for various viewing conditions and applications that depart from the CIE94 reference conditions. It is also worth noting that when CIE94 color differences are averaged across the color space, they are significantly smaller in magnitude than CIELAB color differences for the same stimuli pairs. Thus the use of CIE94 color differences to report overall performance in applications such as the colorimetric characterization of imaging devices results in the appearance of significantly improved performance if the numbers are mistakenly considered equivalent to CIELAB color differences. The same is true for CMC color differences.

The CIE (1995) established a set of reference conditions for the use of the CIE94 color-difference equations, as follows:

Illumination: CIE illuminant D65 simulator
Illuminance: 1000 lux
Observer: Normal color vision
Background: Uniform, achromatic, $L^* = 50$
Viewing Mode: Object
Sample Size: Greater than 4° visual angle
Sample Separation: Direct edge contact
Sample Color-difference Magnitude: 0 to 5 CIELAB units
Sample Structure: No visually apparent pattern or nonuniformity

The tight specification of reference conditions for the CIE94 equations highlights the vast amount of research that remains to be performed in order to generate a universally useful process for color-difference specification. These issues are some of the same that must be tackled in the development of color appearance models. It should also be noted that for sample sizes greater than 4°, use of the CIE 1964 supplementary standard colorimetric observer is recommended.

3.10   The Next Step

This chapter reviewed the fundamentals of basic colorimetry (and began to touch on advanced colorimetry) as illustrated conceptually in Figure 3-1. While these techniques are well established and have been successfully used for decades, much more information needs to be added to the triangle of color in

Figure 3-1 in order to extend basic colorimetry toward the specification of the color appearance of stimuli under a wide variety of viewing conditions. Other issues that must be considered include the following:

- Chromatic adaptation
- Light adaptation
- Luminance level
- Background color
- Surround color

These issues are explored in the remaining chapters of this book.

CHAPTER 4

Color-appearance Terminology

4.1   Importance of Definitions
4.2   Color
4.3   Hue
4.4   Brightness and Lightness
4.5   Colorfulness and Chroma
4.6   Saturation
4.7   Unrelated and Related Colors
4.8   Definitions in Equations
4.9   Brightness-Colorfulness Versus Lightness-Chroma

IN ANY SCIENTIFIC FIELD, a large portion of the knowledge contained therein is in the definitions of terms used by its practitioners. Newcomers to a field often realize that although they understand the fundamental scientific concepts, they must first learn the language in order to communicate accurately, precisely, and effectively in the new discipline. Nowhere is this more true, or more important, than in the study of color appearance. Hunt (1978) showed concern that color scientists and technologists might be taking an attitude like Humpty-Dumpty, who is quoted as saying, "When I use a word, it means just what I choose it to mean—neither more nor less." And we all know what happened to Humpty-Dumpty.

The careful definition of color-appearance terms presented in this chapter is intended to put everyone on a level playing field and help ensure that the concepts, data, and models discussed in this book are presented and interpreted consistently. As can be seen throughout this book, consistent use of terminology has not been the historical norm and continues to be one of the challenges of color-appearance research and application.

4.1   Importance of Definitions

Why should it be particularly difficult to agree upon consistent terminology in the field of color appearance? Perhaps the answer lies in the very nature of the subject. Almost everyone *knows* what color is. After all, they have had first-hand experience of it since shortly after birth. However, very few can precisely describe their color experiences or even precisely define color. This innate knowledge, along with the imprecise use of color terms (e.g., warmer, cooler, brighter, cleaner, fresher), leads to a subject that everyone knows about, but few can discuss precisely. If color appearance is to be described in a systematic, mathematical way, definitions of the phenomena being described need to be precise and universally agreed upon.

Since color appearance modeling remains an area of active research, the required definitions have not been set in stone for decades. The definitions presented in this chapter have been culled from three sources. The first, and authoritative, source is the *International Lighting Vocabulary* published by CIE (CIE, 1987). This source includes the definitions of approximately 950 terms and quantities related to light and color "to promote international standardization in the use of quantities, units, symbols, and terminology." The other two sources, both articles by Hunt (1977, 1978), provide descriptions of some of the work that led to the latest revision of the *International Lighting Vocabulary*. It should be noted that the *International Lighting Vocabulary* is currently under revision; there is also a relevant ASTM standard on appearance terminology (ASTM, 1995).

Keep in mind that the definitions given here are of perceptual terms, that is, terms that define our perceptions of colored stimuli. These are not definitions of specific colorimetric quantities. In the construction and use of color appearance models, the objective is to develop and use physically measurable quantities that correlate with the perceptual attributes of color appearance defined in this chapter.

4.2   Color

The definition of the word *color* itself provides some interesting challenges and difficulties. While most of us know what color is, it is an interesting challenge to try to write a definition of the word that does not contain an example. As can be seen next, even the brightest and most dedicated color scientists who set out to write the *International Lighting Vocabulary* could not meet this challenge.

> **Color:** Attribute of visual perception consisting of any combination of chromatic and achromatic content. This attribute can be described by chromatic color names such as yellow, orange, brown, red, pink, green, blue, purple, etc., or by achromatic color names such as white, gray, black, etc., and qualified by bright, dim, light, dark, etc., or by combinations of such names.

The authors of this definition were also well aware that the perception of color was not a simple matter and added a note that captures the essence of why color appearance models are needed:

> **Note:** Perceived color depends on the spectral distribution of the color stimulus, on the size, shape, structure, and surround of the stimulus area, on the state of adaptation of the observer's visual system, and on the observer's experience of the prevailing and similar situations of observations.

This note opens the door for the vast array of physical, physiological, psychological, and cognitive variables that influence color appearance—many of which are discussed in this book.

While this definition might not be very satisfying due to its circularity, any definitions that avoid the circularity seem to be equally dissatisfying. One such example would be to define color as those attributes of a visual stimulus that are independent of spatial and temporal variations. Even this definition is flawed, since in the absence of all temporal and spatial variation, even the perception of color vanishes. Despite the difficulty in defining color, the various attributes of color can be defined much more precisely. Those are the terms that are of utmost importance in color appearance modeling.

4.3   Hue

Hue: Attribute of a visual sensation according to which an area appears to be similar to one of the perceived colors: red, yellow, green, and blue, or to a combination of two of them.

Achromatic Color: Perceived color devoid of hue.

Chromatic Color: Perceived color possessing a hue.

Once again, it is difficult, if not impossible, to define hue without using examples. This is due, in part, to the nature of the hue perception. It is a natural interval scale as illustrated by the traditional description of a "hue circle." There is no natural "zero" hue. Color without hue can be described, but there is no perception that corresponds to a meaningful hue of zero units. Thus the color appearance models described in later chapters never aspire to describe hue with more than an interval scale.

The "circular" nature of the definition of hue can be observed in Figure 5-2, which illustrates the hue dimension in the *Munsell Book of Color*. The hue circle in this figure also illustrates how all of the hues can be described using the terms red, yellow, green, blue, or combinations thereof as predicted by Hering's opponent-colors theory of color vision. Other examples of hue include the variation of color witnessed in a projected visible spectrum or a rainbow. Three of the rendered cubes in Figure 4-1 are of three different hues: red, green, and blue. The fourth is white and thus achromatic—that is, possessing no hue.

Figure 4-1.   A computer-graphics rendering of four solid blocks illuminated by two light sources of differing intensities and angles of illumination. To be used for demonstration of various color appearance attributes.

4.4   Brightness and Lightness

> **Brightness:** Attribute of a visual sensation according to which an area appears to emit more or less light.

> **Lightness:** The brightness of an area judged relative to the brightness of a similarly illuminated area that appears to be white or highly transmitting.

> **Note:** Only related colors (see Section 4.7) exhibit lightness.

The definitions of brightness and lightness are straightforward and rather intuitive. The important distinction is that brightness refers to the absolute level of the perception, while lightness can be thought of as relative brightness, normalized for changes in the illumination and viewing conditions.

A classic example is to think about a piece of paper, such as this book page. If this page was viewed in a typical office environment, the paper would have some brightness and a fairly high lightness (perhaps it is the lightest stimulus in the field of vision and therefore white). If the book was viewed outside on a sunny summer day, there would be significantly more energy reflected from the page and the paper would appear brighter. However, the page would still likely be the lightest stimulus in the field of vision and so would retain its high lightness, approximately the same lightness it exhibited in office illumination. In other words, the paper still appears white, even though it is brighter outdoors. This is an example of approximate lightness constancy.

Figure 4-1 illustrates four rendered cubes that are illuminated by two light sources of different intensities, but of the same color. Imagine you are actually viewing the cubes in their illuminated environment. In this case, it would be clear that the different sides of the cubes are illuminated differently and exhibit different brightnesses. However, if you were asked to judge the lightness of the cubes, you could give one answer for all of the visible sides, since you would interpret lightness as their brightness relative to the brightness of a similarly illuminated white object.

4.5   Colorfulness and Chroma

Colorfulness: Attribute of a visual sensation according to which the perceived color of an area appears to be more or less chromatic.

Note: For a color stimulus of a given chromaticity and, in the case of related colors, of a given luminance factor, this attribute usually increases as the luminance is raised, except when the brightness is very high.

Chroma: Colorfulness of an area judged as a proportion of the brightness of a similarly illuminated area that appears white or highly transmitting.

Note: For given viewing conditions and at luminance levels within the range of photopic vision, a color stimulus perceived as a related color, of a given chromaticity, and from a surface having a given luminance factor, exhibits approximately constant chroma for all levels of luminance except when the brightness is very high. In the same circumstances, at a given level of illuminance, if the luminance factor increases, the chroma usually increases.

As was discussed in Chapters 1 and 3, color perception is generally thought of as being three-dimensional. Two of those dimensions (hue and brightness/lightness) have already been defined. Colorfulness and chroma define the remaining dimension of color. Colorfulness is to chroma as brightness is to lightness. It is appropriate to think of chroma as relative colorfulness just as lightness can be thought of as relative brightness. Colorfulness describes the intensity of the hue in a given color stimulus. Thus achromatic colors exhibit zero colorfulness and chroma, and as the amount of color content increases (with constant brightness/lightness and hue), colorfulness and chroma increase.

Like lightness, chroma is approximately constant across changes in luminance level. Note, however, that chroma is likely to change if the color of the illumination is varied. Colorfulness, on the other hand, increases for a given object as the luminance level increases, since it is an absolute perceptual quantity. Figure 4-1 illustrates the difference between colorfulness and chroma. Again, imagine you are in the illuminated environment viewing the cubes in Figure 4-1. Since different sides of each cube are illuminated with differing amounts of the same color energy, they vary in colorfulness. On the other hand, if you were to judge the chroma of the cubes, you could provide one answer for each of the cubes. This is because you would be judging each side relative to a similarly illuminated white object. The sides of the cubes with greater illumination exhibit greater colorfulness, but the chroma is roughly constant within each cube.

4.6   Saturation

Saturation: Colorfulness of an area judged in proportion to its brightness.

Note: For given viewing conditions and at luminance levels within the range of photopic vision, a color stimulus of a given chromaticity exhibits approximately constant saturation for all luminance levels, except when brightness is very high.

Saturation is a unique perceptual experience separate from chroma. Like chroma, saturation can be thought of as relative colorfulness. However, saturation is the colorfulness of a stimulus relative to its own brightness, while chroma is colorfulness relative to the brightness of a similarly illuminated area that appears white. For a stimulus to have chroma, it must be judged in relation to other colors, while a stimulus seen completely in isolation can have saturation. An example of a stimulus that exhibits saturation, but not chroma, is traffic signal lights viewed in isolation on a dark night. The lights, typically red, yellow, or green, are quite saturated and can be compared to the color appearance of oncoming headlights, whose saturation is very nearly zero (since they typically appear white).

Saturation is sometimes described as a shadow series. This refers to the range of colors observed when a single object has a shadow cast upon it. As the object falls into deeper shadow, it becomes darker, but saturation remains constant. This can be observed in Figure 4-1 by assuming that the rendered environment is illuminated by a single light source. The various sides of the cubes will all be of approximately constant saturation.

4.7   Unrelated and Related Colors

Unrelated Color: Color perceived to belong to an area or object seen in isolation from other colors.

Related Color: Color perceived to belong to an area or object seen in relation to other colors.

The distinction between related and unrelated colors is critical for a firm understanding of color appearance. The definitions are simple enough; related colors are viewed in relation to other color stimuli, while unrelated colors are viewed completely in isolation. Almost every color-appearance application of interest deals with the perception of related colors; they are the main focus of this book. However, it is important to keep in mind that many of the visual experiments that provide the foundations for understanding color vision and color appearance were performed with isolated stimuli, that is, unrelated colors. It also is important to keep the distinction in mind and not try to predict phenomena that occur only with related colors using models defined for unrelated colors and vice versa.

At times, related colors are thought of as object colors and unrelated colors are thought of as self-luminous colors. There is no correlation between the two concepts. An object color can be seen in isolation and thus be unrelated. Also, self-luminous stimuli (such as those presented on CRT displays) can be seen in relation to one another and thus be related colors.

There are various phenomena, discussed throughout this book, that occur only for related or unrelated colors. One interesting example of this is the perception of colors described by certain color names such as gray and brown. It is not possible to see unrelated colors that appear either gray or brown. Gray is an achromatic color with a lightness significantly lower than white. Brown is an orange color with low lightness. Both of these color name definitions require specific lightness levels. Since lightness and chroma require judgments relative to other stimuli that are similarly illuminated, they cannot possibly be perceived as unrelated stimuli. To convince yourself, search for a light that can be viewed in isolation (i.e., a completely dark environment) and that appears either gray or brown.

A nice demonstration of these related colors can be made by taking a spot of light that appears either white or orange and surrounding it with increasingly higher luminances of white light. As the luminance of the background light increases, the original stimuli will change in appearance from white and orange to gray and brown. If the background luminance is increased far enough, the original stimuli can be made to appear black. For interesting discussions on the color brown, see Bartleson (1976), Fuld et al. (1983), and Mausfeld et al. (1993).

The perceptual color terms defined previously are applied differently to related and unrelated colors. Unrelated colors exhibit only the perceptual attributes of hue, brightness, colorfulness, and saturation. The attributes that require judgment relative to a similarly illuminated white object cannot be perceived with unrelated colors. On the other hand, related colors exhibit all of the perceptual attributes of hue, brightness, lightness, colorfulness, chroma, and saturation.

4.8   Definitions in Equations

The various terms used to carefully describe color appearance can be confusing at times. To keep the definitions straight, think of them in terms of simple equations. These equations, while not strictly true in a mathematical sense, provide a first-order description of the relationships between the various color percepts. In fact, an understanding of the definitions in terms of the following equations provides the first building block toward understanding the construction of the various color appearance models.

Chroma can be thought of as colorfulness relative to the brightness of a similarly illuminated white, as shown in Equation 4-1:

$$\text{Chroma} = \frac{\text{Colorfulness}}{\text{Brightness (White)}}$$

$$(4\text{-}1)$$

Saturation can be described as the colorfulness of a stimulus relative to its own brightness, as illustrated in Equation 4-2:

$$\text{Saturation} = \frac{\text{Colorfulness}}{\text{Brightness}}$$

$$(4\text{-}2)$$

Finally, lightness can be expressed as the ratio of the brightness of a stimulus to the brightness of a similarly illuminated white stimulus, as given in Equation 4-3:

$$\text{Lightness} = \frac{\text{Brightness}}{\text{Brightness (White)}}$$

$$(4\text{-}3)$$

The utility of these simple definitions in terms of equations is illustrated by a derivation of the fact that an alternative definition of saturation (used in some color appearance models) is given by the ratio of chroma and lightness in Equation 4-4:

$$\text{Saturation} = \frac{\text{Chroma}}{\text{Lightness}} \tag{4-4}$$

This can be proven by first substituting the definitions of chroma and lightness from Equations 4-1 and 4-3 into Equation 4-4, arriving at Equation 4-5:

$$\text{Saturation} = \frac{\text{Colorfulness}}{\text{Brightness (White)}} \bullet \frac{\text{Brightness (White)}}{\text{Brightness}} \tag{4-5}$$

Completing the algebraic exercise by canceling out the brightness of the white terms in Equation 4-5 results in saturation being expressed as the ratio of colorfulness to brightness, as shown in Equation 4-6, which is identical to the original definition in Equation 4-2:

$$\text{Saturation} = \frac{\text{Colorfulness}}{\text{Brightness}} \tag{4-6}$$

4.9   Brightness-Colorfulness Versus Lightness-Chroma

While color is typically thought of as three-dimensional and color matches can be specified by just three numbers, it turns out that three dimensions are not enough to completely specify color appearance. In fact, five perceptual dimensions, or attributes, are required for a complete specification of color appearance:

- Brightness
- Lightness
- Colorfulness
- Chroma
- Hue

Saturation is redundant, since it is known if these five attributes are known. However, in many practical color-appearance applications it is not necessary to know all five. Typically, related colors are of most interest and only the relative appearance attributes are of significant importance. Thus it is often sufficient to be concerned with only the relative attributes of lightness, chroma, and hue.

There might seem to be some redundancy in using all five attributes to describe a color appearance. However, this is not the case, as was elegantly described by Nayatani et al. (1990). That article illustrated both theoretically and experimentally the distinction between brightness-colorfulness appearance matches and lightness-chroma appearance matches and showed that in most viewing conditions, the two types of matches are distinct. Imagine viewing a yellow school bus outside on a sunny day. The yellow bus will exhibit its typical appearance attributes of hue (yellow), brightness (high), lightness (high), colorfulness (high), and chroma (high). Now imagine viewing a printed photographic reproduction of the school bus in the relatively subdued lighting of an office or home. The image of the bus could be a perfect match to the original object in hue (yellow), lightness (high), and chroma (high). However, the brightness and colorfulness of the print viewed in subdued lighting could never equal that of the original school bus viewed in bright sunlight. This is simply because of the lack of energy reflecting off of the print relative to that reflecting off of the original object. If that same print was carried outside into the bright sunlight that the original bus was viewed under, it is then possible that the reproduction could match the original object in all five attributes.

So which is more important, the matching (i.e., reproduction) of brightness and colorfulness or the matching of lightness and chroma? (Note that hue is defined the same way in either circumstance.) The answer depends on the application, but it is safe to say that much more often it is lightness and chroma that are of greater importance. As illustrated by the previous example, it is typically possible and desirable in color-reproduction applications to aspire to lightness-chroma matching. Imagine trying to make reproductions with a brightness-colorfulness matching objective. To reproduce the sunlight-illuminated school bus in office lighting, one would have to make a print that was literally glowing in order to reproduce brightness and colorfulness. This is not physically possible. Going in the other direction (subdued lighting to bright lighting) is possible, but is it desirable? Imagine taking a photograph of a person at a candlelight dinner and making a reproduction to be viewed under bright sunlight. It would be easy to reproduce the brightness and colorfulness of the original scene, but the print would be extremely dark (essentially black everywhere) and it would be considered a very poor print. Customers of such color reproductions expect lightness-chroma reproduction.

There are a few situations in which brightness and colorfulness might be more important than lightness and chroma. Nayatani et al. (1990) suggest a few such situations. One of those is in the specification of the color-rendering properties of light sources. In such an application, it might be more important to know how bright and colorful objects appear under a given light source, rather than just lightness and chroma. Another situation might be in the judgment of image quality for certain types of reproductions. For example, in comparing the quality of projected transparencies in a darkened (or not so darkened) room, observers might be more interested in the brightness and colorfulness of an image than in the lightness and chroma. In fact, the lightness and chroma of image elements might remain nearly constant as the luminance of the projector is decreased, while it is fairly intuitive that the perceived image quality would be decreasing. It is very rare for observers to comment that they wish the image from a slide or overhead projector was not so bright unless a bright overhead projector image is displayed adjacent to a dim 35mm-projector image!

CHAPTER 5

Color-order Systems

5.1 Overview and Requirements
5.2 The *Munsell Book of Color*
5.3 The Swedish Natural Color System
5.4 The Colorcurve System
5.5 Other Color-order Systems
5.6 Uses of Color-order Systems
5.7 Color-naming Systems

SINCE COLOR APPEARANCE is a basic perception, the most direct method to measure it is through psychophysical techniques designed to elucidate perceptually uniform scales of the various color-appearance attributes defined in Chapter 4. When such experiments are performed, it is possible to specify stimuli using basic colorimetry that embody the perceptual color-appearance attributes. A collection of such stimuli, appropriately specified and denoted, forms a color-order system. Such a system does allow a fairly unambiguous specification of color appearance. However, there is no reason to expect that the specified appearances will generalize to other viewing conditions or be mathematically related to physical measurements in any straightforward way. Thus color-order systems provide data and a technique for specifying color appearance, but they do not provide a mathematical framework to allow extension of those data to novel viewing conditions. So color-order systems are of significant interest in the development and testing of color appearance models, but they cannot serve as a replacement for such models.

This chapter provides an overview of some color-order systems that are of particular interest in color appearance modeling and device-independent color imaging. Their importance and application will become self-evident in this and later chapters. Additional details on color-order systems can be found in Hunt's text on color measurement (Hunt, 1991a), Wyszecki and Stiles's reference volume on color science (Wyszecki and Stiles, 1982), Wyszecki's review chapter on color appearance (Wyszecki, 1986), and Derefeldt's review on color-appearance systems (Derefeldt, 1991).

5.1   Overview and Requirements

Many definitions of color-order systems have been suggested. It is probably most useful to adopt some combination of all of them. Wyszecki (1986) points out that color-order systems fall into three broad groups:

- One based on the principles of additive mixtures of color stimuli. A well-known example is the Ostwald system.
- One consisting of systems based on the principles of colorant mixtures. The Lovibond Tintometer provides an example of a color-specification system based on the subtractive mixture of colorants.

- One consisting of those based on the principles of color perception or color appearance.

In fact, Derefeldt (1991) suggests that color-appearance systems are the only systems appropriate for general use. She goes on to state that color-appearance systems are defined by perceptual color coordinates or scales and by uniform or equal visual spacing of colors according to these scales.

This chapter focuses on color-appearance systems such as the Natural Color System (NCS) and the Munsell system. Hunt (1991a) adds a useful constraint to the definition of color-order systems by stating that it must be possible to interpolate between samples in the system in an unambiguous way. Finally, a practical restriction on color-order systems is that they be physically embodied with stable samples that are produced to tight tolerances.

To summarize, color-order systems are constrained in that they must

- Be an orderly (and continuous) arrangement of colors.
- Include a logical system of denotation.
- Incorporate perceptually meaningful dimensions.
- Be embodied with stable, accurate, and precise samples.

A further objective for a generally useful color-order system is that the perceptual scales represent perceived magnitudes uniformly or that differences on the scales be of equal, perceived magnitude.

This definition excludes some color systems that have been found useful in practical applications. For example, the *Pantone Color Formula Guide* is a useful color-specification system for inks, but it is not a color-order system because it does not include continuous scales or an appropriate embodiment. It is more appropriately considered a *color-naming system*. It is discussed briefly later in this chapter. Swatches used to specify paint colors also fall into this category.

There are a variety of applications of color-order systems in the study of color appearance. They provide independent data on the perceptual scaling of various appearance attributes such as lightness, hue, and chroma that can be used to evaluate mathematical models. Their embodiments provide reliable sample stimuli that can be used unambiguously in psychophysical experiments on color appearance. Their nomenclature provides a useful system for the specification and communication of color appearances. They also provide a useful educational tool for the explanation of various color-appearance attributes and phenomena. The uses of color-order systems in color appearance modeling are discussed in more detail in Section 5.6.

5.2   The *Munsell Book of Color*

One of the most widely used color-order systems, particularly in the United States, is the Munsell system, embodied in the *Munsell Book of Color*. The history of the Munsell system has been reviewed by Nickerson (1940, 1976a, b, c). Also, interesting insight can be obtained by reviewing the visual experiments leading to the renotation of the Munsell colors in the 1940s (Newhall, 1940).

The system was developed by artist Albert H. Munsell in the early part of the twentieth century. Munsell was particularly interested in developing a system that would aid in the education of children. Basically, color appearance is specified according to three attributes:

- Value (V)
- Hue (H)
- Chroma (C)

The definitions of the three Munsell attributes match the current definitions of the corresponding appearance attributes, with *Munsell Value* referring to lightness. Munsell's objective was to specify colors (both psychophysically and physically) that have equal visual increments along each of the three perceptual dimensions.

Value Attribute

The Munsell Value scale is the anchor of the system. There are ten main steps in the Munsell Value scale, with white given a notation of 10, black a 0, and intermediate grays given notations ranging between 0 and 10. The design of the Munsell Value scale is such that an intermediate gray with a Munsell Value of 5 (denoted N5 for a neutral sample with Value 5) is perceptually halfway between an ideal white (N10) and an ideal black (N0). Also, the perceived lightness difference between N3 and N4 samples is equivalent to the lightness difference between N6 and N7 samples or any other samples varying by one step in Munsell Value. Lightness perceptions falling in between two Munsell Value steps are denoted with decimals. For example, Munsell Value 4.5 falls perceptually halfway between Munsell Values 4 and 5. It is important to note that the relationship between Munsell Value, V, and relative luminance, Y, is nonlinear. In fact, it is specified by the 5th-order polynomial given in Equation 5-1 and plotted in Figure 5-1:

$$Y = 1.2219V - 0.23111V^2 + .23951V^3 - 0.021009V^4 + 0.0008404V^5 \quad (5\text{-}1)$$

As can be seen in Figure 5-1, a sample that is perceived to be a middle gray (N5) has a relative luminance (or luminous reflectance factor) of about

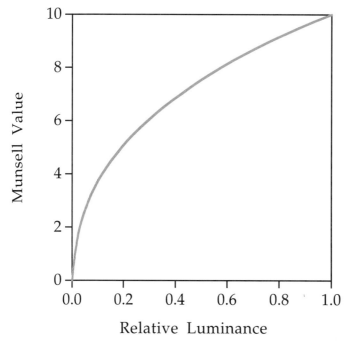

Figure 5-1.   Munsell Value as a function of relative luminance.

20%. The Munsell Value of any color (independent of hue or chroma) is defined by the same univariate relationship with relative luminance. Thus, if the Munsell Value of a sample is known, so is its relative luminance, CIE Y, and vice versa. Unfortunately, the 5th-order polynomial in Equation 5-1 cannot be analytically inverted for practical application. Since the CIE lightness scale, L^*, was designed to model the Munsell system, it provides a very good computational approximation to the Munsell Value. As a very useful and accurate general rule, the Munsell Value of a stimulus can be obtained from its CIE L^* (Ill. C, 2° Obs.) by simply dividing by 10.

Hue Attribute

The next dimension of the Munsell system is hue. The hue circle in the Munsell system is divided into five principal hues (purple, blue, green, yellow, and red, denoted 5P, 5B, 5G, 5Y, and 5R, respectively) and is designed to divide the complete hue circle into equal perceptual intervals. Five intermediate hues are also designated in the Munsell system as 5PB, 5BG, 5GY, 5YR, and 5RP, for a total of ten hue names. For each of the ten hues, there are ten integral hues with notations as illustrated by the range between 5P and 5PB and consisting of 6P,

7P, 8P, 9P, 10P, 1PB, 2PB, 3PB, and 4PB. This type of sequence continues around the entire hue circle, resulting in a hundred integer hue designations that are intended to be equal, perceived hue intervals. Hues intermediate to the integer designations are denoted with decimal values (e.g., 7.5PB).

Chroma Attribute

The third attribute of the Munsell system is chroma. The chroma scale is designed to have equal visual increments from a chroma of zero for neutral samples to increasing chromas for samples with stronger hue content. There is no set maximum for the chroma scale. The highest chromas achieved depend on the hue and the Munsell Value of the samples and on the colorants used to produce them. For example, there are no high chroma samples with a yellow hue and a low Munsell Value or a purple hue and a high Munsell Value. Such stimuli cannot be physically produced due to the nature of the human visual response.

Figure 5-2 illustrates the three-dimensional arrangement of the Munsell system in terms of a constant Munsell Value plane [Figure 5-2(a)] and a constant hue plane [Figure 5-2(b)]. Figure 5-3 illustrates similar planes and a three-dimensional perspective of the Munsell system generated using a computer-graphics model of the system. The Munsell system is used to denote a specific colored stimulus using its Munsell hue, Value, and chroma designations in a triplet arranged with the hue designation followed by the Value, a forward slash (/), and then the chroma. For example, a red stimulus of medium lightness and fairly high chroma would be designated 7.5R5/10 (hue Value/chroma).

The Munsell system is embodied in the *Munsell Book of Color*. This book consists of about 1,500 samples arranged on 40 pages of constant hue. Each hue page is arranged in order of increasing lightness (bottom to top) and chroma (center of book to edge). The samples consist of painted paper and are available in both gloss and matte surfaces. Larger-sized Munsell samples can also be purchased for special applications such as visual experiments or the construction of test targets for imaging systems. Munsell samples are produced to colorimetric aim points that were specified by the experiments leading up to the Munsell renotation (Newhall, 1940). The chromaticity coordinates and luminance factors for each Munsell sample (including many that cannot be easily produced) can be found in Wyszecki and Stiles (1982). The colorimetric specifications utilize CIE illuminant C and the CIE 1931 Standard Colorimetric Observer (2°). These specifications should be kept in mind when viewing any embodiment of the Munsell system. The perceptual uniformity of the system is valid only under source C, on a uniform middle gray (N5) background, with a sufficiently high illuminance level (e.g., greater than 500 lux). Viewing the samples in the *Munsell Book of Color* under any other viewing conditions does not represent an embodiment of the Munsell system.

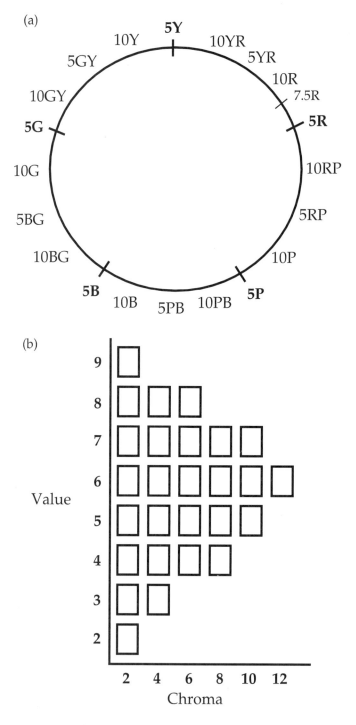

Figure 5-2. A graphical representation of (a) the hue circle and (b) a Value/chroma plane of constant hue in the Munsell system.

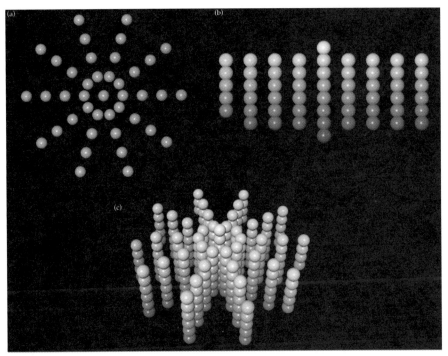

Figure 5-3. A color rendering of samples from the Munsell system in (a) a constant Value plane, (b) a pair of constant-hue planes, and (c) a three-dimensional perspective.

5.3   The Swedish Natural Color System

More recently, the Natural Color System (NCS) has been developed in Sweden (Hard and Sivik, 1981) and adopted as a national standard in Sweden (SS 01 91 02 and SS 01 91 03) and a few other European countries. The NCS is based on Hering's opponent-colors theory. The hue circle is broken up into four quadrants defined by the unique hues red, yellow, green, and blue, as illustrated in Figure 5-4(a). The four unique hues are arranged orthogonally with equal numbers of steps between them. Thus, while the NCS hues are spaced with equal, perceived intervals between each unique hue, the intervals are of different magnitude within each of the four quadrants. This is because there are, for example, more visually distinct hues between unique red and unique blue than between unique yellow and unique green. Perceived hues that fall between the unique hues are given notations representing the relative perceptual composition of the two neighboring unique hues. For example, an orange hue that is perceived to be midway between unique red and unique yellow would be given the notation Y50R.

(a)

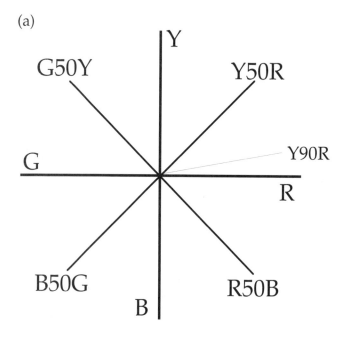

(b)

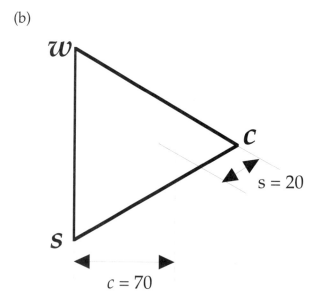

Figure 5-4. A graphical representation of (a) the hue circle and (b) a blackness/ chromaticness plane of constant hue in the NCS.

Once the NCS hue notation is established, the remaining two dimensions of relative color appearance are specified on trilinear axes, as illustrated in Figure 5-4(b). The three corners of the triangle represent colors of maximal blackness (S), whiteness (W), and chromaticness (C). For any stimulus, the whiteness, blackness, and chromaticness must sum to 100. Thus the sample of maximum blackness is denoted as $s=100$, $w=0$, and $c=0$. The sample of maximum whiteness is denoted as $s=0$, $w=100$, and $c=0$, and the sample of maximum chromaticness is denoted $s=0$, $w=0$, and $c=100$. Since the three numbers must sum to 100, only two are required for a complete specification (along with the hue notation). Typically, blackness and chromaticness are used. For example, an intermediate sample might be denoted $s=20$ and $c=70$, implying a whiteness of $w=10$.

The maximum chromaticness for each hue is defined using a mental anchor of the maximally chromatic sample that could be perceived for that hue. Thus there is no direct relationship between Munsell chroma and NCS chromaticness. Likewise, there is no simple relationship between the Munsell Value and NCS blackness.

It is also important to note that the samples of maximum chromaticness are of different relative luminance and lightness for the various hues. The Munsell system and NCS represent two different ways to specify perceptual color appearance. It is not possible to say that one is better than the other; it can only be stated that the two differ. This was recently reaffirmed in the report of CIE TC1-31 (CIE, 1996a), which was requested by ISO to recommend a single color-order system as an international standard along with techniques to convert from one to another.

A color is denoted in the NCS by its blackness *(s)*, chromaticness *(c)*, and hue. For example, the stimulus described in the previous section with a Munsell designation of 7.5YR 5/10 has an NCS designation of 20, 70, Y90R. This suggests that the sample is nearly a unique red, with only 10% yellow content. It is further described as being highly chromatic (70%) with only a small amount of blackness (20%). Note that even though this sample is of medium Munsell Value, it is of substantially lower blackness (or higher whiteness) in the NCS. This illustrates the fundamental difference between the Munsell Value scale and the NCS whiteness-blackness-chromaticness scale.

Like the Munsell system, the NCS is embodied in an atlas and specified by CIE tristimulus values based on extensive visual observations. The NCS atlas includes 40 different hues and samples in steps of 10 along the blackness and chromaticness scales. Since it is not possible to produce all of the possible samples, due to limitations in pigments, there are approximately 1,500 samples in the atlas. The NCS atlas should be viewed under daylight illumination with appropriate luminance levels and background.

NCS samples are also available in various sizes for different specialized applications. As a Swedish national standard, the NCS is taught at a young age and used widely in color communication in Sweden, thereby enabling an enviable level of precision in everyday color communication.

5.4   The Colorcurve System

A recently developed color-order system is the Colorcurve system (Stanziola, 1992), which was designed as a color communication system that represents a combination of a color-appearance system and a color-mixture system. The system is designed such that colors can be specified within the system and then spectral reflectance data for each sample can be used to formulate matching samples in various materials or media. Thus each sample in the system is specified not only by its colorimetric coordinates, but also by its spectral reflectance properties.

The Colorcurve system uses the CIELAB color space as a starting point. Eighteen different L^* levels were chosen at which to construct constant-lightness planes in the system. The L^* levels range from 30 to 95 in steps of five units, with a few extra levels incorporated at the higher lightness levels that are particularly important in design (e.g., light colors are popular for wall paint). At each lightness level, nine starting points were selected. These consisted of one gray ($a^*=0$, $b^*=0$) and eight chromatic colors with chroma, C^*, of 60. The chromatic starting points were red (60, 0), orange (42.5, 42.5), yellow (0, 60), yellow/green (-42.5, 42.5), green (-60, 0), blue/green (-42.5, -42.5), blue (0, -60), and purple (42.5, -42.5). Thus the starting points were defined using principles of a color-appearance space.

The remainder of the system was constructed using additive color mixing. Each quadrant of the CIELAB a^*b^* plane was filled with a rectangular sampling of additive mixtures of the gray and three chromatic starting points in that quadrant. Equal steps in the Colorcurve designations represent equal additive mixtures among the four starting points. These principles were used to define all of the aim points for the Colorcurve system. The samples were then formulated with real pigments such that the system could be embodied along with the desired spectral reflectance curve specifications.

The system is embodied in two atlases made up of samples of nitrocellulose lacquer coated on paper. The Master Atlas consists of about 1,200 samples at the 18 different lightness levels. There is also a Gray and Pastel Atlas made up of 956 additional samples that more finely sample the regions of color space that are near grays or pastels. Since the Colorcurve system is specified by the spectral reflectance characteristics of the samples, the viewing illumination is not critical as long as a spectral match is made to the color curve sample. If a spectral match is made, the sample produced will match the Colorcurve sample under all light sources. This is not possible with other color-order systems. Like the other color-order systems, Colorcurve samples can be obtained in a variety of forms and sizes for different applications.

One unique attribute of the Colorcurve system is of particular interest. The samples in the atlases are circular rather than square as found in most systems. The circular samples avoid two difficulties with color atlases. The first is the contrast illusion of dark spots that appear at the corners between square samples (the Hermann grid illusion). The second is that it is impossible to mount a circular sample crooked!

5.5   Other Color-order Systems

The Munsell and Natural Color systems are the most important color-order systems in the study of color appearance models. The Colorcurve system provides an interesting combination of color-appearance and color-mixture systems that could provide a useful source of samples for research in color appearance and reproduction. However, many other color-order systems have been created for a variety of purposes. Derefeldt (1991) and Wyszecki and Stiles (1982) provide more details, but there are a few systems that warrant mention here. These include the OSA Uniform Color Scales, the DIN system, and the Ostwald system and are discussed in the following sections.

Uniform Color Scales System

The Optical Society of America (OSA) set up a committee on Uniform Color Scales in 1947. The ultimate results of this committee's work were described by MacAdam (1974, 1978) as the OSA Uniform Color Scales system, or OSA UCS. The OSA UCS is a color-appearance system, but it differs significantly from either the Munsell system or NCS.

OSA UCS is designed such that a given sample is equal in perceptual color difference from each of its neighbors in three-dimensional color space (not simply one dimension at a time, as in the Munsell system). The OSA UCS space is designed in a three-dimensional Euclidean geometry, with L, j, and g axes representing lightness, yellowness-blueness, and redness-greenness, respectively. To make each sample equally spaced from each of its neighbors, a regular rhombohedral sampling of the three-dimensional space is required in which each sample has 12 nearest neighbors, all at an equal distance from each other. If the 12 points of the nearest neighbors to a sample are connected, they form a polyhedron called a *cubo-octohedron*. Such sampling allows rectangularly sampled planes of the color space to be viewed from a variety of directions. Figure 5-5(a) shows a computer-graphics representation of two adjacent constant lightness planes in the OSA UCS illustrating the sampling scheme. Figure 5-5(b) illustrates a three-dimensional representation of the OSA system. The objective of equal color differences in all directions results in a very different type of color-order system. Perhaps due to its complex geometry (and the lack of a useful embodiment), the OSA UCS is not very popular. However, it does provide another set of data that could be used in the evaluation of color appearance and color-difference models. The OSA UCS space was also specified in terms of equations to transform from CIE coordinates to the OSA UCS's L, j, and g coordinates. Unfortunately, the equations are not invertible; this limits their practical utility. The equations and sample point specifications for the OSA UCS can be found in Wyszecki and Stiles (1982).

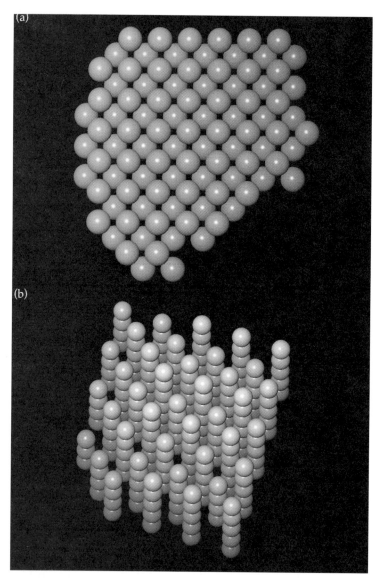

Figure 5-5.   A color rendering of samples from the OSA UCS system in (a) a pair of adjacent constant lightness planes and (b) a three-dimensional projection.

Deusches Institut für Normung System

The Deusches Institut für Normung (DIN) system was developed in Germany with the perceptual variables of hue, saturation, and darkness. An historical overview of the DIN system was presented by Richter and Witt (1986).

The specification of colors in the DIN system is closely related to colorimetric specification on a chromaticity diagram. Colors of equal hue in the DIN system fall on lines of constant dominant (or complementary) wavelength on the chromaticity diagram (i.e., straight lines radiating from the white point). Colors of constant DIN saturation represent constant chromaticities. The sampling of DIN hue and saturation is designed to be perceptually uniform. DIN darkness is related to the luminous reflectance of the sample relative to an ideal sample (a sample that either reflects all or none of the incident energy at each wavelength) of the same chromaticity. The result is a darkness scale that is similar to NCS blackness rather than the Munsell Value. The DIN system is embodied in the DIN Color Chart, which includes constant hue pages with a rectangular sampling of darkness and saturation. Thus columns on a DIN page represent constant chromaticity (DIN saturation) and appear as a shadow series (a single object illuminated at various levels of the same illuminant). The DIN charts also illustrate that chromaticity differences become less distinguishable as darkness increases. The bottom row of any given DIN page appears uniformly black.

Ostwald System

The Ostwald system has been widely used in art and design and is therefore of substantial historical interest (Derefeldt, 1991). Like the NCS, the Ostwald system is based on Hering's opponent-colors theory. However, the Ostwald system, much like the Colorcurve system, represents a combination of a color-appearance system and a color-mixture system.

Ostwald used Hering's four unique hues to set up a hue circle. However, rather than placing the perceptually opponent hues opposite one another, he used colorimetric complements (chromaticities connected by a straight line through the white point on a chromaticity diagram) in the opposite positions on the hue circle.

The Ostwald system also includes a trilinear representation of white content, black content, and full-color content on each constant-hue plane. In the NCS, these planes are defined according to perceptual color scales. However, in the Ostwald system they were defined by additive color mixtures of the three maxima located at the corners of the triangles. Thus the Ostwald system was set up with color appearance in mind, but the samples were filled in using additive color mixing. (This is exactly analogous to the much more recent formulation of the Colorcurve system based on CIELAB.)

5.6   Uses of Color-order Systems

Color-order systems have a variety of applications in the study of color appearance and related areas. These include use as samples in experiments, color design, communication, education, model testing, test targets, and others in which physical samples are helpful.

Color-order Systems in Visual Experiments

Often in visual experiments aimed at studying color appearance, one must view and/or match a variety of colored stimuli under different viewing conditions. Color-order systems provide a useful source of samples for such experiments. For example, an experimenter might select a collection of Munsell, NCS, or Colorcurve samples to scale in a color-appearance experiment. These samples will have well-known characteristics. When researchers publish the notations of the samples used, they provide a useful definition of the stimuli that can be used by others to replicate the experiments. Note that using actual samples from the color-order systems, and not just their designations on arbitrary samples, has the advantage that the reflectance characteristics of the samples are also defined.

A related use of color-order systems in appearance experiments involves teaching the system to observers and then asking them to assign designations to samples viewed under a variety of conditions (Munsell and NCS are particularly useful in this type of experiment). This allows a specification of the change in appearance caused by various changes in viewing conditions. This specification can be used along with the colorimetric specifications of each sample in each viewing condition to formulate and test color appearance models.

Color-order Systems in Art and Design

Color-order systems are often used in art and design. Their very nature as an orderly arrangement of colors allows designers to easily select samples with various color relationships. For example, with the Munsell system it is simple to select a range of colors of constant lightness or hue or to select hues that complement one another in various ways. Color-mixing systems provide this utility in addition to providing some insight for artists to help them actually produce the colors in various media. The color-order systems not only provide a design tool. They also incorporate a communication tool in their designations that allows the chosen colors to be communicated to those producing the materials to be incorporated into a design.

Color-order Systems in Communication

Clearly, precise communication of color appearance is an application for color-order systems. This is effective as long as the people on both ends of the communication link are viewing the systems in properly controlled environments. While colorimetric coordinates have the potential to provide much more precise, accurate, and useful specifications of colors, the perceptual meaning is not so readily apparent to various users. A color-order system can provide a more readily accessible communication tool. It can also be used to describe a color appearance to someone who is familiar with the system, but who may not have an atlas. An interesting example of this type of communication can be found in the ANSI specifications for viewing of color images (ANSI, 1989) in which the backgrounds are specified in terms of the Munsell Value when a reflectance factor alone is sufficient and potentially more precise.

Color-order Systems in Education

Color-order systems are immensely useful in education regarding color appearance (as well as many other aspects of color).

For example, examination of the Munsell system allows a visual definition of the color-appearance attributes of lightness, chroma, and hue. Moving pages in the *Munsell Book of Color* from a low luminance level to a high luminance level allows for a nice demonstration of how brightness and colorfulness increase substantially, while lightness and chroma remain nearly constant.

The DIN system is useful to illustrate the difference between chroma and saturation and how saturation is related to a shadow series (a single object illuminated by decreasing illuminance levels of the same spectral power distribution).

Color-order systems can also be educational in their limitations. For example, in the Munsell system, a constant Value is defined as constant relative luminance. However, it is well known that as samples increase in chroma at constant relative luminance, they appear lighter (see the Helmholtz-Kohlrausch effect described in Chapter 6). One needs only to examine a series of Munsell samples of constant Value and varying chroma to see that indeed there is a large and systematic variation in lightness.

Lastly, systems like the NCS can be a great aid in education regarding the opponent-colors theory of color vision, particularly in their hue designations, which closely follow the physiological encoding of color.

Color-order Systems to Evaluate Mathematical Color Appearance Models

Since color-order systems such as Munsell and NCS are based on perceptual scaling of color appearance, they provide readily available data that can be used to evaluate mathematical color appearance models.

For example, the Munsell system includes planes of constant lightness and hue and cylindrical surfaces of constant chroma. The tristimulus specifications of the Munsell system can be converted into the appropriate color-appearance predictors for a given color appearance model in order to see how well it predicts the constant hue, lightness, and chroma contours. Such an evaluation provides a useful, widely understood technique for the intercomparison of various color appearance models.

Intercomparison of the predictions of Munsell and NCS contours in various models also allows further study to understand the fundamental differences between the two systems.

Color-order Systems and Imaging Systems

Color-order systems can also be used as sources for test targets for imaging systems or other measurement devices. For example, the Macbeth Color Checker Chart (McCamy et al., 1976) is a commonly used test target for imaging systems that is partially based on samples from the Munsell system. Despite its common use, the chart incorporates only a small sample of colors (24) and an incomplete sampling of color space.

Targets of greater practical utility could fairly easily be constructed. Samples from various color-order systems can be used to develop custom test targets that can be reliably specified and replicated elsewhere.

Limitations of Color-order Systems

While color-order systems have a variety of useful applications in color appearance, they are not a substitute for a color appearance model. In general, they suffer from two significant limitations in this regard. First, they are not specified mathematically in relation to physically measurable values. While both the Munsell system and NCS have colorimetric specifications for each sample in the system, there are no equations to relate the colorimetric coordinates to the perceptual coordinates of the color-order systems. Approximate equations have been derived by statistical fitting and neural network modeling, but the only reliable technique for transformation from CIE colorimetry to color-order system coordinates remains look-up table interpolation. So, the lack of mathematical definitions in the forward direction precludes the possibility of the analytical inverse models for the required reverse direction.

Second, color-order systems have been established as perceptual scales of color appearance for a single viewing condition. They provide no data with respect to the changes in color appearance induced by changes in viewing conditions.

5.7   Color-naming Systems

There are a variety of color-specification systems available that do not meet the requirements for consideration as true color-order systems, but they are useful for some practical applications. Generally such systems, more properly called *color-naming systems,* are not arranged in a perceptually ordered manner (although some are arranged in order according to some imaging process) and are not presented or specified for controlled viewing conditions. In addition, the physical embodiments of these systems are not controlled with the colorimetric accuracy required for precise color communication. Examples of such systems include the Pantone, Trumatch, Toyo, and Focoltone systems.

The Pantone System

The main component of the Pantone system is the *Pantone Color Formula Guide.* This guide is a swatch book containing 1,012 Pantone spot-color ink mixtures on coated and uncoated stock. Each swatch has a numerical identifier that can be used to communicate the desired color to a printer. The printer mixes the spot-color ink using the prescribed Pantone formula. The resulting printed color should be a reasonable approximation of the color in the swatch book. This system is the prevalent tool for the specification of spot color in the United States.

The Pantone system also includes the *Pantone Process Color Imaging Guide.* This guide has 942 color swatches that show the Pantone spot colors that can be reasonably well simulated with a four-color (CMYK) printing process. The swatch book includes a patch of the Pantone spot color adjacent to the process-color simulation to indicate the variance that can be expected.

The Trumatch System

The *Trumatch Colorfinder* is a swatch book that has more than 2,000 process-color samples. These samples are arranged in an order that is slightly more perceptually based than the Pantone system. Such a system allows computer users to select CMYK color specifications according to the appearance of printed swatches rather than relying on the approximate color represented on a CRT display for a given CMYK specification.

The user finds the desired color in the swatch book and sets the particular area in an image to those CMYK values. The user then proceeds to ignore the often inappropriate appearance of the computer display, confident that the final printed color will be a fairly close approximation to the color selected in the swatch book.

Other Systems

In addition to the Pantone and Trumatch systems, process color guides that can be used as shortcuts to specification of colors that are ultimately to be printed, there is the *PostScript Process Color Guide* published by Agfa that includes over 16,000 examples of process colors representing a complete sampling of CMY combinations from 0% to 100% (dot coverage) in 5% increments, with additional samples incorporating four different levels of black ink coverage. The samples are presented on both coated and uncoated stock.

Given the variability in printing inks, papers, and processes, all of these systems can be considered only as approximate guides. They are known to be not very stable, and replacement of swatch books every six months or so is recommended. Regardless, using them is far better than working with no guides and uncalibrated/uncharacterized imaging systems. On the other hand, a system in which all of the imaging devices have been carefully calibrated and characterized, and in which viewing conditions are carefully controlled, will be capable of easily producing superior color accuracy and precision for within-gamut colors. For out-of-gamut colors (such as metallic inks that cannot be simulated on a two-dimensional computer graphics display), a swatch-book system might still prove invaluable.

CHAPTER 6

Color-appearance Phenomena

6.1   What Are Color-appearance Phenomena?
6.2   Simultaneous Contrast, Crispening and Spreading
6.3   Bezold-Brücke Hue Shift (Hue Changes with Luminance)
6.4   Abney Effect (Hue Changes with Colorimetric Purity)
6.5   Helmholtz-Kohlrausch Effect (Brightness Depends on Luminance and Chromaticity)
6.6   Hunt Effect (Colorfulness Increases with Luminance)
6.7   Stevens Effect (Contrast Increases with Luminance)
6.8   Helson-Judd Effect (Hue of Nonselective Samples)
6.9   Bartleson-Breneman Equations (Image Contrast Changes with Surround)
6.10   Discounting-the-Illuminant
6.11   Other Context and Structural Effects
6.12   Color Constancy?

CHAPTER 3 DESCRIBES the fundamental concepts of basic colorimetry. While the CIE system of colorimetry has proven to be extremely useful, it is important to remember that it has limitations. Most of its limitations are inherent in the design of a system of tristimulus values based on color matching. Such a system can accurately predict color matches for an average observer, but it incorporates none of the information necessary for specifying the color appearance of those matching stimuli. Such is the realm of color appearance models. Tristimulus values can be considered as a nominal (or at best ordinal) scale of color. They can be used to state whether two stimuli match. The specification of color differences requires interval scales, and the description of color appearance requires interval scales (for hue) and ratio scales (for brightness, lightness, colorfulness, and chroma). Additional information is needed, in conjunction with tristimulus values, to derive these more sophisticated scales.

Where is this additional information found? Why is it necessary? What causes tristimulus colorimetry to "fail"? These questions can be answered through examination of various color-appearance phenomena, several of which are described in this chapter. These phenomena represent instances in which one of the necessary requirements for the success of tristimulus colorimetry is violated. Understanding what causes these violations and the nature of the discrepancies enables the construction of color appearance models.

6.1 What Are Color-appearance Phenomena?

Two stimuli with identical CIE XYZ tristimulus values will match in color for an average observer as long as certain constraints are followed. These constraints include the retinal locus of stimulation, the angular subtense, and the luminance level. In addition, the two stimuli must be viewed with identical surrounds, backgrounds, size, shape, surface characteristics, illumination geometry, and so on. If any of these constraints are violated, it is likely that the color match will no longer hold. However, in many practical applications, the constraints necessary for successful color-match prediction using simple tristimulus colorimetry cannot be met. It is these applications that require colorimetry to be enhanced to include the influences of these variables. Such enhancements are color appearance models. The various phenomena that "break" the simple XYZ tristimulus system are the topics of the following sections of this chapter.

Figure 6-1 illustrates a simple example of one color-appearance phenomenon: *simultaneous contrast*, or *induction*. In Figure 6-1(a), the two gray patches with identical *XYZ* tristimulus values match in color because they are viewed under identical conditions (both on the same gray background). If one of the gray patches is placed on a white background and the other on a black background, as in Figure 6-1(b), the two patches no longer match in appearance. Their tristimulus values, however, remain equal. Since the constraint that the stimuli are viewed in identical conditions is violated in Figure 6-1(b), tristimulus colorimetry can no longer predict a match. Instead, a model that includes the effect of background luminance factor on the appearance of the patches would be required.

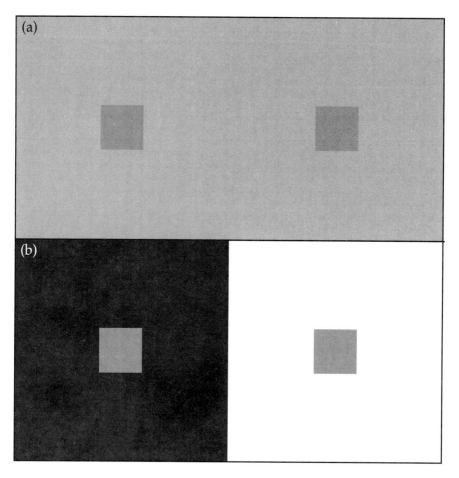

Figure 6-1.   An example of simultaneous contrast. The gray patches on the gray background (a) are physically identical to those on the white and black backgrounds (b).

Simultaneous contrast is just one of the many color-appearance phenomena described in this chapter. Discussion of other phenomena will address the effects of changes in surround, luminance level, illumination color, cognitive interpretation, and other viewing parameters. These phenomena justify the need to develop color appearance models and define the required input data and output predictions.

6.2   Simultaneous Contrast, Crispening, and Spreading

Simultaneous contrast, crispening, and spreading are three color-appearance phenomena that are directly related to the spatial structure of the stimuli.

Simultaneous Contrast

Figure 6-1 illustrates simultaneous contrast. The two identical gray patches presented on different backgrounds appear distinct. The black background causes the gray patch to appear lighter, while the white background causes the gray patch to appear darker. *Simultaneous contrast* causes a stimulus to shift in color appearance when the color of its background is changed. These apparent color shifts follow the opponent-colors theory of color vision in a contrasting sense along the opponent dimensions. In other words, a light background induces a stimulus to appear darker, a dark background induces a lighter appearance, red induces green, green induces red, yellow induces blue, and blue induces yellow. Josef Albers (1963), in his classic study *Interaction of Color,* explores various aspects of simultaneous contrast and teaches artists and designers how to avoid the pitfalls and take advantage of the effects. More-complete explorations of the effect are available in classic color-vision texts such as Hurvich (1981), Boynton (1979), and Evans (1948). Cornelissen and Brenner (1991) explore the relationship between adaptation and chromatic induction based on the concept that induction can be at least partially explained by localized chromatic adaptation. Blackwell and Buchsbaum (1988a) describe some of the spatial and chromatic factors that influence the degree of induction.

Robertson (1996) presented an interesting example, reproduced in Figure 6-2, of chromatic induction that highlights the complex spatial nature of this phenomenon. The red squares in Figure 6-2(a) or the cyan squares in Figure 6-2(b), are all surrounded by the same chromatic edges (two yellow edges and two blue edges for each square). If chromatic induction were strictly determined by the colors at the edges, then all of the red squares and all of the cyan squares should appear similar. However, it is clear from Figure 6-2 that the squares that appear to be falling on the yellow stripes are subject to induction from the yellow and thus appear darker and bluer. On the other hand, the

(a)

(b)

Figure 6-2. Stimuli patterns that illustrate the complexity of simultaneous contrast. The local contrasts for the left and right sets of squares are identical. However, simultaneous contrast is apparently driven by the stripes on which the square patches appear to rest. Tilt the page to one side and view the figure at a grazing angle to see an even larger effect.

136

squares falling on the blue stripes appear lighter and yellower. Clearly, the simultaneous contrast for these stimuli depends more on the spatial structure than simply the local edges.

Crispening

A related phenomenon is crispening. *Crispening* is the increase in perceived magnitude of color differences when the background on which the two stimuli are compared is similar in color to the stimuli themselves. Figure 6-3 illustrates crispening for a pair of gray samples. The two gray stimuli appear to be of greater lightness difference on the gray background than on either the white or black background. Similar effects occur for color differences. Semmelroth (1970) published a comprehensive study on the crispening effect along with a model for its prediction.

Figure 6-3.   An example of crispening. The pairs of gray patches are physically identical on all three backgrounds.

Spreading

When the stimuli increase in spatial frequency, or become smaller, the simultaneous contrast effect disappears and is replaced with a spreading effect. *Spreading* is the apparent mixture of a color stimulus with its surround. This effect is complete at the point of spatial fusion when the stimuli are no longer viewed as discrete but fuse into a single stimulus (such as when a halftone image is viewed at a sufficient distance such that the individual dots cannot be resolved). Spreading, however, occurs at spatial frequencies above those at which fusion occurs. Thus while the stimuli are still observed as distinct from the background, their colors begin to blend.

Classic studies by Chevreul (1839) explored the importance of spreading and contrast in the design of tapestries. In such cases, it was often desired to preserve the color appearance of design elements despite changes in spatial configuration and background color. Thus the tapestry designers were required to physically change the colors used throughout the tapestry in order to preserve color appearance. A related, although more complex, phenomenon known as neon spreading is discussed by Bressan (1993). Neon spreading is an interesting combination of perceptions of spreading and transparency.

Figure 6-4 illustrates both simultaneous contrast and spreading along a color dimension. Colorimetrically achromatic stimuli patches of various spatial frequency are presented on a red background. For the low-frequency (large) patches, simultaneous contrast takes place and the patches appear slightly greenish. However, at higher spatial frequencies (small patches), spreading occurs and the patches appear pinkish. The dependency on spatial

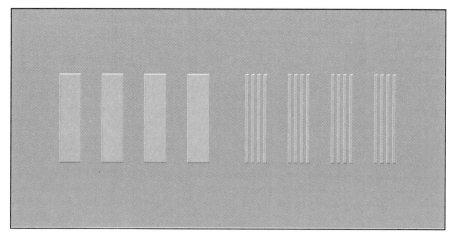

Figure 6-4. A demonstration of the difference between simultaneous contrast and spreading. The gray areas are physically identical but appear different depending on their spatial scale with respect to the red background.

frequency can be explored by examining Figure 6-4 from various viewing distances. Simultaneous contrast and spreading point to lateral interactions and adaptation effects in the human visual system.

6.3   Bezold-Brücke Hue Shift (Hue Changes with Luminance)

It is often assumed that hue can be specified by the wavelength of a monochromatic light. Unfortunately, this is not the case, as illustrated by phenomena such as the Bezold-Brücke hue shift. This hue shift occurs when one observes the hue of a monochromatic stimulus while changing its luminance. The hue will not remain constant.

Some typical experimental results on the Bezold-Brücke hue shift have been reported by Purdy (1931). Figure 6-5 represents some of the results from

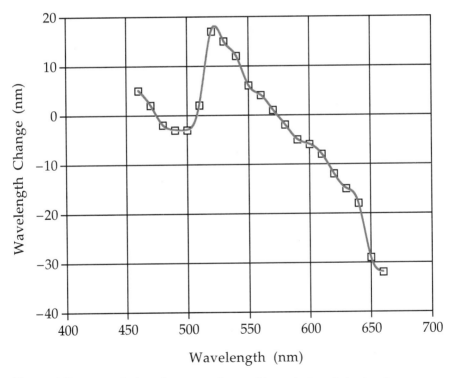

Figure 6-5.   Example data illustrating the Bezold-Brücke hue shift. Plot shows the wavelength shift required to maintain constant hue across a 10X reduction in luminance.

the Purdy (1931) research. The data in the figure indicate the change in wavelength required to preserve a constant hue appearance across a reduction in luminance by a factor of 10. For example, to match the hue of 650-nm light at a given luminance would require a light of 620 nm at one-tenth the luminance level (-30-nm shift). Recall that a given monochromatic light will have the same relative tristimulus values no matter what the luminance level (since absolute luminance level is usually not considered in tristimulus colorimetry). Thus tristimulus values alone would predict that the color of a monochromatic light should remain constant at all luminance levels. Purdy's results clearly disprove that hypothesis and point to the need to consider the absolute luminance level in order to predict color appearance.

The Bezold-Brücke hue shift suggests that there are nonlinear processes in the visual system after the point of energy absorption in the cones, but prior to the point that judgments of hue are made. Hunt (1989) has shown that the Bezold-Brücke hue shift does not occur for related colors.

6.4   Abney Effect (Hue Changes with Colorimetric Purity)

If one were to additively mix white light with a monochromatic light of a given wavelength, the mixture would vary in colorimetric purity while retaining a constant dominant wavelength. Perhaps it is reasonable that a collection of such mixtures, falling on a straight line between the white point and the monochromatic stimulus on a chromaticity diagram, would be of constant perceived hue. However, as the Bezold-Brücke hue shift illustrated, the wavelength of a monochromatic stimulus is not a good physical descriptor of perceived hue. Mixing a monochromatic light with white light also does not preserve constant hue. This phenomenon is called the *Abney effect*.

The Abney effect can be illustrated by plotting lines of constant perceived hue for mixtures of monochromatic and white stimuli. Such results, from a study by Robertson (1970), are illustrated in Figure 6-6. That figure shows several lines of constant perceived hue based on psychophysical results from three observers. The curvature of lines of constant perceived hue in chromaticity diagrams holds up for other types of stimuli as well. This can be illustrated for object-color stimuli (related stimuli) by examining lines of constant Munsell hue from the Munsell renotation studies published by Newhall (1940), an example of which is illustrated in Figure 6-7.

To summarize, the Abney effect points out that straight lines radiating from the white point in a chromaticity diagram are not lines of constant hue. Like the Bezold-Brücke effect, the Abney effect suggests nonlinearities in the visual system between the stages of cone excitation and hue perception. This was discussed by Purdy (1931). Recent experimental data on the Bezold-Brücke hue shift and Abney effect have been published by Ayama et al. (1987).

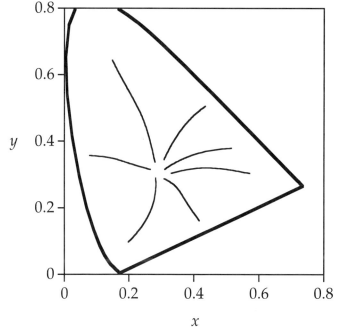

Figure 6-6.   Contours of constant hue in the CIE 1931
chromaticity diagram illustrating the Abney effect.

6.5   Helmholtz-Kohlrausch Effect (Brightness Depends on Luminance and Chromaticity)

In the CIE system of colorimetry, the Y tristimulus value defines the luminance, or luminance factor, of a stimulus. Since luminance is intended to represent the effectiveness of the various stimulus wavelengths in evoking the perception of brightness, it is often erroneously assumed that the Y tristimulus value produces a direct estimate of perceived brightness.

One phenomenon that confirms this error is the Helmholtz-Kohlrausch effect. This effect is best illustrated by examining contours of the constant brightness-to-luminance ratio, as shown in Figure 6-8, which is adapted from Wyszecki and Stiles (1982). The contours represent chromaticity loci of constant perceived brightness at a constant luminance. The labels on the contours represent the brightness of those chromaticities relative to the white point, again with constant luminance. These contours illustrate that, at constant luminance, perceived brightness increases with increasing saturation. They also illustrate that the effect depends on hue.

Various approaches have been taken to model the Helmholtz-Kohlrausch effect. One involves using the Ware and Cowan equations (Hunt, 1991a).

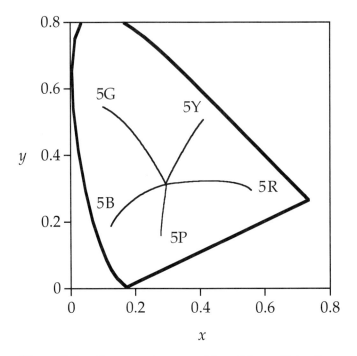

Figure 6-7.   Contours of constant Munsell hue at Value 5 plotted in the CIE 1931 chromaticity diagram.

These equations rely on the calculation of a correction factor that depends on chromaticity, as shown in Equation 6-1:

$$F = 0.256 - 0.184y - 2.527xy + 4.656x^3y + 4.657xy^4 \qquad (6\text{-}1)$$

Correction factors are calculated for all of the stimuli in question, and two stimuli are deemed to be equally bright if the equality in Equation 6-2 holds:

$$\log(L_1) + F_1 = \log(L_2) + F_2 \qquad (6\text{-}2)$$

In Equation 6-2, L is luminance and F is the correction factor given by Equation 6-1.

The Ware and Cowan equations were derived for unrelated colors. Similar experiments have shown that the Helmholtz-Kohlrausch effect also holds for related colors. A review of some of this research and a derivation of a simple predictive equation was published by Fairchild and Pirrotta (1991). In this work, a correction to the CIELAB lightness predictor, L^*, was derived as a function of CIELAB chroma, C^*_{ab}, and hue angle, h_{ab}. The predictor of chromatic lightness, L^{**}, had the form of Equation 6-3:

$$L^{**} = L^* + f_2(L^*)f_1(h_{ab})C^*_{ab} \qquad (6\text{-}3)$$

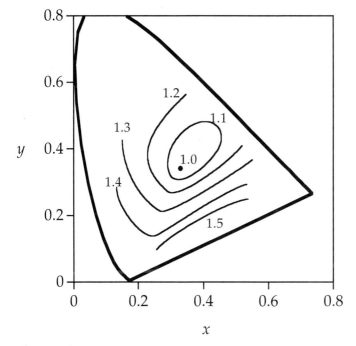

Figure 6-8. Contours of constant brightness-to-luminance ratio illustrating the Helmholtz-Kohlrausch effect.

Equation 6-3 describes the Helmholtz-Kohlrausch effect by adjusting the luminance-based predictor of lightness, L^*, with an additive factor of the chroma, C^*_{ab}, that depends on the lightness and hue of the stimulus. Details of this predictor of lightness can be found in Fairchild and Pirrotta (1991).

An example of the Helmholtz-Kohlrausch effect can be found in the samples of the *Munsell Book of Color*. Samples of constant Munsell Value have been defined to also have a constant luminance factor. Thus as one examines Munsell chips of a given hue and Value, the luminance factor is constant while chroma is changing. An examination of such sets of chips illustrates that the higher chroma chips do appear brighter and that the magnitude of the effect depends on the particular hue and Value being examined.

The Helmholtz-Kohlrausch effect illustrates that perceived brightness (and thus lightness) cannot be strictly considered a one-dimensional function of stimulus luminance (or relative luminance). As a stimulus becomes more chromatic, at constant luminance, it appears brighter. The differences between spectral luminous efficiency measured by flicker photometry [as in the $V(\lambda)$ curve] and heterochromatic brightness matching (as described by the Helmholtz-Kohlrausch effect) as a function of observer age have been examined by Kraft and Werner (1994).

6.6   Hunt Effect (Colorfulness Increases with Luminance)

Careful observation of the visual world shows that the color appearances of objects change significantly when the overall luminance level changes. Objects appear vivid and contrasty on a bright summer afternoon and more subdued at dusk. The Hunt effect and the Stevens effect (Section 6.7) describe these attributes of appearance.

The Hunt effect obtains its name from a study on the effects of light and dark adaptation on color perception published by Hunt (1952). In that study, Hunt collected corresponding colors data via haploscopic matching, in which each eye was adapted to different viewing conditions and matches were made between stimuli presented in each eye. Figure 6-9 shows a schematic representation of Hunt's results. The data points represent corresponding colors for various levels of adaptation. What these results show is that a stimulus of low

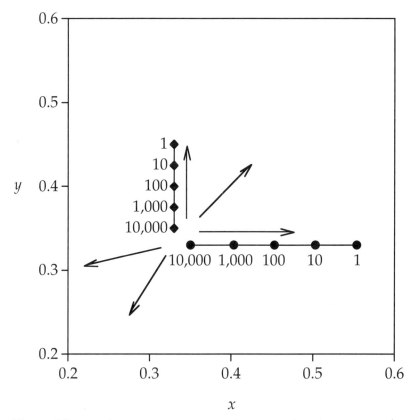

Figure 6-9.   A schematic representation of corresponding chromaticities across changes in luminance showing the Hunt effect. Points are labeled with luminance levels.

colorimetric purity viewed at 10,000 cd/m² is required to match a stimulus of high colorimetric purity viewed at 1 cd/m². Stated more directly, as the luminance of a given color stimulus is increased, its perceived colorfulness also increases.

The Hunt effect can be illustrated by viewing Figure 4-1 and imagining you are in the illuminated environment along with the rendered cubes. Notice that the sides of the cubes with more illumination falling on them appear more colorful. The Hunt effect can also be witnessed by taking a color image, such as that in Figure 4-1, and changing the level of illumination under which it is viewed. When the image is viewed under a low level of illumination, the colorfulness of the various image elements will be quite low. If the image is then moved to a significantly brighter viewing environment (e.g., a viewing booth or bright sunlight), the image elements will appear significantly more colorful.

In summary, according to the Hunt effect the colorfulness of a given stimulus increases with luminance level. This effect highlights the importance of considering the absolute luminance level in color appearance models—something that traditional colorimetry does not do.

6.7   Stevens Effect (Contrast Increases with Luminance)

The Stevens effect is a close relative of the Hunt effect. While the Hunt effect refers to an increase in chromatic contrast (colorfulness) with luminance, the Stevens effect refers to an increase in brightness (or lightness) contrast with increasing luminance. For the purposes of understanding these effects, contrast should be thought of as the rate of change of perceived brightness (or lightness) with respect to luminance. For a more complete discussion on contrast, refer to Fairchild (1995b).

Like the Hunt effect, the Stevens effect draws its name from a classic psychophysical study (Stevens and Stevens, 1963). In this study, observers were asked to perform magnitude estimations on the brightness of stimuli across various adapting conditions. The results illustrated that the relationship between perceived brightness and measured luminance tended to follow a power function. This power function is sometimes called, in psychophysics, the Stevens Power law. A relationship that follows a power function when plotted on linear coordinates becomes a straight line (with slope equal to the exponent of the power function) on log-log coordinates. Typical results from Stevens and Stevens (1963) experiments are plotted on logarithmic axes in Figure 6-10. That figure shows average relative brightness magnitude estimations as a function of relative luminance for four different adaptation levels. The figure shows that the slope of this relationship (and thus the exponent of the power function) increases with increasing adapting luminance.

The Stevens effect indicates that as the luminance level increases, dark colors will appear darker and light colors will appear lighter. While this pre-

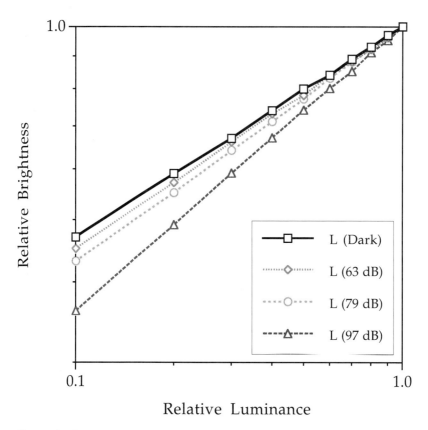

Figure 6-10.   Changes in lightness contrast as a function of adapting luminance according to the Stevens effect.

diction might seem somewhat counterintuitive, it is indeed the case. The Stevens effect can be demonstrated by viewing an image at high and low luminance levels. A black-and-white image is particularly effective for this demonstration. At a low luminance level, the image will appear to have rather low contrast. White areas will not appear very bright and, perhaps surprisingly, dark areas will not appear very dark. If the image is then moved to a significantly higher level of illumination, white areas appear substantially brighter and dark areas appear darker—the perceived contrast has increased.

6.8   Helson-Judd Effect (Hue of Nonselective Samples)

The Helson-Judd effect is elusive and perhaps cannot even be observed in normal viewing conditions. It is probably unimportant in practical situations.

However, its description is included here because two color appearance models (Hunt and Nayatani et al.) make rather strong predictions of this effect. Thus it is important to understand its definition and consider its importance when implementing those models. The experimental data first describing the Helson-Judd effect were presented by Helson (1938).

In Helson's experiment, observers were placed in a light booth (effectively a closet) that was illuminated with nearly monochromatic light. They were then asked to assign Munsell designations (after a training period) to various nonselective (Neutral Munsell patches) samples. Typical results are illustrated in Figure 6-11 for a background of Munsell Value 5/. Similar trends were observed on black-and-white backgrounds. Figure 6-11 shows the perceived chroma (in Munsell units) for nonselective samples of various Munsell Values. The results indicate that these nonselective samples did not appear neutral under strongly chromatic illumination. Samples lighter than the

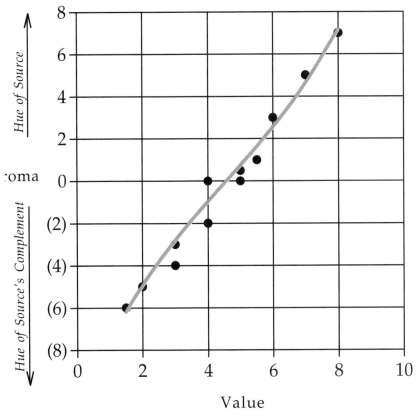

Figure 6-11.   A representation of some of the original Helson (1938) results illustrating the Helson-Judd effect. Munsell hue and chroma scaling of nonselective samples under a green source on a gray background.

background exhibited chroma of the same hue as the source, while samples darker than the background exhibited chroma of the hue of the source's complement. It is important to note that this effect occurred only for nearly monochromatic illumination. Helson (1938) stated that the effect completely disappeared if as little as 5% white light was added to the monochromatic light. Thus the effect is of little practical importance, since colored stimuli should never be evaluated under monochromatic illumination.

However, the effect is predicted by some color appearance models and has been observed in one recent experiment (Mori et al., 1991). The Mori experiment was performed with haploscopic viewing (each eye adapted differently), which might increase the chance of observing the Helson-Judd effect. However, it is not possible to observe or demonstrate the Helson-Judd effect under normal viewing conditions. This raises an interesting question with respect to the original Helson (1938) study. Why was the effect so large? While this question cannot be directly answered, perhaps the effect was caused by incomplete chromatic adaptation (explaining the hue of the light samples) and chromatic induction (explaining the hue of the dark samples). In normal viewing situations, cognitive mechanisms are thought to "discount-the-illuminant" and thus result in the preservation of achromatic appearance of nonselective samples. Perhaps Helson's monochromatic chamber and more recent haploscopic viewing experiments did not allow these cognitive mechanisms to fully function.

Another difficulty with the Helson results is that observers scaled chroma as high as 6–8 Munsell units for samples with values less than 2. Such perceptions are not possible in the surface mode, since value 2 is nearly black and an object cannot appear both that dark and highly chromatic at the same time. It seems that observers see a highly chromatic "glowing light" superimposed on the dark objects under these viewing conditions.

Recent attempts to demonstrate the Helson-Judd effect in the Munsell Color Science Laboratory have verified the unique nature of the percept. The effect cannot be observed with complex stimuli. It is observed only when individual nonselective patches are viewed on a uniform background. (Even a step tablet of nonselective stimuli of various reflectances is too complex to produce the effect.) Also, nearly monochromatic light is required, as originally reported by Helson (1938). Under these conditions, observers do report a "glowing light" of the hue complementary to the light source superimposed on the samples that are darker than the background. Interestingly, only about 50% of observers report seeing any effect at all.

While the practical importance of the Helson-Judd effect might be questionable, it does raise some interesting questions and warrants consideration, since it influences the predictions of some color appearance models.

To review, the Helson-Judd effect suggests that nonselective samples, viewed under highly chromatic illumination, take on the hue of the light source if they are lighter than the background and take on the complementary hue if they are darker than the background.

6.9 Bartleson-Breneman Equations (Image Contrast Changes with Surround)

While Stevens and Stevens (1963) showed that perceived contrast increased with increasing luminance level, Bartleson and Breneman (1967) were interested in the perceived contrast of elements in complex stimuli (images) and how it varied with luminance level and surround. They observed results similar to those described by the Stevens effect with respect to luminance changes. However, they also observed some interesting results with respect to changes in the relative luminance of an image's surround.

Their experimental results, obtained through matching and scaling experiments, showed that the perceived contrast of images increased when the image surround was changed from dark to dim to light. This effect occurs because the dark surround of an image causes dark areas to appear lighter, while having little effect on light areas (white areas still appear white despite changes in surround). Thus, since there is more of a perceived change in the dark areas of an image than in the light areas, there is a resultant change in perceived contrast.

These results are consistent with the historical requirements for optimum image tone reproduction. Photographic prints viewed in an average surround are reproduced with a one-to-one relationship between relative luminances in the original scene and the print. Photographic transparencies intended for projection in a dark surround are reproduced with a system transfer function that is a power function with an exponent of approximately 1.5 (roughly, a photographic gamma of 1.5 for the complete system). This is one reason transparencies are produced with a physically higher contrast in order to counteract the reduction in perceived contrast caused by the dark surround. Similarly, television images, typically viewed in a dim surround, are reproduced using a power function with an exponent of about 1.25 (roughly, the gamma for the complete television system). More details regarding the history and importance of surround compensation in image reproduction can be found in Hunt (1995) and Fairchild (1995a).

Bartleson and Breneman (1967) published equations that predict their experimental results quite well. Bartleson (1975), in a paper on optimum image tone reproduction, published a simplified set of equations that are of more practical value. Figure 6-12 illustrates predictions of perceived lightness as a function of relative luminance for various surround conditions according to results of the type described by Bartleson and Breneman. This plot is virtually identical to the results of Stevens and Stevens given in Figure 6-10 on logarithmic axes. The straight lines of various slopes on logarithmic axes transform into power functions with various exponents on linear axes, such as those in Figure 6-12. Color appearance models such as Hunt's and RLAB include predictions of the surround effects on perceived contrast of images.

Often, when working at a computer workstation, users turn off the room lights in order to make the CRT display appear of higher contrast. This produces a darker surround that should perceptually lower the contrast of the

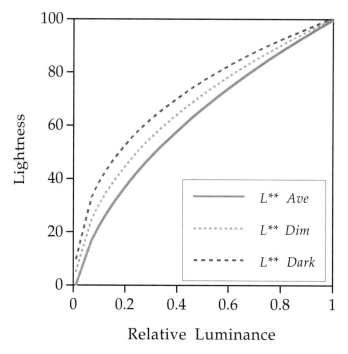

Figure 6-12.   Changes in lightness contrast as a function of surround relative luminance according to the results of Bartleson and Breneman (Bartleson, 1975).

display. The predictions of Bartleson and Breneman are counter to everyday experience in this situation. The reason for this is that the room lights are usually introducing a significant amount of reflection off the face of the monitor and thus reducing the physical contrast of the displayed images. If the surround of the display can be illuminated without introducing reflection off the face of the display (e.g., by placing a light source behind the monitor that illuminates the surrounding area), the perceived contrast of the display will actually be higher than when it is viewed in a completely darkened room.

6.10   Discounting-the-Illuminant

Mechanisms of chromatic adaptation can be classified as sensory or cognitive. It is well established (Hunt and Winter, 1975; Fairchild, 1992b, 1993a) that sensory mechanisms are not capable of complete chromatic adaptation. However, under most typical viewing conditions, observers perceive colored objects as if adaptation to the color of the illumination were complete (i.e., a white object appears white under tungsten light, fluorescent light, and daylight). Since the sensory mechanisms are incapable of mediating such perceptions, it can be shown that cognitive mechanisms (based on knowledge about objects, illumination, and the viewing environment) take over to complete the job. Further details of these different mechanisms of adaptation are presented in Chapter 8.

Discounting-the-illuminant is the cognitive ability of observers to interpret the colors of objects based on the illuminated environment in which they are viewed. This allows observers to perceive the colors of objects more independently of changes in the illumination and is consistent with the typical notion that color somehow "belongs" to an object. Discounting-the-illuminant is important to understand and has been allowed for in some color appearance models (e.g., Hunt and RLAB). It is of importance in imaging applications in which comparisons are made across various media. For example, when viewing prints observers will be able to discount the illumination color. However, when viewing a CRT display there are no illuminated objects and discounting-the-illuminant does not occur. Thus, in some situations, it might be necessary to model this change in viewing mode.

There is a significant amount of recent research that addresses issues related to changes in color appearance induced by complex stimulus structure and observer interpretation. Examples of relevant references include Gilchrist (1980), Arend and Reeves (1986), Arend and Goldstein (1987, 1990), Schirillo et al. (1990), Arend et al. (1991), Arend (1993), Schirillo and Shevell (1993, 1996), Schirillo and Arend (1995), and Cornelissen and Brenner (1995). Recent work by Craven and Foster (1992) and Speigle and Brainard (1996) addresses the ability of observers to detect changes in illumination separate from changes in object colors.

6.11   Other Context and Structural Effects

A wide variety of color-appearance effects depend on the structure and/or context of the stimuli. Some of these fall into the category of optical illusions. Others present interesting challenges to traditional colorimetry and, at times, color appearance modeling.

There are many interesting optical illusions and almost every good text on color or vision includes a number of interesting illusions (e.g., Wandell, 1995; Barlow and Mollon, 1982; Hurvich, 1981). Thus they will not be repeated here. However, a few examples do help to illustrate the importance of context and structural effects on color appearance. Figure 6-13 shows a structural illusion that has little to do with color, but it does illustrate the importance of surround. The two central circles in the figure are of physically identical diameter. However, the one surrounded by larger circles appears

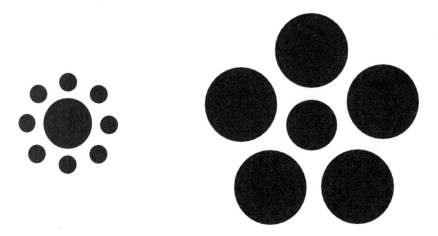

Figure 6-13. An example of simultaneous contrast in shape. The central circles in the two patterns are identical in size.

smaller than the other. While this effect does not specifically address color issues, it does show how spatial variables can influence appearance, and there is certainly an interaction between spatial and chromatic perceptions.

Various transparency effects help to illustrate the interaction of spatial and chromatic perceptions. One such demonstration has been constructed by Adelson (1993). Figure 6-14(a) shows two rows of identical gray diamonds. In Figure 6-14(b), the same diamonds have been partially placed on two different backgrounds. Since parts of the diamonds overlap the two backgrounds, the change in the appearance of the two rows is minimal. In Figure 6-14(c), tips of two different gray levels have been added to the diamonds, again with little or no effect on the appearance of the diamonds themselves. However, in Figure 6-14(d), a more complete picture has been put together in which both the backgrounds and the tips have been added to the diamonds. Now there is a significant perceptual difference between the two rows of diamonds, since the difference between them can be cognitively interpreted as either a transparency or shadow effect.

The same demonstration can be completed in color by using, for example, yellow and blue backgrounds and tips. This demonstration illustrates that it is not just the spatial configuration of the stimuli, but their perceptual interpretation that influences appearance. Additional recent psychophysical data related to stimulus structure and appearance can be found in the work of Logvinenko and Menshikova (1994) and Taya et al. (1995).

Other evidence of important structural effects on color appearance has been reported by Shevell (1993). In these experiments, simple spatial structures were added to the surrounds of colored stimuli with profound effects that could not be explained by the usual theories of simultaneous contrast and

(a)

(b)

Figure 6-14. An apparent contrast effect that depends on interpretation of spatial structure: (a) Two rows of identical gray diamonds; (b) the same diamonds with tips added that have little influence on their appearance; (c) the same diamonds on two backgrounds that have little influence on appearance, since the diamonds overlap both backgrounds; (d) the combination of the tips and backgrounds on the same diamonds. In (d), there is a striking appearance change, since the lower row of diamonds can be interpreted as light objects that are partially in a shadow or behind a filter.

(c)

(d)

153

adaptation. Such results highlight the importance of considering spatial and color variables in conjunction with one another, not as separate entities. While various color appearance models do include spatial variables in a simple way, more-complex approaches along the lines of those suggested by Poirson and Wandell (1993) need to be explored further.

Other interesting color demonstrations and effects rely to some degree on cognitive interpretation of the structure and context of the stimuli. Classic experiments on memory color that are often described in sensation and perception textbooks fall into this category. Memory color refers to the idea that observers remember prototypical colors for familiar objects. In image reproduction, objects such as sky, skin, and foliage are often mentioned (Hunt, 1995; Bartleson, 1960; Hunt et al., 1974). Other examples include simple experiments (which are quite repeatable!) in which, for example, cutouts in the shape of a tomato and a banana are made from orange construction paper and observers are asked to scale the color appearance of the two objects. The orange cutout in the shape of the banana will typically be perceived as slightly more yellow than an arbitrary shape cut out of the same paper, and the cutout in the shape of a tomato will be perceived as more red. These effects are small, but consistent, and they reiterate the importance of observers' interpretations of stimuli. Additional experimental results on the characteristics of color memory have been published by Nilsson and Nelson (1981) and Jin and Shevell (1996).

Two-color Projections

Partially related to memory color are the somewhat famous two-color projections that were demonstrated by Land (1959) and that apparently included a full range of color appearances despite not having the three primaries required by conventional colorimetric theory.

Figure 6-15 illustrates the process of the two-color projection. The original color image [Figure 6-15(a)] is separated into three black-and-white positive transparencies [Figure 6-15(b)], representing the red, green, and blue information, as done in Maxwell's (1858–62) original color photographic process. Normally, the three transparencies would be projected through red, green, and blue filters and superimposed to produce an accurate reproduction of the original [as in Figure 6-15(a)]. In the Land two-color projection, however, the red separation is projected through a red filter, the green separation is projected with white light, and the blue separation is not projected at all [Figure 6-15(c)]. While one might expect such a projection could produce only a pinkish image, the result does indeed appear to be fairly colorful (although not nearly as colorful as a three-color projection, to which Land apparently never demonstrated a direct comparison!). The quality of the two-color projection depends on the subject matter, since the effect is strengthened if memory color can be applied. The remainder of the colors appearing in the

Figure 6-15. Example of a two-color image reproduction:
(a) original full-color image; (b) red, green, and blue separations of the image; (c) combination of the red separation "projected" through a red filter and the green separation "projected" with white light.

two-color projection can be quite easily explained by simultaneous contrast and chromatic adaptation (Judd, 1960; Valberg and Lange-Malecki, 1990).

Cognitive aspects of color appearance and recognition are of significant interest but largely outside the scope of this book. An interesting monograph on the subject has been published by Davidoff (1991). It is noteworthy that the cognitive model proposed by Davidoff is consistent with the various interpretations of color appearance necessary to explain the phenomena and models described in this book. The topic of cognitive aspects of color appearance cannot be fully considered without reference to the classic work of Katz (1935), which provides fascinating insight into the topic.

6.12   Color Constancy?

Color constancy is another phenomenon that is often discussed. Typically color constancy is defined as the apparent invariance in the color appearance of objects upon changes in illumination. This definition is somewhat misleading. The main reason for this is because color constancy does not exist in humans! The data presented in the previous sections of this chapter and the discussion of chromatic adaptation in Chapter 8 should make this point abundantly clear.

An interesting thought experiment also points out the difficulty with the term "color constancy." If the colors of objects were indeed constant, then one would not have to include the light source in colorimetric calculations in order to predict color matches. In fact, color appearance models would not be necessary, as CIE XYZ colorimetry would define color appearance for all viewing conditions. Clearly, this is not the case—as demonstrated, sometimes painfully, by metameric object color matches. Such objects match in color under one light source but mismatch under others. Thus both objects in a metameric pair cannot be color constant.

Then why does the term "color constancy" exist? Perhaps a quote from Evans (1943) answers that question best: "... in everyday life we are accustomed to thinking of most colors as not changing at all. This is due to the tendency to remember colors rather than to look at them closely." When colors are closely examined, the lack of color constancy becomes extremely clear. The study of color appearance and the derivation of color appearance models, in fact, aim to quantify and predict the failure of color constancy. Examples of recent data are presented in the research of Blackwell and Buchsbaum (1988b), Foster and Nascimento (1994), and Kuriki and Uchikawa (1996).

There still remains a great deal of interest in the concept of color constancy. At first this might seem strange, since it is known not to exist in human observers. However, the study of color constancy can potentially lead to theories describing how the human visual system might strive for approximate color constancy and the fundamental limitations preventing color constancy in the real world. Such studies take place in the arena of computational color constancy with applications in machine vision (see, e.g., Maloney and Wandell, 1986; Drew and Finlayson, 1994; Finlayson et al., 1994a, b).

Jameson and Hurvich (1989) discussed some interesting concepts in regard to color constancy, the lack thereof in humans, and the utility in not being color constant that are suitable to end this chapter. They pointed out the value of having multiple mechanisms of chromatic adaptation, thus producing imperfect color constancy and retaining information about the illumination, to provide important information both about changes, such as weather, light, and time of day, and about the constant physical properties of objects in the scene.

CHAPTER 7

Viewing Conditions

7.1   Configuration of the Viewing Field
7.2   Colorimetric Specification of the Viewing Field
7.3   Modes of Viewing
7.4   Unrelated and Related Colors Revisited

SOME OF THE VARIOUS color appearance phenomena that produce the need for extensions to basic colorimetry were presented in Chapter 6. It is clear from these phenomena that various aspects of the visual field impact the color appearance of a stimulus. In this chapter, practical definitions and descriptions of the components of the viewing field that allow the development of reasonable color appearance models are given along with the required colorimetric measurements for these components. Accurate use of color appearance models requires accurate definition and measurement of the various components of the viewing field.

Different configurations of the viewing field will result in different cognitive interpretations of a stimulus and, in turn, different color perceptions. The last part of this chapter includes explanations of some of these phenomena and definitions of the various modes of viewing that can be observed for colored stimuli. Understanding these modes of viewing can help explain why seemingly physically identical stimuli can appear significantly different in color.

Related to the specification of viewing fields for color appearance models are the various definitions of standard viewing conditions used in different industries. These attempt to minimize difficulties with color appearance by defining appropriate viewing field configurations. One example of such a standard is the ANSI (1989) standard that defines viewing conditions for prints and transparencies. This document is currently under revision.

7.1 Configuration of the Viewing Field

The color appearance of a stimulus depends on the stimulus itself as well as on other stimuli that are nearby in either space or time. Temporal effects, while important, are generally not encountered in typical color-appearance applications. They are dealt with by ensuring that observers have had adequate time to adapt to the viewing environment and presenting stimuli that do not vary in time. (However, there are several recent applications, such as digital video, that will push color-appearance studies toward the domain of temporal variation.) The spatial configuration of the viewing field is always of critical importance. (Since the eyes are constantly in motion, it is impossible, in practical situations, to separate spatial and temporal effects.) The ideal spatial representation of the visual field would be to have a fully specified image of the scene. Such an image would have to have a spatial resolution greater than the visual acuity of the fovea and each pixel would be represented by a complete spectral power distribution. With such a representation of the entire visual field, one would have almost all of the information necessary to specify the color appearance of any element of the scene; however, the cognitive experience of the observer and temporal information would still be missing. Some interesting recent data on the impact of the spatial configuration of stimulus and surround were published by Abramov et al. (1992).

Such a specification of the viewing field is not practical for several reasons. First, the extensive data required are difficult to obtain accurately, even in a laboratory setting. It is not plausible to require such data in practical applications. Second, even if the data could be obtained, its huge volume would make its use quite difficult. Third, assuming these technical issues were overcome, one would then require a color appearance model capable of utilizing all of that data. Such a model does not exist and is not likely to be developed in the foreseeable future. When the inter-observer variability in color-appearance judgments is considered, such a detailed model would certainly be unnecessarily complex.

Given these limitations, the situation is simplified by defining a minimum number of important components of the viewing field. The various color appearance models use different subsets of these viewing-field components. The most extensive set is the one presented by Hunt (1991b, 1995) for use with his color appearance model. Since Hunt's definition of the viewing field includes a superset of the components required by all other models, his definitions are presented in this chapter. The viewing field is divided into four components:

1. Stimulus
2. Proximal field
3. Background
4. Surround

Figure 7-1 schematically represents these components of the visual field.

Stimulus

Stimulus is the color element for which a measure of color appearance is desired. Typically the stimulus is taken to be a uniform patch of about 2° angular subtense. Figure 7-1 illustrates a 2° stimulus when viewed from 20 cm. A stimulus of approximately 2° subtense is assumed to correspond to the visual field appropriate for use of the CIE 1931 standard colorimetric observer. The 1931 observer is considered valid for stimuli ranging from 1° to 4° in angular subtense (CIE, 1986). Trichromatic vision breaks down for substantially smaller stimuli. The CIE 1964 supplementary standard colorimetric observer should be considered for use with larger stimuli (10° or greater angular subtense).

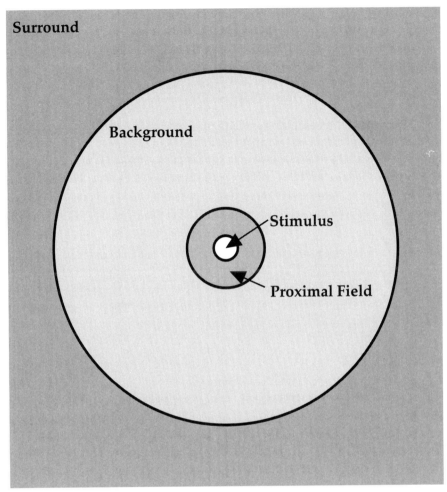

Figure 7-1.   Specification of components of the viewing field. When the figure is viewed from a distance of 20 cm, the angular subtenses are correct (i.e., the 2° stimulus area will actually subtend a visual angle of 2°).

The inhomogeneity of the retina with respect to color responsivity is a fundamental theoretical limitation to this definition of the stimulus. However, it is a practical necessity that has served basic colorimetry well since 1931. A more practical limitation, especially in imaging applications, is that the angular subtense of image elements is often substantially smaller than 2° and rarely as large as 10°. Fortunately, such limitations are often nulled, since in color reproduction, the objective is to reproduce a nearly identical spatial configuration of colors (i.e., the image). Thus any assumptions that are not completely valid are equally violated for both the original and the reproduction. Care should be taken, however, when reproducing images with significant size changes or when trying to reproduce a color from one scene in a completely different visual context (e.g., spot color or sampling color from an image).

When viewing real scenes, observers often consider an entire object as a "uniform" stimulus. For example, one might ask, What color is that car? Even though different areas of the car will produce widely different color appearances, most observers would reply with a single answer. Thus the stimulus is not a 2° field, but rather the entire object. This occurs to a limited extent in images. However, it is more conceivable for observers to break an image apart into smaller image elements.

Proximal Field

A proximal field is the immediate environment of the stimulus extending for about 2° from the edge of the stimulus in all, or most, directions. The definition of the proximal field is useful for modeling local contrast effects such as lightness or chromatic induction, crispening, or spreading. Of current models, only Hunt's (1991b) distinguishes the proximal field from the background.

While knowledge of the proximal field is necessary for detailed appearance modeling, specifying it precisely is often impractical. For example, in an image, the proximal field for any given element would be defined by its surrounding pixels. While these data are available (at least in digital imaging applications), utilizing them should require the parameters of the color appearance model to be recalculated for each spatial location within the image. Currently, such computations are prohibitive, and probably of little practical value. In cases in which the proximal field is not known, it is normally specified to be equal to the background.

Background

The background is defined as the environment of the stimulus, extending for about 10° from the edge of the stimulus (or proximal field, if defined) in all, or most, directions. Specification of the background is absolutely necessary for modeling simultaneous contrast. If the proximal field is different, its specification can be used for more complex modeling.

As with the proximal field, defining the background in imaging applications is difficult. For a given image element, the background usually consists of the surrounding image areas, the exact specification of which will change with image content and by location in the image. Thus precise specification of the background in images would require point-wise recalculation of appearance-model parameters. Since this is impractical in any typical applications, the background is usually assumed to be constant and of some medium chromaticity and luminance factor (e.g., a neutral gray with 20% luminance factor).

Alternatively, the background can be defined as the area immediately adjacent to the image. However, such a definition tends to attribute more importance to this area than is warranted. The difficulty of such definitions of background and the impact on image reproduction are discussed by Braun and Fairchild (1995, 1997a).

Fortunately, the need for precise definition of the background is minimized in most imaging applications, since the same spatial configuration of colors can be found in the original and in the reproduction. However, careful consideration of the background is critical for spot color applications, in which it is desired to reproduce the same color appearance in various spatial configurations (e.g., application of the Pantone system).

Surround

A surround is the field outside the background. In practical situations, the surround can be considered to be the entire room or the environment in which the image (or other stimulus) is viewed. For example, printed images are usually viewed in an illuminated (average) surround, projected slides in a dark surround, and video displays in a dim surround. Thus, even in imaging applications, it is easy to specify the surround. It is the area outside the image display filling the rest of the visual field.

Specification of the surround is important for modeling long-range induction, flare (stimulus and within the eye), and overall image contrast effects (e.g., Bartleson and Breneman, 1967; Fairchild, 1995b). Practical difficulties arise in specifying the surround precisely when typical situations are encountered, particularly those involving a wide range of surround relative luminances and inhomogeneous spatial configurations.

7.2   Colorimetric Specification of the Viewing Field

Various color appearance models utilize more or less colorimetric information on each component of the visual field. Essentially, it is necessary to know absolute (luminance or illuminance units) tristimulus values for each component of the field of view. However, some models require or utilize more or less data. In addition to the components of the visual field discussed in the previous section, a specification of the "adapting stimulus" is often required to implement color appearance models. The adapting stimulus is sometimes considered to be the background; at other times (or in other models), it is considered to be a measure of the light source itself. Thus it becomes necessary to specify absolute tristimulus values for the illumination or a white object under the given illumination.

When measuring the absolute tristimulus values for each of the visual fields, one must consider the standard observer used (usually the CIE 1931 (2°) standard colorimetric observer) and the geometry of measurement and viewing. It is ideal to make colorimetric measurements using the precise geometry under which the images will be viewed. Often this is not possible, and a compromise must be made. It is important to remember that this compromise, if necessary, has been made and that it can influence the correlation between model predictions and visual evaluation.

When dealing with self-luminous displays (such as CRT monitors), one can determine absolute tristimulus values in a straightforward manner by measuring the display with a colorimeter or spectroradiometer. However, when one is dealing with reflective or transmissive media, the situation becomes more complex. Normally, such materials are characterized by their spectral reflectance, or transmittance, distributions as measured with a spectrophotometer. Colorimetric coordinates (such as CIE tristimulus values) are then calculated using a standard colorimetric observer and, normally, one of the defined illuminants (e.g., CIE illuminant D65, D50, A, F2, and so on). Such measurements and calculations are usually adequate in basic colorimetric applications. However, it is an extremely rare case in which images, or other colored objects, are actually viewed under light sources that closely approximate one of the CIE illuminants (Hunt, 1992). The difference in color between that calculated using a CIE illuminant and that observed under a real source intended to simulate such an illuminant can be quite significant.

A conservative example of the differences encountered is given in Table 7-1. The spectral reflectances for seven different colors produced on a digital photographic printer were measured and used to calculate CIELAB coordinates using CIE illuminants D50 and F8. Illuminant F8 is specified as a typical fluorescent illuminant with a correlated color temperature of 5,000 K and can

Table 7-1.   CIELAB coordinates and ΔE^*_{ab} values for photographic samples evaluated using CIE Illuminants D50 and F8.

Sample	D50			F8			
	L^*	a^*	b^*	L^*	a^*	b^*	ΔE^*_{ab}
Gray	53.7	−2.6	−9.7	53.6	−3.2	−9.8	0.6
Red	39.1	41.0	20.4	39.4	41.5	21.0	0.8
Green	43.2	−41.4	22.8	42.9	−40.6	22.0	1.2
Blue	26.5	11.2	−28.8	26.4	9.2	−28.5	2.0
Cyan	64.2	−38.7	−29.4	63.7	−40.8	−30.6	2.5
Magenta	54.7	57.3	−24.8	55.0	56.3	−23.6	1.6
Yellow	85.3	0.5	63.2	85.5	2.2	63.0	1.7
Average							1.49

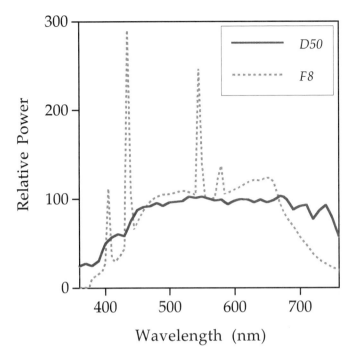

Figure 7-2. The relative spectral power distributions of CIE illuminants D50 and F8 (each normalized to 100.0 at 560 nm).

be thought of as an extremely high-quality illuminant D50 simulator. A real fluorescent lamp, intended to simulate illuminant D50 as found in typical viewing booths, would most likely have a spectral power distribution that deviates more from illuminant D50 than from F8. The spectral power distributions of illuminants D50 and F8 are illustrated (for 10-nm increments) in Figure 7-2.

The color differences in Table 7-1 are as large as 2.5, a magnitude that would be perceptible in an image and easily perceptible in simple patches. These differences, while perceptible, are probably small enough to be insignificant. However, more typical light sources will produce substantially larger errors.

A common example is encountered when colorimetric values (e.g., CIELAB coordinates) are used to color balance an imaging system. When the colorimetric coordinates indicate that printed samples should be neutral (i.e., $a^*=b^*=0.0$), significant chromatic content is often observed. This result can be traced to two causes. The first is differences between the standard illuminant used in the calculation and the light source used for observation. The second is differences between the CIE standard observer and the particular human observer making the evaluation. Differences between individual observers can be siginficant. For color reproduction stimuli, the average CIELAB ΔE^*_{ab}

between colors deemed to be matches by an individual observer is approximately 2.5, with maxima ranging up to 20 units (Fairchild and Alfvin, 1995; Alfvin and Fairchild, 1997). The former cause can be corrected by using the actual spectral power distribution of the observing condition in the colorimetric calculation. The latter is a fundamental limitation of colorimetry (indeed a limitation of any mean value) that cannot be corrected but only understood.

To summarize, it is critical to use the actual spectral power distribution of illuminating sources, rather than CIE standard illuminants, when precise estimates of color appearance are required. When this is not feasible, viewing booths with light sources that are excellent approximations of the CIE illuminants should be used.

While it would be ideal to have absolute spectral power distributions (and thus absolute tristimulus values) for each component of the viewing field, such detailed information is not needed for each model. The minimum data required for each subfield are described later in this chapter. Some models require even less data because they do not consider each of the components of the visual field. The adapting field must be specified by at least its absolute tristimulus values. (An alternative and equivalent specification is to have relative tristimulus values and the absolute luminance or illuminance.) The stimulus must also be specified by absolute tristimulus values (preferably calculated with the actual light source). Similar data are also required for the proximal field and the background, although often the background is assumed to be achromatic and can then be specified using only its luminance factor.

The color of the surround is not considered in any appearance model; it is sufficient to know the relative luminance of the surround with respect to the image (or stimulus) areas. Often, this is even more detail on the surround than is necessary and the surround can be specified sufficiently with an adjective such as dark, dim, and average. As a practical definition of surround relative luminance, dark can be taken to be 0%, dim between 0% and 20%, and average between 20% and 100% of the luminance of the scene, or image, white.

7.3   Modes of Viewing

Clearly, the mode of appearance, and thus apparent color, depends upon the stimulus configuration and cognitive interpretation. This can often be hard to accept, especially by those with an affinity for the physical sciences and engineering. It is most clearly illustrated by example and most clearly understood (and believed!) upon personal experience.

One example has been observed by the author (and others) when viewing a familiar house that was painted a light yellow color and that had a front door of the same yellow color. One time in the late evening, the door of this yellow house appeared to be painted blue. A blue door on a yellow house is a noteworthy perception! However, upon closer examination, it was determined that the door was still its normal yellow color. The illumination was such that the house was illuminated directly by the sunlight from the setting sun (quite yellow), while a small brick wall (at first unnoticed) cast a shadow that just covered the door. Thus the door was illuminated only by skylight (and no direct sunlight) and therefore appeared substantially bluer than the rest of the house. On first sight, the scene was interpreted as a yellow house and blue door under uniform illumination. However, once it was understood that the illumination was not uniform (and the door was actually illuminated with blue light), the door actually changed in appearance from blue to yellow. The change in color appearance of the door was based completely on cognitive information about the illumination and could not be reversed once the illumination was known. A simulation of this effect is illustrated in Figure 7-3, which is an ambiguous figure that can be geometrically interpreted in two ways. In one interpretation, the darker area looks like a shadow and the entire object seems to be of one yellowish color. In the other geometric interpretation, the darker area cannot

Figure 7-3. An ambiguous figure illustrating the concept of discounting the illuminant. In one spatial interpretation, the gray area looks like a shadow, while in the other, it appears to be paint on the object.

be a shadow and is interpreted as a piece of the object "painted" a different color. Turning the figure upside down can sometimes enhance the effect. This effect is similar to the transparency effect observed in Figure 6-14.

In another example, a young child was watching black-and-white photographic prints being developed in a darkroom under an amber safelight. Of course, the prints were completely achromatic and the illumination was of a single, highly chromatic color. However, the child insisted she could recognize the colors of familiar objects in a print as it was being developed. Only when the parent took the black-and-white print out into an illuminated room did the child believe that the "known" colors of the familiar objects were not present on the print. This is another example of when knowledge of the object produced a color perception. This phenomenon is completely compatible with the cognitive model of color recognition proposed by Davidoff (1991). An interesting discussion of the apparent brightness of snow, in the context of viewing modes, was presented by Koenderink and Richards (1992).

Other similar phenomena and a systematic description of the modes of viewing that produce them have been described nicely in Chapter 5 of the OSA (1963) publication, *The Science of Color*. Following are the five modes of viewing defined in the OSA chapter:

1. Illuminant
2. Illumination
3. Surface
4. Volume
5. Film

These are defined and described in the following sections. Table 7-2 summarizes the color appearance attributes (see also Chapter 4) that are most commonly associated with each mode of viewing.

Table 7-2.   Color-appearance attributes most commonly associated with the various modes of appearance. Those in parentheses are possible, although less likely.

Attribute	Illuminant (glow)	Illumination (fills space)	Surface (object)	Volume (object)	Film (aperture)
Brightness	* * *	* * *			* * *
Lightness			* * *	* * *	(* * *)
Colorfulness	* * *	* * *			* * *
Chroma			* * *	* * *	(* * *)
Hue	* * *	* * *	* * *	* * *	* * *

In addition to the normal color-appearance attributes given in the table, other attributes can be considered, such as duration, size, shape, location, texture, gloss, transparency, fluctuation, insistence, and pronouncedness (as defined by OSA, 1963). The modes of viewing described with respect to the interpretation of color appearance are strikingly similar to the types of "objects" of which efforts are made to produce realistic renderings in the field of computer graphics (Foley et al., 1990, Chapter 16). This similarity points to a fundamental link between the perception and interpretation of stimuli and their synthesis.

- Illuminant

  The *illuminant* mode of appearance is color perceived as belonging to a source of light. Illuminant-color perceptions are likely to involve the brightest perceived colors in the field of view. Thus objects much lighter than the surroundings can sometimes give rise to the illuminant-mode perception. This mode is an "object mode" (i.e., the color belongs to an object) and can be typified as "glow."

- Illumination

  The *illumination* mode of appearance is color attributed to properties of the prevailing illumination rather than to objects. The "blue door" example described earlier in this section is an example of a mode change between illuminant and surface. OSA (1963) gave an example of a perception of irregular splotches of highly chromatic yellow paint (surface mode) on the dark shadowed side of a railroad coach. When the observer came closer, noticed the angle of the setting sun, and could see penumbral gradations, he realized that the yellow patches were due to sunlight masked off by obstructions (illumination mode). The illumination mode of perception is a "non-object" mode and is mediated by the presence of illuminated objects that reflect light and cast shadows (sometimes particles in the atmosphere).

- Surface

  The *surface* mode of appearance is color perceived as belonging to a surface. Any recognizable, illuminated object provides an example of the surface mode. This mode requires the presence of a physical surface and light being reflected from the surface. It is an "object mode."

- Volume

  The *volume* mode of appearance is color perceived as belonging to the bulk of a more or less uniform and transparent substance. For example, as the number of air bubbles in a block of ice increases, the lightness of the block increases toward white, while the transparency decreases toward zero. Thus a volume color transforms into a surface color. The volume mode of perception is an "object mode" and requires transparency and a three-dimensional structure.

- Film

  The *film* mode of appearance (also referred to as *aperture mode*) is color perceived in an aperture with no connection to an object. For example, failure to focus on a surface can cause a mode shift from surface to film. An aperture screen accomplishes this, since the observer tends to focus on the plane of the aperture. The film mode of perception is a "nonobject" mode. All other modes of appearance can be reduced to film mode.

7.4   Unrelated and Related Colors Revisited

Unrelated and related colors were defined in Chapter 4. However, a revisit is warranted because of their fundamental importance in color-appearance specification, their simplifying and unifying theme with respect to modes of appearance, and their important relation to the specification of brightness-colorfulness or lightness-chroma appearance matches. The distinction between related and unrelated colors is the single most important viewing-mode concept to understand.

> **Unrelated Color:**   Color perceived to belong to an area or object seen in isolation from other colors.

> **Related Color:**   Color perceived to belong to an area or object seen in relation to other colors.

Unrelated colors exhibit only the perceptual attributes of hue, brightness, colorfulness, and saturation. The attributes that require judgment relative to a similarly illuminated white object cannot be perceived with unrelated colors. On the other hand, related colors exhibit all of the perceptual attributes of hue, brightness, lightness, colorfulness, chroma, and saturation.

Recall that five perceptual dimensions are required for a complete specification of the color appearance of related colors: brightness, lightness, colorfulness, chroma, and hue. However, in most practical color-appearance applications it is not necessary to know all five of these attributes. Typically, for related colors, only the three relative appearance attributes are of significant importance: lightness, chroma, and hue. See Chapter 4 for a discussion of the distinction between brightness-colorfulness matching and lightness-chroma matching and their relative importance.

CHAPTER 8

Chromatic Adaptation

8.1 Light, Dark, and Chromatic Adaptation
8.2 Physiology
8.3 Sensory and Cognitive Mechanisms
8.4 Corresponding-colors Data
8.5 Models
8.6 Computational Color Constancy

VARIOUS COLOR-APPEARANCE phenomena were discussed in Chapter 6. These phenomena illustrated cases in which simple tristimulus colorimetry was not capable of adequately describing appearance. Many of those phenomena could be considered second-order effects. The topic of this chapter, chromatic adaptation, is clearly the most important first-order color-appearance phenomenon. Tristimulus colorimetry tells us when two stimuli match for an average observer when viewed under identical conditions. Interestingly enough, such visual matches persist when the stimuli are viewed (as a pair) under an extremely wide range of viewing conditions. While the match persists, the color appearance of the two stimuli might be changing drastically. Changes in chromatic adaptation are one instance in which matches persist but appearance changes. It is this change in appearance that must be understood in order to construct a color-appearance model.

Chromatic adaptation is the human visual system's capability to adjust to widely varying colors of illumination in order to approximately preserve the appearance of object colors. Perhaps it is best illustrated by considering a system that does not have the capacity for chromatic adaptation—photographic transparency film. Most transparency film is designed for exposure under daylight sources. If such film is used to make photographs of objects under incandescent illumination, the resulting transparencies have an unacceptable yellow-orange cast. This is because the film cannot adjust the relative responsivities of its red, green, and blue imaging layers in the way the human visual system adjusts the responsivities of its color mechanisms. Humans perceive relatively little change in the colors of objects when the illumination is changed from daylight to incandescent.

This chapter reviews some of the basic concepts of chromatic adaptation. Issues related to chromatic adaptation have been studied for much of modern history. The topic is even discussed by Aristotle (Wandell, 1995).

> In woven and embroidered stuffs the appearance of colors is profoundly affected by their juxtaposition with one another (purple, for instance, appears different on white than on black wool), and also by differences of illumination. Thus embroiderers say that they often make mistakes in their colors when they work by lamplight, and use the wrong ones.
>
> Aristotle, *Meteorologica*

There are many excellent discussions of chromatic adaptation available in books (e.g., Barlow and Mollon, 1982; Wyszecki and Stiles, 1982; Spillman and Werner, 1990; Wandell, 1995) and journals (e.g., Terstiege, 1972; Hunt, 1976; Bartleson, 1978; Wright, 1981a; Lennie and D'Zmura, 1988). The interested reader is encouraged to explore this extensive and fascinating literature.

8.1   Light, Dark, and Chromatic Adaptation

Adaptation is the ability of an organism to change its sensitivity to a stimulus in response to changes in the conditions of stimulation. The general concept of adaptation applies to all domains of perception. The various mechanisms of adaptation can act over extremely short durations (on the order of milliseconds) or very long durations (weeks, months, or years!). In general, adaptation mechanisms make the observer less sensitive to a stimulus when the physical intensity of the stimulus is greater. For example, one might be keenly aware of the ticking of a clock in the middle of a quiet night, but completely unable to perceive the same ticking during a busy cocktail party. In the realm of vision, three types of adaptation become important—light, dark, and chromatic.

Light Adaptation

Light adaptation is the decrease in visual sensitivity upon increases in the overall level of illumination. For example, it is easy to see millions of stars on a clear night. An equivalent number and variety of stars are present in the sky on a clear day; however, we are unable to perceive them. This is because the overall luminance level of the sky is several orders of magnitude higher in the daytime than at night. This causes visual sensitivity to changes in luminance to be reduced in the daytime relative to night. Thus the luminance change that served to produce the perception of millions of stars at night is inadequate to allow their perception during the day.

As another example, imagine waking up in the middle of the night and switching on a bright room light. At first, your visual system is dazzled, you are unable to see much of anything, and you might even feel a little pain. Then, after tens of seconds, you begin to be able to view objects normally in the illuminated room. What has happened is that the mechanisms of vision were at their most sensitive in the dark room. When the light was first switched on, they were overloaded due to their high sensitivity. After a short period, they light-adapted, thus decreasing their sensitivity and allowing normal vision.

Dark Adaptation

Dark adaptation is similar to light adaptation, except that dark adaptation is changes in the opposite direction. Thus dark adaptation is the increase in visual sensitivity experienced upon decreases in luminance level. While the phenomena associated with light and dark adaptation are similar, it is useful to distinguish the two because they are mediated by different mechanisms and exhibit different visual performance.

For example, light adaptation takes place much more quickly than dark adaptation. One can experience dark adaptation when entering a dark movie theater after being outdoors in bright sunlight. At first, the theater will seem completely dark. Often people stop walking immediately upon entering a darkened room because they cannot see anything. However, after a short period objects in the room (theater seats, other people, and so on) begin to become visible. After several minutes, objects will become quite visible and there is little difficulty identifying other people, finding better seats, and so on. All of this happens because the mechanisms of dark adaptation are gradually increasing the overall sensitivity of the visual system. Light and dark adaptation in the visual system can be thought of as analogous to automatic exposure controls in cameras.

Chromatic Adaptation

The processes of light and dark adaptation have profound impacts on the color appearance of stimuli. Thus they will be considered in various color appearance models. However, chromatic adaptation, a third type of visual adaptation, is far more important and must be included in all color appearance models. *Chromatic adaptation* is the largely independent sensitivity regulation of the mechanisms of color vision. Often it is considered to be only the independent changes in responsivity of the three types of cone photoreceptors (while light and dark adaptation refer to overall responsivity changes in all of the receptors). However, it is important to keep in mind that there are other mechanisms of color vision (e.g., at the opponent level and even at the object recognition level) that are capable of changes in sensitivity that can be considered mechanisms of chromatic adaptation.

To see an example of chromatic adaptation, consider a piece of white paper illuminated by daylight. When the paper is in a room with incandescent light, it still appears white despite the fact that the energy reflected from the paper has changed from predominantly blue to predominantly yellow (this is the change in illumination to which the transparency film discussed in the introduction to this chapter couldn't adjust). Figure 8-1 illustrates such a change in illumination. Figure 8-1(a) illustrates a typical scene under daylight illumination. Figure 8-1(b) shows what the scene would look like under incandescent illumination when viewed by a visual system that is incapable of chromatic adaptation. Figure 8-1(c) illustrates the same scene viewed under incandescent illumination by a visual system capable of adaptation similar to that observed in the human visual system.

Figure 8-1. Illustration of (a) a scene illuminated by daylight; (b) the same scene illuminated by tungsten light as perceived by a visual system incapable of chromatic adaptation; and (c) the scene illuminated by tungsten light as perceived by a visual system with typical von Kries-type chromatic adaptation (similar to the human visual system). Original lighthouse image from Kodak Photo Sampler PhotoCD.

178

AFTERIMAGES

A second example of chromatic adaptation is *afterimages*. Examples of afterimages are shown in Figure 8-2. Stare at the black dot in the center of Figure 8-2 and memorize the positions of the various colors. After approximately 30 seconds, move your gaze to an illuminated white area such as a white wall or blank piece of white paper. Notice the various colors and their locations. These afterimages are the result of independent sensitivity changes of the color mechanisms. For example, the retinal areas exposed to the red area in Figure 8-2 became less sensitive to red energy during the adapting exposure, thus resulting in the cyan appearance of the afterimage when viewing a white area. This is caused by the lack of red response in this area that would normally be expected when viewing a white stimulus. Similar explanations hold for the other colors observed in the afterimage. While light adaptation can be thought of

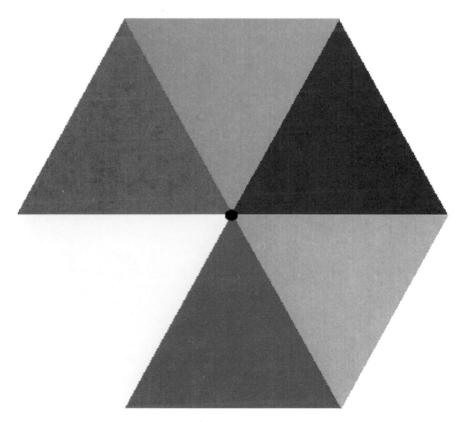

Figure 8-2. An example of afterimages produced by local retinal adaptation. Fixate the black spot in the colored pattern for about 30 seconds and then move your gaze to a uniform white area. Note the colors of the afterimages with respect to the original colors of the pattern.

as analogous to an automatic exposure control, chromatic adaptation can be thought of as analogous to an automatic white-balance feature on a video camera.

8.2   Physiology

While the various phenomena of adaptation are interesting in their own right, one must understand something of the physiological mechanisms of adaptation in order to model them properly. There are various adaptation mechanisms, ranging from strictly sensory, reflex-like responses to purely cognitive responses. All of these mechanisms are not fully understood, but it is instructive to examine them in order to later understand how they are incorporated into various models. The mechanisms discussed here are the following:

- Pupil dilation/constriction
- Rod-cone transition
- Receptor gain control
- Subtractive mechanisms
- High-level adaptation

Pupil Dilation/Constriction

The most apparent mechanism of light and dark adaptation is dilation and constriction of the pupil. In ordinary viewing situations, the pupil diameter can range from about 3 mm to 7 mm. This represents a change in pupil area of approximately a factor of 5. Thus the change in pupil size could explain light and dark adaptation over a 5X range of luminances. While this might seem significant, the range of luminance levels over which the human visual system can comfortably operate spans about ten orders of magnitude. So, while the pupil provides one mechanism of adaptation, it is insufficient to explain observed visual capabilities. There must be additional adaptational mechanisms embedded in the physiological mechanisms of the retina and beyond.

Role of the Rods and Cones

There are two classes of photoreceptors in the human retina: rods and cones. The cones are less sensitive and respond to higher levels of illumination, while the rods are more sensitive, responding to lower levels of illumination. Thus the transition from cone vision to rod vision (which occurs at luminances on the order of 0.1–1.0 cd/m²) provides an additional mechanism for light and dark adaptation.

The decrease in responsivity of the cones upon exposure to increased luminance levels (light adaptation) takes place fairly rapidly, requiring a few minutes at most. In contrast, the increase in sensitivity of the rods upon exposure to decreased luminance levels requires more time. This can be illustrated with a classic dark-adaptation curve showing the recovery of threshold after exposure to an extremely bright adapting stimulus, as illustrated in Figure 8-3. The first phase of the curve shows the recovery of sensitivity of the cones, which levels off after a couple of minutes. Then, after about 10 minutes, the rods have recovered enough sensitivity to become more sensitive than the cones and the curve takes another drop. After about a total of 20 minutes, the rods have reached their maximal sensitivity and the dark-adaptation curve levels off. This curve explains the perceptions observed over time after entering a darkened movie theater.

The rod-cone transition, in addition to providing a mechanism for light and dark adaptation, has a profound impact on color appearance. Recall that there are three types of cones to serve the requirements of color vision, but only

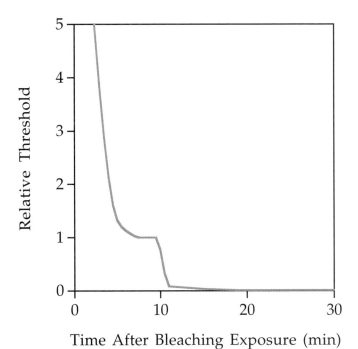

Time After Bleaching Exposure (min)

Figure 8-3. A typical dark adaptation curve showing the recovery of threshold after a strong exposure.

one type of rod. When luminance is reduced to levels at which only the rods are active, humans become effectively color-blind, seeing the world only in shades of gray. Thus the rod-cone transition is of limited interest in color appearance and chromatic-adaptation models. Other mechanisms must be considered. (*Note:* The influence of rods on color appearance can be important at low luminance levels, and it is incorporated in Hunt's color appearance model.)

Receptor Gain Control

Perhaps the most important mechanism of chromatic adaptation is independent sensitivity changes in the photoreceptors, sometimes called *receptor gain control*. It is possible to imagine a gain control that varies the relationship between the number of photons incident on a photoreceptor and the electrochemical signal it produces in response to those photons. Chromatic adaptation would be served by turning down the gain when there were many photons (high levels of excitation for the particular cone type) and turning up the gain when photons were less readily available. The key to chromatic adaptation is that these gain controls are independent in each of the three cone types. (Gain control is certainly a mechanism of light adaptation as well, but light adaptation could be served by a single gain control for all three cone types. It is overly well served by independent mechanisms of chromatic adaptation.)

Physiologically, changes in photoreceptor gain can be explained by pigment depletion at higher luminance levels. Light breaks down molecules of visual pigment (part of the process of phototransduction) and thus decreases the number of molecules available to produce further visual response. Therefore, at higher stimulus intensities, there is less photopigment available and the photoreceptors exhibit a decreased responsivity.

While pigment depletion provides a nice explanation, there is evidence that the visual system adapts in a similar way at luminance levels for which there is insignificant pigment depletion. This adaptation is thought to be caused by gain-control mechanisms at the level of the horizontal, bipolar, and ganglion cells in the retina. Gain control in retinal cells beyond the photoreceptors helps to explain some of the spatially low-pass characteristics of chromatic adaptation.

Subtractive Mechanisms

There is also psychophysical evidence for subtractive mechanisms of chromatic adaptation in addition to gain control mechanisms (e.g., Walraven, 1976; Shevell, 1978). Physiological mechanisms for such subtractive adaptation can be found by examining the temporal impulse response of the cone photoreceptors, which is biphasic and thus enhances transients and suppresses steady signals. Similar processes are found in lateral inhibitory mechanisms in the retina that produce its spatially antagonistic impulse response that enhances spatial transients and suppresses spatially uniform stimuli.

Physiological models of adaptation that require both multiplicative (gain) and subtractive mechanisms (e.g., Hayhoe et al., 1987, 1989) can be made completely compatible with models typically proposed in the field of color appearance (see Chapter 9) that include only gain controls. This can be done by assuming that the subtractive mechanism takes place after a compressive nonlinearity. If the nonlinearity is taken to be logarithmic, then a subtractive change after a logarithmic transformation is identical to a multiplicative change before the nonlinearity. This bit of mathematical manipulation serves to provide consistency between the results of threshold psychophysics, physiology, and color appearance. It also highlights the importance of compressive nonlinearities as mechanisms of adaptation.

Figure 8-4 illustrates a nonlinear response function typical of the human visual system (or any imaging system). The function exhibits a threshold level below which the response is constant and a saturation level above which the response is also constant. The three sets of inputs with 100:1 ratios at different adapting levels are illustrated. It can be seen in the figure that a 100:1 range of

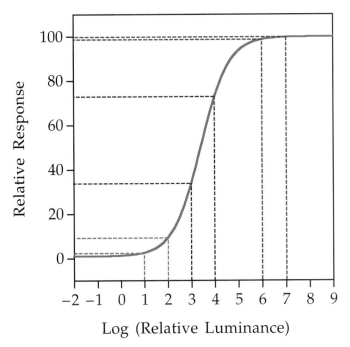

Figure 8-4. A prototypical response function for the human visual system illustrating response compression at low and high levels of the input signal.

input stimuli produces a small output range at low- and high-adapting luminance levels and a large output range at intermediate levels. The decrease in response at low levels has to do with the fundamental limitation of the
mechanism's sensitivity, while the response compression at high levels can be
considered a form of adaptation (decreased responsivity with increased input
signal). Nonlinear response functions such as the one illustrated in Figure 8-4
are required in color appearance models to predict phenomena such as the
Stevens and Hunt effects described in Chapter 6.

High-level Adaptation Mechanisms

Thus far, the mechanisms discussed have been at the front end of the visual system. These are low-level mechanisms that respond and adapt to very simple
stimulus configurations. Webster and Mollon (1994) present interesting results
that illustrate the relationship between spatial contrast, color appearance, and
higher-level visual mechanisms. There are also numerous examples of visual
adaptation that must take place at higher levels in the system (i.e., in the visual
cortex). Examples of such cortical adaptation include

- The McCollough effect
- Spatial frequency adaptation
- Motion adaptation

It is useful to consider these examples as illustrations of the potential for other
types of high-level adaptation not yet considered.

MCCOLLOUGH EFFECT

A nice example of the McCollough effect can be found in Barlow and Mollon
(1982). To experience the McCollough effect, one must intermittently view a
pattern of red and black strips in one orientation, say horizontal, and another
pattern of green and black strips of a second orientation, say vertical.
Observers view each pattern for several seconds and then switch to the other,
thus ensuring that no simple afterimages are formed. After continuing this
adaptation process for about 4 minutes, they then turn their attention to patterns of black and white strips of spatial frequency similar to the adapting patterns. What will be observed is that black-and-white patterns of a vertical
orientation will appear black and pink, and black-and-white patterns in a horizontal pattern will appear black and pale green.

This effect is contingent upon the color and orientation of the adapting
stimuli and cannot be explained as a simple afterimage. It suggests adaptation
at a cortical level in the visual system, where neurons that respond to particular orientations and spatial frequencies are first observed. The effect is also
very persistent, sometimes lasting for several days or longer!

SPATIAL FREQUENCY ADAPTATION

Spatial frequency adaptation can be observed by examining Figure 8-5. One can see this at work by doing the following. First, adapt to Figure 8-5(a) by gazing at the black bar in the center for 1 to 2 minutes. To avoid producing simple afterimages, do not fixate on a single point, but rather let your gaze move back and forth along the black bar. Next, after the adaptation period, fixate on the black dot in the middle of Figure 8-5(b). The pattern on the left in Figure 8-5(b) should look like it is of a higher spatial frequency than the pattern on the right. The two patterns in Figure 8-5(b) are identical. The difference in appearance after adaptation to Figure 8-5(a) is caused by the adaptation of mechanisms sensitive to various spatial frequencies.

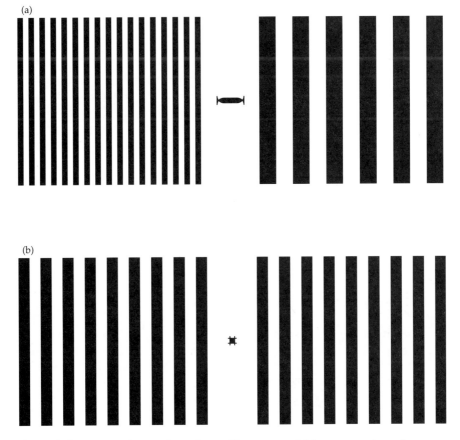

Figure 8-5.   A stimulus configuration to illustrate spatial frequency adaptation. Gaze at the black bar in the middle of (a) for about 60 seconds and then fixate on the black point in the middle of (b). Note the perceived relative spatial frequencies in the two patterns in (b) after this adaptation period.

As this example illustrates, when one is adapting to a high spatial frequency, other patterns appear to be of lower spatial frequency and vice versa. Once again, this adaptation is attributed to cortical cells that selectively respond to various spatial frequencies.

MOTION ADAPTATION

Motion adaptation provides in the temporal domain evidence similar to spatial frequency adapation.

An example of motion adaptation can be observed when one views the credits at the end of a motion picture (or text scrolling up a computer terminal). If the credits are scrolling up the screen (and being observed) for several minutes, then when the final credit is stationary on the screen, it appears to be moving downward—and going nowhere at the same time! This occurs because the cortical mechanisms selective for upward motion have become adapted while viewing the moving credits. Once the motion stops, the response of the upward and downward selective mechanisms should be nulled out, but the adapted (i.e., fatigued) upward mechanisms are not responding as strongly as they should and the stationary text appears to be moving downward.

Motion adaptation can also be observed sometimes after one drives a car on a highway for long periods of time. The visual system adapts to the motion toward the observer. Then, when the car is stopped, it can sometimes appear as if the outside world is moving away from the observer even though there is no physical motion.

These examples of cortical adaptation lead to the next logical step. That is, if there are adaptive mechanisms at such a high level in the visual system, is it possible that there are also cognitive mechanisms of adaptation? This issue is discussed in the following section.

8.3   Sensory and Cognitive Mechanisms

It is tempting to assume that chromatic adaptation can be considered a sensory mechanism that is some sort of automatic response to changes in the stimulus configuration. However, there is clear evidence for mechanisms of chromatic adaptation that depend on knowledge of the objects and their illuminated environment (Fairchild, 1992a, b, 1993a). These are cognitive mechanisms of adaptation.

Chromatic-adaptation mechanisms can be classified into two groups:

- Sensory—those that respond automatically to the stimulus energy
- Cognitive—those that respond based upon an observer's knowledge of scene content

Sensory Mechanisms

Sensory chromatic-adaptation mechanisms are well-known and have been widely discussed in the vision and color-science literature. The physiological locus of such mechanisms is generally believed to be sensitivity control in the photoreceptors and neurons in the first few stages of the visual system, as discussed previously. Most modern theories and models of sensory chromatic adaptation trace their roots to the work of von Kries (1902), who wrote:

> . . . the individual components present in the organ of vision are completely independent of one another and each is fatigued or adapted exclusively according to its own function. [trans. MacAdam, 1970]

This is not precisely correct today, but the concept is accurate and provides useful insight. To this day, the idea that chromatic adaptation takes place through normalization of cone signals is known as the *von Kries coefficient law* and serves as the basis of all modern models of chromatic adaptation and color appearance.

Cognitive Mechanisms

Cognitive mechanisms have also been long recognized in color-science literature. However, perhaps because of the difficulty of quantifying cognitive effects, they are usually discussed briefly and are not as widely recognized or understood. To help understand the idea of cognitive chromatic-adaptation mechanisms, it might be best to quote some of those who have mentioned them in the past two centuries.

Helmholtz (1866), in his treatise on physiological optics, discussed object color appearance:

> We learn to judge how such an object would look in white light, and since our interest lies entirely in the object color, we become unconscious of the sensations on which the judgement rests. [trans. Woodworth, 1938]

Hering (1920), who is known for hypothesizing the opponent-colors theory of color vision, discussed the concept of memory color:

> All objects that are already known to us from experience, or that we regard as familiar by their color, we see through the spectacles of memory color. [trans. Hurvich and Jameson, 1964]

Judd (1940), who made innumerable contributions to the field of color science, referred to two types of chromatic-adaptation mechanisms:

The processes by means of which the observer adapts to the illuminant or discounts most of the effect of a nondaylight illuminant are complicated; they are known to be partly retinal and partly cortical.

Evans (1943), who wrote and lectured on many aspects of color photography and color perception, discussed why the colors in photographs look acceptable:

> . . . in everyday life we are accustomed to thinking of most colors as not changing at all. This is in large part due to the tendency to remember colors rather than to look at them closely.

Also, Jameson and Hurvich (1989) discussed the value of having multiple mechanisms of chromatic adaptation to provide important information both about changes such as weather, light, and time of day and about constant physical properties of objects in the scene. Finally, Davidoff (1991) published a monograph on the cognitive aspects of color and object recognition.

Hard-copy Versus Soft-copy Output

While it is clear that chromatic adaptation is complicated and relies on both sensory and cognitive mechanisms, it is less clear how important it is to distinguish between the two types of mechanisms when viewing image displays. If an image is being reproduced in the same medium as the original and is viewed under similar conditions, one can safely assume that the same chromatic-adaptation mechanisms are active when viewing both the original and the reproduction. But what happens when the original is presented in one medium, such as a soft-copy display output, such as CRT display, and the reproduction is viewed in a second medium, such as a hard-copy output, such as on a piece of paper? A series of experiments have been described (Fairchild, 1992b, 1993a) that quantify some of the characteristics of chromatic-adaptation mechanisms. They indicate that the same mechanisms are not active when soft-copy output is viewed as are active when a hard-copy display or an original scene is viewed.

When a hard-copy image is being viewed, the image is perceived as an object that is illuminated by the prevailing illumination. Thus both sensory mechanisms, which respond to the spectral energy distribution of the stimulus, and cognitive mechanisms, which discount the "known" color of the light source, are active. When a soft-copy output is being viewed, it cannot easily be interpreted as an illuminated object. So there is no "known" illuminant color, and only sensory mechanisms are active. This can be demonstrated by viewing a piece of white paper under incandescent illumination and comparing its appearance to that of a CRT display of a uniform field with exactly the same chromaticity and luminance viewed in a darkened room. The paper will appear white or just slightly yellowish. The CRT display will appear relatively high-chroma yellow. In fact, a white piece of paper illuminated by that CRT display will appear white while the display itself retains a yellow appearance! Color appearance models such as RLAB and the Hunt model include provisions for various degrees of cognitive "discounting-the-illuminant."

The Time-course of Adaptation

Another important feature of chromatic adaptation mechanisms is their time-course. The time-course of chromatic adaptation for color appearance judgments has been explored in detail (Fairchild and Lennie, 1992; Fairchild and Reniff, 1995). The results of these studies suggest that the sensory mechanisms of chromatic adaptation are about 90% complete after 60 seconds for changes in adapting chromaticity at constant luminance. Sixty seconds can be considered a good general rule for the minimum duration for which observers should adapt to a given viewing environment prior to making critical judgments. Adaptation is slightly slower when significant luminance changes are also involved (Hunt, 1950).

Cognitive mechanisms of adaptation rely on knowledge and interpretation of the stimulus configuration. Thus they can be thought of as effectively instantaneous once such knowledge is obtained. However, in some unusual viewing situations, the time required to interpret the scene can be quite lengthy, if not indefinite.

8.4   Corresponding-colors Data

Corresponding-colors data are the most extensively available visual data on chromatic adaptation. *Corresponding colors* are two stimuli, viewed under differing viewing conditions, that match in color appearance. For example, a stimulus specified by the tristimulus values, XYZ_1, viewed in one set of viewing conditions, might appear the same as a second stimulus specified by the tristimulus values, XYZ_2, viewed in a second set of viewing conditions. XYZ_1 and XYZ_2, together with specifications of their respective viewing conditions, represent a pair of corresponding colors. It is important to note, however, that XYZ_1 and XYZ_2 are rarely numerically identical.

Corresponding-colors data have been obtained through a wide variety of experimental techniques. Wright (1981a) provides an historical review of how and why chromatic adaptation has been studied. Some of the techniques, along with studies that have used them, are briefly described here.

- Asymmetric matching

  Since the collection of corresponding-colors data requires a visual match to be determined across a change in viewing conditions, the experiments are sometimes referred to as *asymmetric matching experiments*. Ideally, color matches are made by direct, side-by-side comparison of the two stimuli. This is technically impossible to accomplish with two sets of viewing conditions unless some simplifying assumptions are made.

  Perhaps the most fascinating example is an experiment reported by MacAdam (1961) in which differential retinal conditioning was used. In this experiment, two different areas of the retina (left and right halves) were exposed to different adapting stimuli and then test and matching stimuli were presented in the two halves of the visual field for color matching. This technique requires the assumption that differential adaptation of the two halves of the retina is similar to adaptation in normal viewing. This assumption is likely false, and the differential retinal-conditioning technique is only of historical interest.

- Haploscopic matching

  In *haploscopic matching,* one eye is adapted to one viewing condition and the other eye is adapted to a second viewing condition. Then a test stimulus presented in one eye is compared and matched with a stimulus presented to the other eye. Haploscopic experiments require the assumption that adaptation takes place independently in the two eyes. This assumption might be valid for sensory mechanisms, but it is certainly not valid for cognitive mechanisms.

  Some of the advantages and disadvantages of haploscopic experiments in color-appearance research have been described by Fairchild et al. (1994). Hunt (1952) provides an example of a classic study using haploscopic viewing. Breneman (1987) described a clever device for haploscopic matching, and an extensive study completed by the Color Science Association of Japan (Mori et al., 1991) used haploscopic viewing with object-color stimuli.

- Memory matching

  To avoid the assumptions of differential retinal conditioning or haploscopic viewing, one must give up the precision of direct color matches in exchange for more realistic viewing conditions. One technique that allows more natural viewing is *memory matching.*

  In memory matching, observers generate a match in one viewing condition to the remembered color of a stimulus in a different viewing condition. Helson, Judd, and Warren (1952) used a variation of memory matching in which observers assigned Munsell coordinates to various color stimuli. In effect, the observers were matching the stimuli to remembered Munsell samples under standard viewing conditions. Wright (1981a) suggested that achromatic memory matching (matching a gray appearance) would be an extremely useful technique for studying chromatic adaptation. Such a technique has been used to derive a variety of corresponding-colors data (Fairchild, 1990, 1991b, 1992b, 1993a).

- Magnitude estimation

   Another technique that allows natural viewing is magnitude estimation. In *magnitude estimation,* observers assign scale values to various attributes of appearance such as lightness, chroma, and hue, or brightness, colorfulness, and hue. Such experiments can provide color-appearance data as well as corresponding-colors data. An extensive series of magnitude-estimation experiments has been reported by Luo et al. (1991a, b) and summarized by Hunt and Luo (1994).

- Cross-media comparisons

   Braun et al. (1996) published an extensive series of experiments aimed at comparing various viewing techniques for cross-media image comparisons. They concluded that a short-term memory-matching technique produced the most reliable results. It is also worthwhile to note that the Braun study showed that the common practice of comparing CRT displays and reflection prints side-by-side produces unpredictable color appearances (or, alternatively, predicted matching images that are unacceptable when viewed individually).

Given all of these experimental techniques for deriving corresponding-colors data, what can one learn from the results? Figure 8-6 illustrates corresponding-colors data from the study of Breneman (1987). The circles represent chromaticities under illuminant D65 adaptation that match the corresponding chromaticities under illuminant A adaptation plotted using triangles. Given these data, one can safely assume that the pairs of corresponding colors represent lightness-chroma matches in color appearance across the change in viewing conditions. This is the case, since lightness and chroma are the appearance parameters most intuitively judged for related colors. With this assumption, one can use the corresponding-colors data to test a color appearance model by taking the set of values for the first viewing condition, using the model to predict lightness-chroma matches for the second viewing condition, and comparing the predictions with the visual results.

This same sort of test can be completed with a simpler form of model—a *chromatic-adaptation transform,* or, more commonly, *chromatic-adaptation*

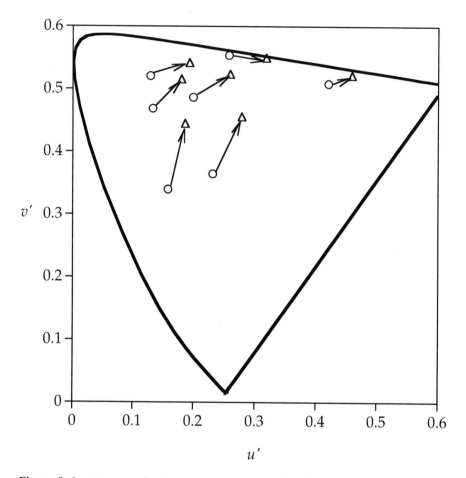

Figure 8-6. An example of corresponding-colors data for a change in chromatic adaptation from the chromaticity of illuminant D65 to that of illuminant A plotted in the *u'v'* chromaticity diagram.

model. A chromatic-adaptation model does not include correlates of appearance attributes such as lightness, chroma, and hue. Instead, it simply provides a transformation from tristimulus values in one viewing condition to matching tristimulus values in a second set of viewing conditions.

8.5   Models

As described in the previous section, a chromatic-adaptation model allows prediction of corresponding-colors data. A general form of a model can be expressed as shown in Equations 8-1 through 8-3.

$$L_a = f(L, L_{white}, \ldots) \tag{8-1}$$

$$M_a = f(M, M_{white}, \ldots) \tag{8-2}$$

$$S_a = f(S, S_{white}, \ldots) \tag{8-3}$$

This generic chromatic-adaptation model is designed to predict three cone signals, L_a, M_a, and S_a, after all of the effects of adaptation have acted upon the initial cone signals, L, M, and S. Such a model requires, as a minimum, the cone excitations for the adapting stimulus, L_{white}, M_{white}, and S_{white}. It is quite likely that an accurate model would require additional information as well (represented by the ellipses).

A chromatic-adaptation model can be converted into a chromatic-adaptation transform by combining the forward model for one set of viewing conditions with the inverse model for a second set. Often such a transform is expressed in terms of CIE tristimulus values, as shown in Equation 8-4:

$$XYZ_2 = f\left(XYZ_1, XYZ_{white1}, XYZ_{white2}, \ldots\right) \tag{8-4}$$

To accurately model the physiological mechanisms of chromatic adaptation, one must express stimuli in terms of cone excitations, *LMS*, rather than CIE tristimulus values, *XYZ*. Fortunately, cone excitations can be reasonably approximated by a linear transformation (3 × 3 matrix) of CIE tristimulus values. Thus a generic chromatic-adaptation transform can be described as shown in the flow chart in Figure 8-7. The complete process is as follows:

1. Begin with CIE tristimulus values $(X_1 Y_1 Z_1)$ for the first viewing condition.
2. Transform these values to cone excitations $(L_1 M_1 S_1)$.
3. Incorporate information about the first set of viewing conditions (VC_1) using the chromatic-adaptation model to predict adapted cone signals $(L_a M_a S_a)$.
4. Reverse the process for the second set of viewing conditions (VC_2) to determine the corresponding color in terms of cone excitations $(L_2 M_2 S_2)$ and ultimately CIE tristimulus values $(X_2 Y_2 Z_2)$.

Examples of specific chromatic-adaptation models are given in Chapter 9.

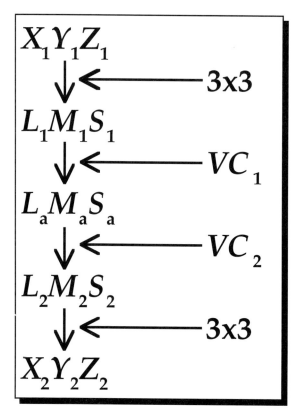

Figure 8-7.   A flow chart of the application of a chromatic-adaptation model to the calculation of corresponding colors.

Chromatic-adaptation models provide predictions of corresponding colors and thus can be used to predict required color reproductions for changes in viewing conditions. If this is the only requirement for a given application, then a chromatic-adaptation model might provide a simpler alternative to a complete color appearance model. Chromatic-adaptation models are also the basic building blocks of all color appearance models. However, they do have some disadvantages. A chromatic-adaptation model does not provide any predictors of appearance attributes such as lightness, chroma, and hue. These attributes might be necessary for some applications, such as image editing and gamut mapping. In these circumstances, a more complete color appearance model is required.

8.6   Computational Color Constancy

There is another field of study that produces mathematical models that are, at times, closely related to chromatic-adaptation models. This is the field of *computational color constancy*. The objective in this approach is to take limited color information available in a typically trichromatic representation of a scene and produce color-constant estimates of the objects. Essentially this reduces to an attempt to estimate signals that depend only on the spectral reflectances of objects and not on the illumination. On the other hand, chromatic-adaptation models aim to predict the failure of color constancy actually observed in humans.

It is simple to prove that precise color constancy is not possible, or desirable, for the human visual system. The examples of metameric object color pairs that cannot both be color constant, along with the need to include illuminants in practical colorimetry, suffice to make this point. All this means is that striving for the most color-constant model is not necessarily a good way to model the human visual system. There are, however, applications in machine vision that could benefit greatly from having the most color-constant sensors possible.

The results in the field of computational color constancy provide some interesting constraints and techniques that could help in modeling human performance. For example, the results of Maloney and Wandell (1986) illustrate limits to the accuracy with which a trichromatic system could possibly estimate surface reflectances in a natural scene. D'Zmura and Lennie (1986) show how a trichromatic visual system can provide color-constant responses for one dimension of color appearance—hue—while sacrificing constancy for the other dimensions. The work of Finlayson et al. (1994) illustrates how optimum sensory spectral responsivities can be derived to utilize the von Kries coefficient rule to obtain near color constancy.

These studies, and many more in the field, provide interesting insights into what the visual system could possibly do at the limits. Such insights can help in the construction, implementation, and testing of color appearance models. They also provide definitive answers to questions pertaining to requirements for the collection of color images (computer vision and other image scanners), the synthesis of realistic images (computer graphics), and the design of colorimetric instrumentation (imaging colorimeters). Brill and West (1986) provide a useful review of the similarities and differences in the studies of chromatic adaptation and color constancy.

CHAPTER 9

Chromatic-adaptation Models

9.1   von Kries Model
9.2   Retinex Theory
9.3   Nayatani et al. Model
9.4   Guth's Model
9.5   Fairchild's Model

CHROMATIC ADAPTATION IS THE single most important property of the human visual system with respect to understanding and modeling color appearance. This importance in vision science means there is significant literature available on various aspects of the topic. Chapter 8 reviewed some of the important properties of adaptation phenomena and mechanisms. It also provided the generic outline of a chromatic-adaptation model for predicting corresponding colors. This chapter builds upon that information by including more-detailed descriptions of a few, specific chromatic-adaptation transformations. It is impossible to cover all of the models that have been published. An attempt has been made to cover a variety of models and show their fundamental relationships to each other. Readers interested in more detail on the models or historical developments should delve into the available literature.

There are several good places to start, including the review papers cited in Chapter 8 (Bartleson, 1978; Terstiege, 1972; Wright, 1981a; Lennie and D'Zmura, 1988). Further details on the early history of chromatic-adaptation models can be found in an interesting overview in a study by Helson, Judd, and Warren (1952). Another excellent review of the entire field of color appearance with significant treatment of chromatic adaptation was written by Wyszecki (1986). Many of the classic papers in the field can be found in the collection edited by MacAdam (1993).

The models described in this chapter allow the computation of corresponding colors, but they are not color appearance models. They include no predictors of appearance attributes such as lightness, chroma, and hue. They are, however, quite useful in predicting color matches across changes in viewing conditions. This is a significant extension of tristimulus colorimetry, all that is necessary in some applications, and the fundamental basis upon which all color appearance models are constructed.

Any physiologically plausible model of chromatic adaptation must act on signals representing the cone responses (or at least relative cone responses). Thus, in applications for which the use of CIE colorimetry is important, it is necessary to first transform from CIE tristimulus values (XYZ) to cone responses (denoted LMS, RGB, or $\rho\gamma\beta$, depending on the model). Fortunately, cone responsivities can be accurately represented using a linear transformation of CIE tristimulus values. An example of such a transformation is illustrated in Figure 9-1. This transformation, or a similar one, is common to all chromatic-adaptation and color appearance models that are compatible with CIE colorimetry. Thus it is not explicitly included in every case in this book. Where the

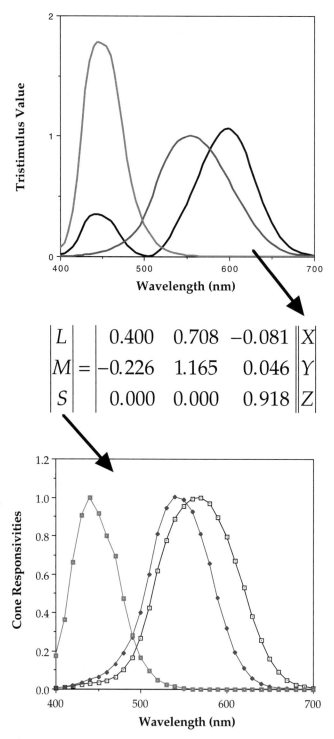

$$\begin{vmatrix} L \\ M \\ S \end{vmatrix} = \begin{vmatrix} 0.400 & 0.708 & -0.081 \\ -0.226 & 1.165 & 0.046 \\ 0.000 & 0.000 & 0.918 \end{vmatrix} \begin{vmatrix} X \\ Y \\ Z \end{vmatrix}$$

Figure 9-1. The process of transformation from XYZ tristimulus values to LMS cone responsivities using an example linear matrix multiplication.

200

particular transformation is of importance to a particular model, it is explicitly included in this and following chapters.

9.1   von Kries Model

All viable modern chromatic-adaptation models can trace their roots, both conceptually and mathematically, to the hypotheses of Johannes von Kries (1902). von Kries laid down some ideas about chromatic adaptation that, to this day, are being "rediscovered." His main idea was to propose a simple model of chromatic adaptation that would serve as a "straw man" for future research. He had fairly low expectations for this and other of his ideas, as illustrated by the following quote from MacAdam's translation of the 1902 paper:

> If some day it becomes possible to distinguish in an objective way the various effects of light by direct observation of the retina, people will perhaps recall with pitying smiles the efforts of previous decades which undertook to seek an understanding of the same phenomena by such lengthy detours.

Over nine decades later, there is no one looking back at von Kries's work with a "pitying smile." Rather, many do so with astonishment at how well it has withstood the test of time.

von Kries (1902) did not outline a specific set of equations as representative of what is today referred to as the von Kries model, the von Kries proportionality law, the von Kries coefficient law, and other similar names. He simply outlined his hypothesis in words and described the potential impact of his ideas. In MacAdam's translation of von Kries's words, he said:

> This can be conceived in the sense that the individual components present in the organ of vision are completely independent of one another and each is fatigued or adapted exclusively according to its own function.

The ideas that von Kries outlined were considered by him to be an extension of Grassmann's laws of additive color mixture to two viewing conditions.

The modern interpretation of the von Kries hypothesis in terms of a chromatic-adaptation model is expressed in Equations 9-1 through 9-3:

$$L_a = k_L L \tag{9-1}$$

$$M_a = k_M M \tag{9-2}$$

$$S_a = k_S S \tag{9-3}$$

L, M, and S represent the initial cone responses; k_L, k_M, and k_S are the coefficients used to scale the initial cone signals (i.e., gain control); and L_a, M_a, and S_a are the post-adaptation cone signals. Equations 9-1 through 9-3 represent a simple gain-control model of chromatic adaptation in which each of the three cone types has a separate gain coefficient. A key aspect of any model is how the particular values of k_L, k_M, and k_S are obtained. In most modern instantiations of the von Kries model, the coefficients are taken to be the inverse of the L, M, and S cone responses for the scene white or maximum stimulus, as illustrated in Equations 9-4 through 9-6:

$$k_L = 1/L_{max} \quad \text{or} \quad k_L = 1/L_{white} \tag{9-4}$$

$$k_M = 1/M_{max} \quad \text{or} \quad k_M = 1/M_{white} \tag{9-5}$$

$$k_s = 1/S_{max} \quad \text{or} \quad k_s = 1/S_{white} \tag{9-6}$$

Equations 9-4 through 9-6 are a mathematical representation of von Kries's statement that "each is fatigued or adapted exclusively according to its own function." Given the previous interpretations of the gain coefficients, the von Kries model can be used to calculate corresponding colors between two viewing conditions by calculating the post-adaptation signals for the first condition, setting them equal to the post-adaptation signals for the second condition, and then reversing the model for the second condition. Performing these steps and completing the algebra result in the transformations given in Equations 9-7 through 9-9 that can be used to calculate corresponding colors:

$$L_2 = \left(L_2 / L_{max1}\right)L_{max2} \tag{9-7}$$

$$M_2 = \left(M_1 / M_{max1}\right)M_{max2} \tag{9-8}$$

$$S_2 = \left(S_1 / S_{max1}\right)S_{max2} \tag{9-9}$$

In some cases, it becomes more convenient to express chromatic-adaptation models in terms of matrix transformations. The interpretation of the von Kries model is expressed in matrix notation in Equation 9-10:

$$\begin{vmatrix} L_a \\ M_a \\ S_a \end{vmatrix} = \begin{vmatrix} 1/L_{max} & 0.0 & 0.0 \\ 0.0 & 1/M_{max} & 0.0 \\ 0.0 & 0.0 & 1/S_{max} \end{vmatrix} \begin{vmatrix} L \\ M \\ S \end{vmatrix} \tag{9-10}$$

The matrix notation can be extended to the calculation of corresponding colors across two viewing conditions and to explicitly include the transformation (matrix $\mathbf{M}$) from CIE tristimulus values (XYZ) to relative cone responses (LMS). This is illustrated in Equation 9-11.

$$\begin{vmatrix} X_2 \\ Y_2 \\ Z_2 \end{vmatrix} = \mathbf{M}^{-1} \begin{vmatrix} L_{max2} & 0.0 & 0.0 \\ 0.0 & M_{max2} & 0.0 \\ 0.0 & 0.0 & S_{max2} \end{vmatrix} \begin{vmatrix} 1/L_{max1} & 0.0 & 0.0 \\ 0.0 & 1/M_{max1} & 0.0 \\ 0.0 & 0.0 & 1/S_{max1} \end{vmatrix} \mathbf{M} \begin{vmatrix} X_1 \\ Y_1 \\ Z_1 \end{vmatrix} \quad (9\text{-}11)$$

The von Kries transformation was used to predict the visual data of Breneman (1987) that were described in Chapter 8. The results are illustrated in a $u'v'$ chromaticity diagram in Figure 9-2. The open symbols represent Breneman's corresponding-colors data, and the filled symbols represent the predictions using a von Kries model. Perfect model predictions would result in the filled triangles' completely coinciding with the open triangles. In this calculation, the chromaticities under daylight adaptation (open circles in Figure

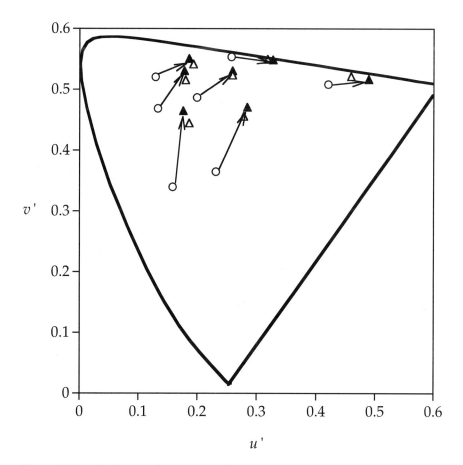

Figure 9-2. Prediction of some example corresponding-colors data using the von Kries model. Open triangles represent visual data, and filled triangles represent model predictions.

9-2) were used to predict the corresponding chromaticities under incandescent adaptation (triangles in Figure 9-2). It is clear in Figure 9-2 that the von Kries hypothesis was indeed a good one and that the modern interpretation as a chromatic-adaptation transformation predicts the data surprisingly well.

Helson, Judd, and Warren (1952) presented an early study in which corresponding colors were derived by memory matching and the von Kries hypothesis was tested and performed quite well. Examples of recent experimental data and analyses that address the utility and limitations of the von Kries hypothesis can be found in the work of Brainard and Wandell (1992) and Chichilnisky and Wandell (1995). There are some discrepancies between these, and other, visual data and the predictions of the von Kries model. Such discrepancies have led investigators down many paths that are described in the remaining sections of this chapter and throughout this book. Perhaps it shouldn't be too surprising to realize that von Kries (1902) himself foresaw this. The next line after his description of what is now referred to as the von Kries model reads:

> But if the real physiological equipment is considered, on which the processes are based, it is permissible to doubt whether things are so simple.

Indeed, things are not so simple. But it is amazing how close such a simple hypothesis comes to explaining the majority of the chromatic-adaptation phenomena.

9.2   Retinex Theory

An often-discussed account of the mechanisms of chromatic adaptation under the rubric of color constancy is the retinex theory developed by Edwin Land and his colleagues (e.g., Land and McCann, 1971; Land, 1977, 1986) The *retinex theory* can be considered an enhanced version of the von Kries model. Various enhancements to the theory have been proposed. However, its key feature is that the retinex algorithm explicitly treats the spatial distribution of colors in a scene in order to better model the visual perceptions that can be observed in complex scenes.

Land's theory was formulated to explain demonstrations of the independence of color appearance on the spectral distribution of reflected light (tristimulus values). He suggested that color appearance is controlled by surface reflectances rather than by the distribution of reflected light. The retinex algorithm, in its most recent form (Land, 1986), is quite simple. Land proposed three color mechanisms with the spectral responsivities of the cone photoreceptors. He called these mechanisms *retinexes* because they are thought to be some combination of retinal and cortical mechanisms. He hypothesized a three-dimensional color-appearance space with the output of the long-, middle-, and short-wavelength sensitive retinexes as the dimensions. The output of a retinex is determined by taking the ratio of the signal at any given point in the scene and normalizing it with an average of the signals in that retinex throughout the scene.

The most interesting feature of this algorithm is that it acknowledges variations in color due to changes in the background of the stimulus. The influence of the background can be varied by changing the spatial distribution of the retinex signals that are used to normalize a given point in the scene. If one takes the normalizing signal to be the scene average for a given retinex, then the retinex algorithm reduces to a typical instantiation of a von Kries-type transformation.

There are some flaws in the physiological implementation of the retinex model (Brainard and Wandell, 1986; Lennie and D'Zmura, 1988), but if one is more interested in the algorithm output than in having a strict physiological model of the visual system (which is also the case for most color appearance models), then the concepts in the retinex theory might prove useful. For example, the retinex algorithm has recently been applied in the development of a digital image-processing algorithm for dynamic range compression and color correction (Jobson et al., 1997). Other applications, challenges, and successes for such a theory have been reviewed by McCann (1993).

The need to consider spatial as well as spectral dimensions in high-level color appearance models is undeniable. Concepts embedded in the retinex theory provide some insight on how this might be accomplished. Other approaches are also under development (e.g., Poirson and Wandell, 1993; Zhang and Wandell, 1996). The retinex theory sets the stage for other developments in chromatic-adaptation models. The general theme is that the von Kries model provides a good foundation, but it needs enhancement in order to address certain adaptation phenomena.

9.3   Nayatani et al. Model

One important enhancement to the von Kries hypothesis is the nonlinear chromatic-adaptation model developed by Nayatani and coworkers. This nonlinear model was developed from a colorimetric background (enhancement to CIE tristimulus colorimetry) within the field of illumination engineering. The early roots of this model can be traced to the work of MacAdam (1961).

MacAdam's Model

MacAdam (1961) described a nonlinear model of chromatic adaptation in which the output of the cones was expressed as a constant plus a multiplicative factor of the cone excitation raised to some power. This nonlinear model represented an empirical fit to MacAdam's (1956) earlier chromatic-adaptation data. Interestingly enough, MacAdam required a visual system with five types of cones in order to explain his data with a linear model! (This is probably because MacAdam used a rather unusual experimental technique in which two halves of the same retina were differentially adapted.) MacAdam's nonlinear model provided a good fit to the data and was the precursor of later nonlinear models.

Nayatani's Model

The nonlinear model of Nayatani et al. (1980, 1981) begins with a gain adjustment followed by a power function with a variable exponent. In this model, the von Kries coefficients are proportional to the maximum long-, middle-, and short-wavelength cone responses and the exponents of the power functions depend on the luminance of the adapting field. The power function nonlinearity was suggested in the classic brightness study by Stevens and Stevens (1963).

Another interesting and important feature of the nonlinear model is that noise terms are added to the cone responses. This helps to model threshold behavior. Equations 9-12 through 9-14 are generalized expressions of the nonlinear model:

$$L_a = a_L \left(\frac{L + L_n}{L_0 + L_n} \right)^{\beta_L} \tag{9-12}$$

$$M_a = a_M \left(\frac{M + M_n}{M_0 + M_n} \right)^{\beta_M} \tag{9-13}$$

$$S_a = a_S \left(\frac{S + S_n}{S_0 + S_n} \right)^{\beta_S} \tag{9-14}$$

L_a, M_a, and S_a are the cone signals after adaptation; L, M, and S are the cone excitations; L_n, M_n, and S_n are the noise terms; and L_0, M_0, and S_0 are the cone excitations for the adapting field. β_L, β_M, and β_S are the exponents and are monotonically increasing functions of the respective cone excitations for the adapting field. a_L, a_M, and a_S are coefficients determined by the principle that exact color constancy holds for a nonselective sample of the same luminance factor as the adapting background.

The formulations for the exponents can be found in Nayatani et al. (1982). Takahama et al. (1984) extended the model to predict corresponding colors for backgrounds of various luminance factors. A version of the model (Nayatani et al., 1987) was accepted for field trial by the CIE. This meant that the CIE, through its technical committee activities, wanted to collect additional data to test the model, possibly improve it, and determine whether it or some other model should be recommended for general use. The results of the field trials were inconclusive, so the CIE did not make a recommendation on this model. Refinements of the model were made during the course of field trials. These are summarized in a CIE technical report (CIE, 1994), which provides full details of the current formulation of the model.

The nonlinear model was used to predict Breneman's (1987) corresponding colors. The results, analogous to those presented in Figure 9-2 for the von Kries model, are illustrated in Figure 9-3. The predictions are quite good but not as good as those of the simple von Kries model (for these particular data).

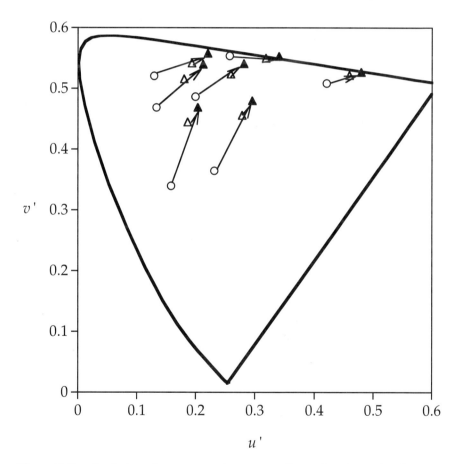

Figure 9-3.   Prediction of some example corresponding-colors data using the nonlinear model of Nayatani et al. Open triangles represent visual data, and filled triangles represent model predictions.

One reason for this is that Breneman's data were collected under viewing conditions for which discounting-the-illuminant could not occur and thus chromatic adaptation was less complete. This is illustrated by the predictions of the Nayatani model, which are all shifted toward the yellow side of the visual data. This indicates that the incandescent adapting field in Breneman's experiment retained some yellowish appearance. Recent enhancements (Nayatani, 1997) have provided techniques to estimate and account for the degree of chromatic adaptation in various experiments.

This nonlinear model is capable of predicting the Hunt (1952) effect (increase in colorfulness with adapting luminance), the Stevens effect (increase in lightness contrast with luminance), and the Helson-Judd effect (hue of nonselective samples under chromatic illumination). It is worth noting that the von Kries chromatic-adaptation transform is luminance-independent and therefore cannot be used to predict appearance phenomena that are functions of luminance. Also, the linear nature of the simple von Kries transform precludes it from predicting the Helson-Judd effect.

Nayatani's nonlinear model is important for several reasons. It provides a relatively simple extension of the von Kries hypothesis that is capable of predicting several additional effects. It also has had a significant historical impact on the CIE work in chromatic-adaptation and color appearance modeling. Further, it provides the basis for one of just two comprehensive color appearance models. The full Nayatani color appearance model is described in Chapter 11.

9.4   Guth's Model

There are many variations of the von Kries hypothesis. One significant variation that is in some ways similar to the Nayatani model, from the field of vision science (rather than colorimetry), is the model described by Guth (1991, 1995). Guth's model is not directly related to CIE tristimulus colorimetry because the cone responsivities used are not linear transformations of the CIE color-matching functions. This produces some practical difficulties in implementing the model. For practical situations, however, it is advisable and certainly not harmful to the predictions to use a set of cone responsivities that can be derived directly from CIE tristimulus values, along with the remainder of Guth's formulation. This model, which is part of the ATD vision model described in Chapter 14, has been developed over many years to predict the results of various vision experiments. Most of these experiments involve classical threshold psychophysics rather than the scaling of the various dimensions of color appearance as defined in Chapter 4.

The general form of Guth's chromatic-adaptation model is given in Equations 9-15 through 9-20:

$$L_a = L_r \left[1 - \left(L_{r0} / (\sigma + L_{r0}) \right) \right] \qquad (9\text{-}15)$$

$$L_r = 0.66 L^{0.7} + 0.002 \qquad (9\text{-}16)$$

$$M_a = M_r \left[1 - \left(M_{r0} / (\sigma + M_{r0}) \right) \right] \qquad (9\text{-}17)$$

$$M_r = 1.0 M^{0.7} + 0.003 \qquad (9\text{-}18)$$

$$S_a = S_r \left[1 - \left(S_{r0} / (\sigma + S_{r0}) \right) \right] \qquad (9\text{-}19)$$

$$S_r = 0.45 S^{0.7} + 0.00135 \qquad (9\text{-}20)$$

L_a, M_a, and S_a are the cone signals after adaptation; L, M, and S are the cone excitations; and L_{r0}, M_{r0}, and S_{r0} are the cone excitations for the adapting field after the nonlinear function. σ is a constant (nominally 300) that can be thought of as representing a noise term. It is also important to note that the cone responses in this model must be expressed in absolute units, since the luminance level does not enter the model elsewhere.

Some algebraic manipulation of the adaptation model as expressed here helps to illustrate its relationship with the von Kries model. If the initial nonlinearity is ignored, a von Kries-type gain control coefficient for the Guth model can be pulled out of Equation 9-15, as shown in Equation 9-21:

$$k_L = 1 - \left(L_{r0} / (\sigma + L_{r0}) \right) \qquad (9\text{-}21)$$

If one uses the algebraic substitutions illustrated in Equations 9-22 through 9-24, the relationship to the traditional von Kries coefficient becomes clear. The difference lies in the σ term, which can be thought of as a noise factor that is more important at low stimulus intensities than at high intensities. Thus, as luminance level increases, the Guth model becomes more and more similar to the nominal von Kries model:

$$k_L = \left((\sigma + L_{r0}) / (\sigma + L_{r0}) \right) - \left(L_{r0} / (\sigma + L_{r0}) \right) \qquad (9\text{-}22)$$

$$k_L = (\sigma + L_{r0} - L_{r0}) / (\sigma + L_{r0}) \qquad (9\text{-}23)$$

$$k_L = \sigma / (\sigma + L_{r0}) \qquad (9\text{-}24)$$

Figure 9-4 shows the Guth model prediction of Breneman's (1987) corresponding-colors data. The calculations were carried out using the nominal, published form of the Guth model. Note that there is a systematic deviation between the observed and predicted results. This discrepancy can be traced to the σ parameter. The Breneman data are fairly well predicted using a simple von Kries model. Thus, if the σ parameter was made smaller, the pre-

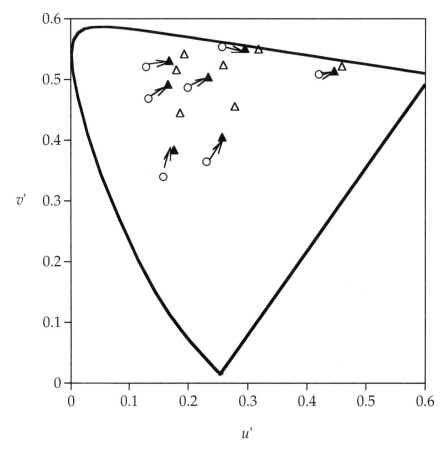

Figure 9-4.   Prediction of some example corresponding-colors data using the Guth model. Open triangles represent visual data, and filled triangles represent model predictions.

diction of the Guth model would improve. This highlights a feature (or a draw-back) of the Guth model. As a framework for a vision model, it is capable of making impressive predictions of available data. However, the model often re-quires a small amount of adjustment in its parameters for any given viewing condition or experiment. This is acceptable when trying to predict various ob-served phenomena. However, it is not practical in many applications such as cross-media color reproduction, where the viewing conditions are often not known until it is time to calculate a predicted image and there is no chance for iterations. Thus, to apply the Guth adaptation model (and the full ATD model described in Chapter 14) to such situations, one must apply some interpreta-tion of how to implement the model.

9.5   Fairchild's Model

The Breneman (1987) results showing incomplete chromatic adaptation inspired a series of experiments (Fairchild, 1990) aimed at measuring the degree of chromatic adaptation to various forms of adapting stimuli. This work led to the development of yet another modification of the von Kries hypothesis that included the ability to predict the degree of adaptation based on the adapting stimulus itself (Fairchild, 1991a, b). This model, like Nayatani's model, is designed to be fully compatible with CIE colorimetry. However, it is more rooted in the field of imaging science than in illumination engineering. It was designed to be a relatively simple model and to include discounting-the-illuminant, the Hunt effect, and incomplete chromatic adaptation.

The model is most clearly formulated as a series of matrix multiplications. The first step is a transformation from CIE tristimulus values, XYZ, to fundamental tristimulus values, LMS, for the first viewing condition, as shown in Equations 9-25 and 9-26. The Hunt-Pointer-Estevez transformation with illuminant D65 normalization is used.

$$\begin{vmatrix} L_1 \\ M_1 \\ S_1 \end{vmatrix} = \mathbf{M} \begin{vmatrix} X_1 \\ Y_1 \\ Z_1 \end{vmatrix} \tag{9-25}$$

$$\mathbf{M} = \begin{vmatrix} 0.4002 & 0.7076 & -0.0808 \\ -0.2263 & 1.1653 & 0.0457 \\ 0.0 & 0.0 & 0.9182 \end{vmatrix} \tag{9-26}$$

The next step is to apply a modified form of the von Kries chromatic-adaptation transform that takes incomplete chromatic adaptation into account, as illustrated in Equations 9-27 through 9-31:

$$\begin{vmatrix} L'_1 \\ M'_1 \\ S'_1 \end{vmatrix} = \mathbf{A}_1 \begin{vmatrix} L_1 \\ M_1 \\ S_1 \end{vmatrix} \tag{9-27}$$

$$A = \begin{vmatrix} a_L & 0.0 & 0.0 \\ 0.0 & a_M & 0.0 \\ 0.0 & 0.0 & a_S \end{vmatrix} \tag{9-28}$$

$$a_M = \frac{p_M}{M_n} \tag{9-29}$$

$$p_M = \frac{\left(1 + Y_n^v + m_E\right)}{\left(1 + Y_n^v + 1/m_E\right)}$$

(9-30)

$$m_E = \frac{3\left(M_n/M_E\right)}{L_n/L_E + M_n/M_E + S_n/S_E}$$

(9-31)

The p and a terms for the short- (S) and long- (L) wavelength sensitive cones are derived in a similar fashion. Y_n is the luminance of the adapting stimulus in cd/m^2. Terms with n subscripts refer to the adapting stimulus, while terms with E subscripts refer to the equal-energy illuminant. The exponent, v, is set equal to 1/3.

The form of these equations for incomplete adaptation is based on those used in the Hunt (1991b) color appearance model, which is described in more detail in Chapter 12. (A separate chromatic-adaptation transformation was never published by Hunt; thus Hunt's model is treated in full in Chapter 12.)

When cognitive discounting-the-illuminant occurs, the p_L, p_M, and p_S terms are all set equal to 1.0. The a terms are modified von Kries coefficients. The p terms represent the proportion of complete von Kries adaptation. They depart from 1.0 as adaptation becomes incomplete. The p values depend on the adapting luminance and color. As the luminance increases, the level of adaptation becomes more complete. As the adapting chromaticity moves farther and farther from a normalizing point (the equal-energy illuminant), adaptation becomes less complete. Equations 9-30 and 9-31 serve to ensure this behavior in the model. These predictions are consistent with the available experimental data (Hunt and Winter, 1975; Breneman, 1987; Fairchild, 1992b).

The final step in the calculation of post-adaptation signals is a transformation that allows luminance-dependent interaction among the three cone types, as shown in Equations 9-32 through 9-34:

$$\begin{vmatrix} L_a \\ M_a \\ S_a \end{vmatrix} = \mathbf{C}_1 \begin{vmatrix} L'_1 \\ M'_1 \\ S'_1 \end{vmatrix}$$

(9-32)

$$\mathbf{C} = \begin{vmatrix} 1.0 & c & c \\ c & 1.0 & c \\ c & c & 1.0 \end{vmatrix}$$

(9-33)

$$c = 0.219 - 0.0784\log_{10}(Y_n)$$

(9-34)

The c term was derived from the work of Takahama et al. (1977) called the *linkage model*. In that model, which the authors later gave up, citing a preference for the nonlinear model (Nayatani et al., 1981), the interaction terms were introduced to predict luminance-dependent effects. This is the same reason the C matrix was included in this model.

To determine corresponding chromaticities for a second adapting condition, one must derive the A and C matrices for that condition, invert them, and apply them as shown in Equations 9-35 through 9-37:

$$\begin{vmatrix} L'_2 \\ M'_2 \\ S'_2 \end{vmatrix} = C_2^{-1} \begin{vmatrix} L_a \\ M_a \\ S_a \end{vmatrix}$$

(9-35)

$$\begin{vmatrix} L_2 \\ M_2 \\ S_2 \end{vmatrix} = A_2^{-1} \begin{vmatrix} L'_2 \\ M'_2 \\ S'_2 \end{vmatrix}$$

(9-36)

$$\begin{vmatrix} X_2 \\ Y_2 \\ Z_2 \end{vmatrix} = M^{-1} \begin{vmatrix} L_2 \\ M_2 \\ S_2 \end{vmatrix}$$

(9-37)

The entire model can be expressed as the single matrix in Equation 9-38:

$$\begin{vmatrix} X_2 \\ Y_2 \\ Z_2 \end{vmatrix} = M^{-1} A_2^{-1} C_2^{-1} C_1 A_1 M \begin{vmatrix} X_1 \\ Y_1 \\ Z_1 \end{vmatrix}$$

(9-38)

Subsequent experiments (e.g., Pirrotta and Fairchild, 1995) showed that the C matrix introduced an unwanted luminance dependency that resulted in an overall shift in lightness with luminance level. This shift did not impact the quality of image reproductions, since the whole image shifted. However, it did introduce significant systematic error in predictions for simple object colors. Thus the model was revised (Fairchild, 1994) by eliminating the C matrix. Doing this improved the predictions for simple colors, while having no effect on the results for images. It also removed the model's capability to predict the Hunt effect. However, in imaging applications, this turns out to be unimportant, since any predictions of the Hunt effect would be counteracted by the process of gamut mapping. These changes, along with some further simplifications in the equations (different normalizations), were compiled and used as the basis for the latest version of the RLAB (Fairchild, 1996) color appearance model presented in Chapter 13.

Figure 9-5 shows predictions of the Breneman (1987) corresponding-colors data using the Fairchild chromatic-adaptation transformation. The pre-

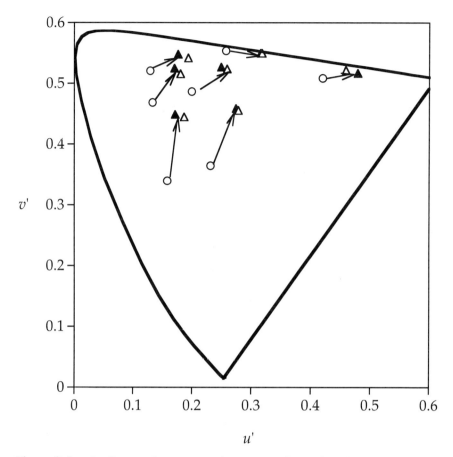

Figure 9-5. Prediction of some example corresponding-colors data using the Fairchild (1991b) model. Open triangles represent visual data, and filled triangles represent model predictions.

dictions are identical for either version (original or simplified) of the model described previously. The predictions are as good as, or better than, each of the models presented thus far. Quantitative analyses of all of Breneman's (1987) data confirm this result (Fairchild, 1991a, b).

CHAPTER 10

Color Appearance
Models

10.1 Definition of Color Appearance Model
10.2 Construction of Color Appearance Models
10.3 CIELAB
10.4 Why Not Use Just CIELAB?
10.5 What About CIELUV?

THE CHROMATIC-ADAPTATION transforms discussed in Chapter 9 go a long way toward extending tristimulus colorimetry toward the prediction of color appearance. However, they are still limited in that they can predict matches only across disparate viewing conditions (i.e., corresponding colors). A chromatic-adaptation transform alone cannot be used to describe the actual color appearance of stimuli. To do this, one must use the appearance parameters defined in Chapter 4—the absolute color-appearance attributes of brightness, colorfulness, and hue and the relative color-appearance attributes of lightness, chroma, saturation, and, again, hue. These terms are used to describe the color appearance of stimuli. Chromatic-adaptation transforms provide no measure of correlates to these perceptual attributes. This is the domain of color appearance models.

10.1 Definition of Color Appearance Model

The world of color measurement is full of various descriptors of color, such as tristimulus values, chromaticity coordinates, uniform chromaticity scales, uniform color spaces, and "just plain-old" color spaces. Sometimes it is difficult to keep all the names and distinctions straight. So just what is it that sets a color appearance model apart from all of these other types of color specification? CIE Technical Committee 1-34 (TC1-34), *Testing Colour Appearance Models*, was given the task of evaluating the performance of various color appearance models and recommending a model for general use. Thus one of the first tasks of this committee became the definition of just what constitutes a color appearance model in order to be included in the tests (Fairchild, 1995a).

TC1-34 agreed on the following definition: A color appearance model is any model that includes predictors of at least the relative color-appearance attributes of lightness, chroma, and hue. For a model to include reasonable predictors of these attributes, it must include at least some form of a chromatic-adaptation transform. Models must be more complex to include predictors of brightness and colorfulness or to model luminance-dependent effects such as the Stevens effect or the Hunt effect.

This definition enables some fairly simple uniform color spaces, such as the CIE 1976 L*a*b* color space (CIELAB) and the CIE 1976 $L^*u^*v^*$ color space (CIELUV), to be considered color appearance models. These color spaces include simple chromatic-adaptation transforms and predictors of lightness, chroma, and hue. The general construction of a color appearance model and then a discussion of CIELAB as a specific example are presented in the following sections.

10.2   Construction of Color Appearance Models

Some general concepts that apply to the construction of all color appearance models are described in this section. All color appearance models for practical applications begin with the specification of the stimulus and viewing conditions in terms of CIE XYZ tristimulus values (along with certain absolute luminances for some models). The first process applied to these data is generally a linear transformation from XYZ tristimulus values to cone responses in order to more accurately model the physiological processes in the human visual system. The importance of beginning with CIE tristimulus values is a matter of practicality. There is a great deal of color-measurement instrumentation that is available to quickly and accurately measure stimuli in terms of CIE tristimulus values. The CIE system is also a well-established, international standard for color specification and communication.

Occasionally, vision-science-based models of color vision and appearance are based upon cone responsivities that are not linear transformations of CIE color-matching functions (e.g., Guth, 1995). The small advantage in performance that such an approach provides is far outweighed by the inconvenience in practical applications.

Given tristimulus values for the stimulus, other data regarding the viewing environment must also be considered in order to predict color appearance. This is illustrated by all of the color-appearance phenomena described in Chapters 6 and 7. As a minimum, the tristimulus values of the adapting stimulus (usually taken to be the light source) are also required. Additional data that might be utilized include the absolute luminance level, colorimetric data on the proximal field, background, surround, and perhaps other spatial or temporal information.

Given some or all of the above data, the first step in a color appearance model is generally a chromatic-adaptation transform such as those described in Chapter 9. The post-adaptation signals are then combined into higher-level signals usually modeled after the opponent-colors theory of color vision and including threshold and/or compressive nonlinearities. These signals are then combined in various ways to produce predictors of the various appearance attributes. Data on the adapting stimulus, background, surround, and so on are incorporated into the model at the chromatic-adaptation stage and at later stages as necessary.

This general process can be seen in all of the color appearance models described in this book. However, each model has been derived with a different approach and various previously given aspects are stressed to a greater or lesser degree. A simple example of a color appearance model that follows most of these construction steps is CIELAB. The interpretation of the CIELAB color space as a color appearance model is described in the next section.

10.3   CIELAB

Those trained in traditional colorimetry usually have a negative reaction when they hear CIELAB described as a color appearance model. This is because the CIE (1986) went to great care to make sure that it was called a *uniform color space* and not an *appearance space*. CIELAB was developed as a color space to be used for the specification of color differences. In the early 1970s, there were as many as 20 different formulas being used to calculate color differences. To promote uniformity of practice pending the development of a better formula, the CIE recommended two color spaces—CIELAB and CIELUV—for use in 1976 (Robertson, 1977, 1990). The Euclidean distance between two points in these spaces is taken to be a measure of their color difference (ΔE^*_{ab} or ΔE^*_{uv}). As an historical note, in 1994 the CIE recommended a single better formula for color-difference measurement based on the CIELAB space, known as ΔE^*_{94} (Berns, 1993a; CIE, 1995). In the process of creating a color-difference formula, the CIE happened to construct a color space with some predictors of color-appearance attributes. Perhaps it is not surprising that the best way to describe the difference in color of two stimuli is to first describe the appearance of each. Thus, with appropriate care, CIELAB can be considered a color appearance model.

Calculating CIELAB Coordinates

To calculate CIELAB coordinates, one must begin with two sets of CIE *XYZ* tristimulus values, those of the stimulus, *XYZ*, and those of the reference white, $X_n Y_n Z_n$. These data are utilized in a modified form of the von Kries chromatic-adaptation transform by normalizing the stimulus tristimulus values to those of the white (i.e., X/X_n, Y/Y_n, and Z/Z_n). Note that the CIE tristimulus values are not first transformed to cone responses as would be necessary for a true von Kries adaptation model.

These adapted signals are then subjected to a compressive nonlinearity represented by a cube root in the CIELAB equations. This nonlinearity is designed to model the compressive response typically found between physical energy measurements and perceptual responses (e.g., Stevens, 1961). These signals are then combined into three response dimensions corresponding to the light-dark, red-green, and yellow-blue responses of the opponent-colors theory of color vision.

Finally, appropriate multiplicative constants are incorporated into the equations to provide the required uniform perceptual spacing and proper relationship among the three dimensions. The full CIELAB equations are given in Equations 10-1 through 10-4:

$$L^* = 116f(Y/Y_n) - 16 \tag{10-1}$$

$$a^* = 500\left[f(X/X_n) - f(Y/Y_n)\right] \tag{10-2}$$

$$b^* = 200\left[f(Y/Y_n) - f(Z/Z_n)\right] \tag{10-3}$$

$$f(\omega) = \begin{cases} (\omega)^{1/3} & \omega > 0.008856 \\ 7.787(\omega) + 16/116 & \omega \le 0.008856 \end{cases} \tag{10-4}$$

The alternative forms for low tristimulus values were introduced by Pauli (1976) to overcome flaws in the original CIELAB equations that limited their application to values of X/X_n, Y/Y_n, and Z/Z_n greater than 0.01. Such low values are not often encountered in color materials, but they sometimes are found in flare-free specifications of imaging systems. It is critical to use the full set of Equations 10-1 through 10-4 in cases for which low values might be encountered.

The L^* measure given in Equation 10-1 is a correlate to perceived lightness ranging from 0.0 for black to 100.0 for a diffuse white (L^* can sometimes exceed 100.0 for stimuli such as specular highlights in images). The a^* and b^* dimensions correlate approximately with red-green and yellow-blue chroma perceptions. They take on both negative and positive values. Both a^* and b^* have values of 0.0 for achromatic stimuli (i.e., white, gray, black). Their maximum values are limited by the physical properties of materials rather than by the equations themselves.

The CIELAB L^*, a^*, and b^* dimensions are combined as Cartesian coordinates to form a three-dimensional color space, as illustrated in Figure 10-1. This color space can also be represented in terms of cylindrical coordinates as shown in Figure 10-2. The cylindrical coordinate system provides predictors of chroma, C^*_{ab}, and hue, h_{ab} (hue angle in degrees) as expressed in Equations 10-5 and 10-6:

$$C^*_{ab} = \sqrt{(a^{*2} + b^{*2})} \tag{10-5}$$

$$h_{ab} = \tan^{-1}(b^*/a^*) \tag{10-6}$$

C^* has the same units as a^* and b^*. Achromatic stimuli have C^* values of 0.0 (i.e., no chroma). Hue angle, h_{ab}, is expressed in positive degrees starting from 0° at the positive a^* axis and progressing in a counterclockwise direction. Figure 10-3 is a full-color, three-dimensional representation of the CIELAB color space sampled along the lightness, chroma, and hue angle dimensions.

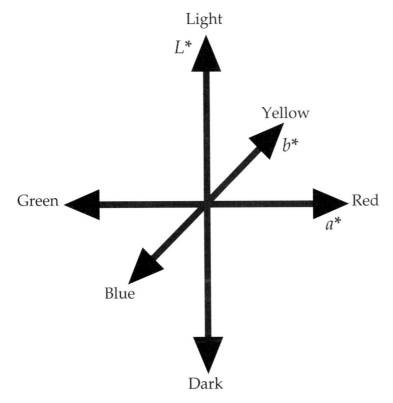

Figure 10-1.   Cartesian representation of the CIELAB color space.

The CIELAB space takes the XYZ tristimulus values of a stimulus and the reference white as input and produces correlates to lightness, L^*, chroma, C^*_{ab}, and hue, h_{ab}, as output. Thus CIELAB is a simple form of a color appearance model. Table 10-1 provides worked examples of CIELAB calculations.

While the CIELAB space provides a simple example of a color appearance model, there are some known limitations. The perceptual uniformity of the CIELAB space can be evaluated by examining plots of constant hue and chroma contours from the *Munsell Book of Color*. Such a plot is illustrated in Figure 10-4. Since the Munsell system is designed to be perceptually uniform in terms of hue and chroma, then to the extent that it realizes this objective, Figure 10-4 should ideally be a set of concentric circles representing the constant chroma contours with straight lines radiating from the center representing constant hue. As can be seen in the figure, the CIELAB space does a respectable job of representing the Munsell system uniformly.

However, further examinations of constant hue contours using a CRT system (capable of achieving higher chroma than generally available in the *Munsell Book of Color*) have illustrated discrepancies between observed and

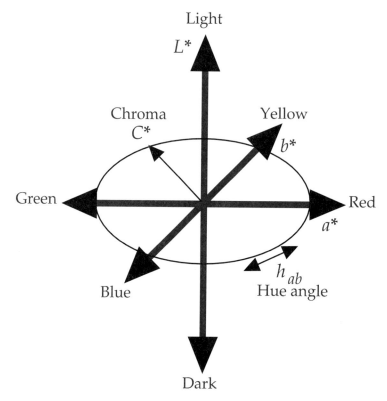

Figure 10-2. Cylindrical representation of the CIELAB color space.

Table 10-1. Example CIELAB calculations.

Quantity	Case 1	Case 2	Case 3	Case 4
X	19.01	57.06	3.53	19.01
Y	20.00	43.06	6.56	20.00
Z	21.78	31.96	2.14	21.78
X_n	95.05	95.05	109.85	109.85
Y_n	100.00	100.00	100.00	100.00
Z_n	108.88	108.88	35.58	35.58
L^*	51.84	71.60	30.78	51.84
a^*	0.00	44.22	−42.69	−13.77
b^*	−0.01	18.11	2.30	−52.86
C^*_{ab}	0.01	47.79	42.75	54.62
h_{ab}	270.0	22.3	176.9	255.4

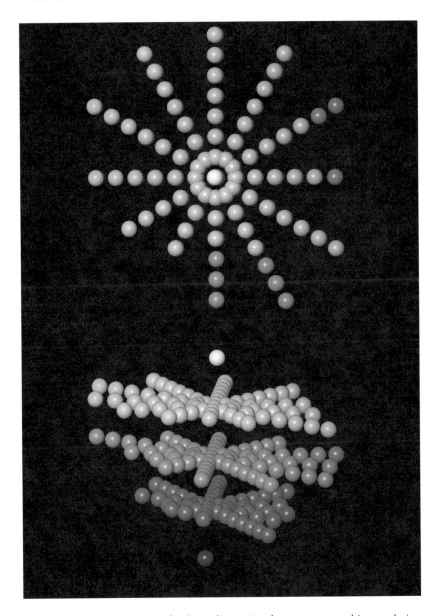

Figure 10-3. Two views of a three-dimensional computer graphics rendering of a sampling of the CIELAB color space along the lightness, chroma, and hue dimensions.

predicted results (e.g., Hung and Berns, 1995). Figure 10-5 shows constant perceived hue lines from Hung and Berns (1995). These lines are curved in the CIELAB space, particularly for red and blue hues.

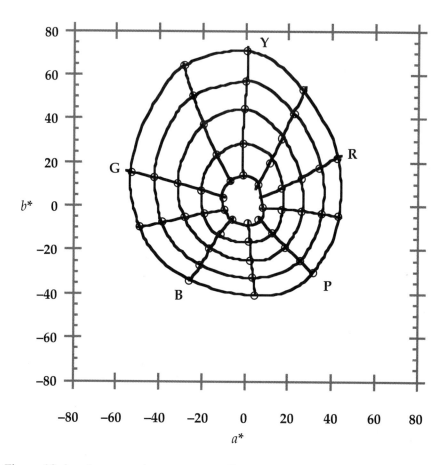

Figure 10-4.   Contours of constant Munsell chroma and hue at Value 5 plotted in the CIELAB a^*b^* plane.

A similar examination of the CIELAB lightness scale can be made by plotting Munsell Value as a function of L^* as shown in Figure 10-6. Clearly, the L^* function predicts lightness, as defined by Munsell Value, quite well. In fact, the L^* function predicts the original Munsell lightness scaling data better than the 5th-order polynomial used to define Munsell Value (Fairchild, 1995b). The result in Figure 10-6 is not surprising, since the historical derivation of the L^* scale is a close approximation of the Munsell Value scale (Robertson, 1990).

It is also worth noting that the perceptual unique hues (red, green, yellow, and blue) do not align directly with the CIELAB a^*b^* axes. The unique hues under daylight illumination lie approximately at hue angles of 24° (red), 90° (yellow), 162° (green), and 246° (blue) (Fairchild, 1996).

Other limitations of CIELAB are caused by the implementation of a von Kries-type chromatic-adaptation transform using CIE XYZ tristimulus values

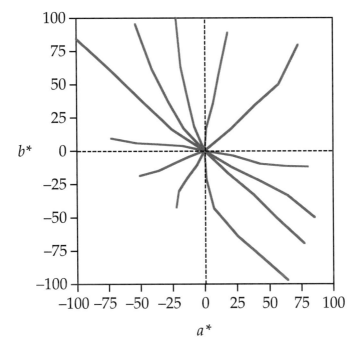

Figure 10-5.   Contours of constant perceived hue from Hung and Berns (1995) plotted in the CIELAB a^*b^* plane.

rather than cone responsivities. This has been called a *wrong von Kries transform* (Terstiege, 1972) and is described in the next section.

Wrong von Kries Transform

Terstiege (1972) has referred to von Kries-type adaptation transforms applied to values other than cone responses (sometimes called fundamental tristimulus values) as *wrong von Kries transforms*. Thus CIELAB incorporates a wrong von Kries transform through its normalization of CIE XYZ tristimulus values to those of the source. It is important to realize that the normalization of XYZ tristimulus values is not equivalent to the process of first transforming (linearly) to cone responses and then performing the normalization. This inequality is illustrated in Equations 10-7 through 10-11, which take a correct von Kries transformation in matrix form and convert it into an operation on CIE tristimulus values.

The wrong von Kries transformation incorporated in CIELAB can be expressed as a diagonal matrix transformation on CIE XYZ tristimulus values. A "right" von Kries transformation is a diagonal matrix transformation of LMS cone responses, as shown in Equation 10-7:

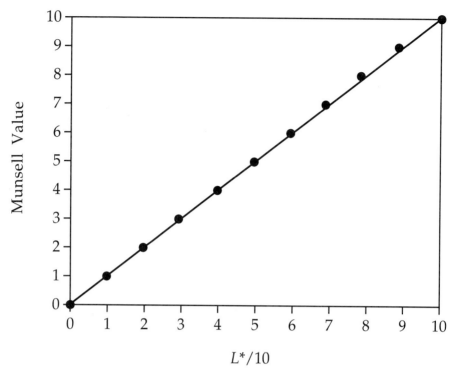

Figure 10-6. Munsell Value plotted as a function of CIELAB L^*. The black line represents a slope of 1.0 and is not fitted to the data.

$$\begin{vmatrix} L_a \\ M_a \\ S_a \end{vmatrix} = \begin{vmatrix} k_1 & 0 & 0 \\ 0 & k_M & 0 \\ 0 & 0 & k_S \end{vmatrix} \begin{vmatrix} L \\ M \\ S \end{vmatrix}$$

(10-7)

Since LMS cone responses can be expressed as linear transformations of CIE XYZ tristimulus values, Equation 10-8 can be derived from Equation 10-7 through a simple substitution. Equation 10-9 then follows through algebraic substitution:

$$\mathbf{M}\begin{vmatrix} X_a \\ Y_a \\ Z_a \end{vmatrix} = \begin{vmatrix} k_1 & 0 & 0 \\ 0 & k_M & 0 \\ 0 & 0 & k_S \end{vmatrix} \mathbf{M}\begin{vmatrix} X \\ Y \\ Z \end{vmatrix}$$

(10-8)

$$\begin{vmatrix} X_a \\ Y_a \\ Z_a \end{vmatrix} = \mathbf{M}^{-1} \begin{vmatrix} k_1 & 0 & 0 \\ 0 & k_M & 0 \\ 0 & 0 & k_S \end{vmatrix} \mathbf{M} \begin{vmatrix} X \\ Y \\ Z \end{vmatrix}$$

(10-9)

The nature of the matrix transformation, $\mathbf{M}$, is critical. $\mathbf{M}$ is never a diagonal matrix. A typical $\mathbf{M}$ matrix is given in Equation 10-10:

$$\mathbf{M} = \begin{vmatrix} 0.390 & 0.689 & -0.079 \\ -0.230 & 1.183 & 0.046 \\ 0 & 0 & 1.000 \end{vmatrix}$$

(10-10)

Evaluating Equation 10-9 using the $\mathbf{M}$ matrix of Equation 10-10 results in Equation 10-11. Since the matrix transformation relating the tristimulus values of the stimulus before and after adaptation is not a diagonal matrix, the wrong von Kries transformation cannot be equal to a correct von Kries transformation applied on cone responses.

$$\begin{vmatrix} X_a \\ Y_a \\ Z_a \end{vmatrix} = \begin{vmatrix} 0.74k_1 + 0.26k_M & 1.32k_1 - 1.32k_M & -0.15k_1 - 0.05k_M + 0.20k_S \\ 0.14k_1 - 0.14k_M & 0.26k_1 + 0.74k_M & -0.03k_1 + 0.03k_M \\ 0 & 0 & k_S \end{vmatrix} \begin{vmatrix} X \\ Y \\ Z \end{vmatrix}$$

(10-11)

An interesting experimental example of the shortcoming of the wrong von Kries transformation embedded in the CIELAB equations has been described by Liu et al. (1995). They studied the perceived hue shift in the color of the gemstone tanzanite upon changes from daylight to incandescent illumination. Some rare examples of tanzanite appear blue under daylight and purple under incandescent light. However, the CIELAB equations predict that the change in hue for these gemstones is from blue toward blue-green upon changing from daylight to incandescent. This prediction is in the opposite direction of the perceived hue change. If the same calculations are performed using a correct von Kries transformation acting on cone responsivities, the correct hue shift is predicted (Liu et al., 1995).

The Breneman (1987) corresponding-colors data that were used to compare chromatic-adaptation models in Chapter 9 were also evaluated using the chromatic-adaptation transform of the CIELAB equations. The predicted and observed results are illustrated using $u'v'$ chromaticity coordinates in Figure 10-7. The errors in the predictions are significantly larger than those found with a normal von Kries transformation (see Figure 9-2). The results indicate particularly large errors in the hue predictions for blue stimuli for this change in adaptation from daylight to incandescent. This is consistent with the errors observed by Liu et al. for tanzanite.

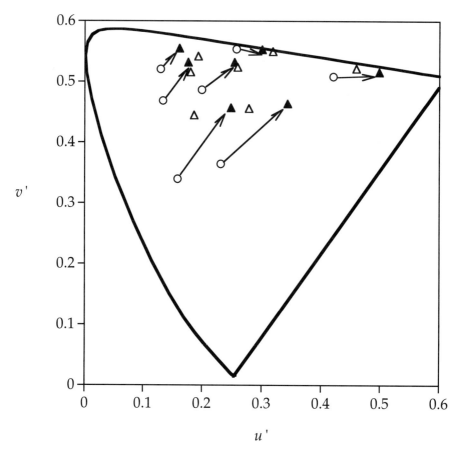

Figure 10-7.   Prediction of some example corresponding-colors data using the CIELAB model. Open triangles represent visual data, and filled triangles represent model predictions.

10.4   Why Not Use Just CIELAB?

Because CIELAB is a well-established, de facto international-standard color space that has been widely used for two decades and that it is capable of color-appearance predictions, why are any other color appearance models necessary? As is shown in Chapter 15, CIELAB performs quite well as a color appearance model in some applications. So why not just quit there and work with CIELAB?

The limitations of CIELAB discussed previously mostly answer these questions. The modified von Kries adaptation transformation incorporated into the CIELAB equations is clearly less accurate than transformations that more closely follow known visual physiology. Also, there are limitations in

CIELAB's ability to predict hue that prompt further work on appearance models.

There are also several aspects of color appearance that CIELAB is incapable of predicting. First, CIELAB incorporates no luminance-level dependency. Thus it is completely incapable of predicting luminance-dependent effects such as the Hunt effect and the Stevens effect. CIELAB also incorporates no background or surround dependency. Therefore it cannot be used to predict simultaneous contrast or the Bartleson-Breneman results showing a change in image contrast with surround relative luminance. Second, CIELAB has no mechanism for modeling cognitive effects, such as discounting-the-illuminant, that can become important in cross-media color-reproduction applications. Lastly, CIELAB does not provide correlates for the absolute appearance attributes of brightness and colorfulness. As a reminder, it is useful to recall note 6 on the CIELAB space from CIE publication 15.2 (CIE, 1986), which states:

> These spaces are intended to apply to comparisons of differences between object colours of the same size and shape, viewed in identical white to middle-grey surroundings, by an observer photopically adapted to a field of chromaticity not too different from that of average daylight.

This long list of limitations seems to indicate that it should be possible to significantly improve upon CIELAB in the development of a color appearance model. Such models are described in the next few chapters. The CIELAB space should be kept in mind as a simple model that can be used as a benchmark to measure whether more sophisticated models are indeed improvements.

10.5   What About CIELUV?

Since CIELAB can be considered a color appearance model, what about the other color space that the CIE recommended in 1976, CIELUV? CIELUV has many of the same properties as CIELAB (e.g., stimulus and source chromaticities as input and lightness, chroma, and hue predictors as output), so it might deserve equal attention.

CIELUV incorporates a different form of chromatic-adaptation transform than CIELAB. It uses a subtractive shift in chromaticity coordinates $(u' - u'_n, v' - v'_n)$ rather than a multiplicative normalization of tristimulus values $(X/X_n, Y/Y_n, Z/Z_n)$. The subtractive-adaptation transform incorporated in CIELUV is even further from physiological reality than the wrong von Kries transform of CIELAB. This subtractive shift can result in predicted corresponding colors being shifted right out of the gamut of realizable colors. (This produces predicted tristimulus values of less than zero, which cannot happen with the CIELAB transformation.) Even if this does not occur, the transform is likely to shift predicted colors outside the gamut of colors producible by any

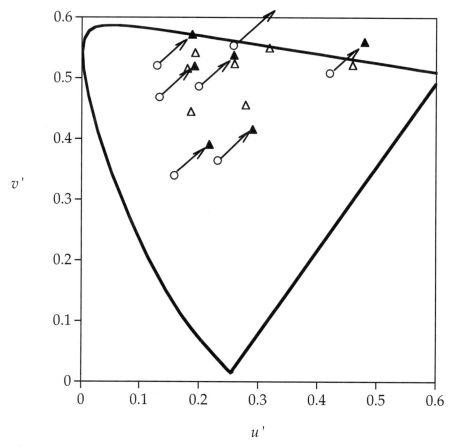

Figure 10-8. Prediction of some example corresponding-colors data using the CIELUV model. Open triangles represent visual data, and filled triangles represent model predictions.

given device. In addition to this problem, the CIELUV adaptation transform is extremely inaccurate with respect to predicting visual data. This is illustrated nicely in Figure 10-8, which shows the CIELUV predictions of the Breneman (1987) corresponding-colors data. Figure 10-8 illustrates how some colors are shifted outside the gamut of realizable colors (outside the spectrum locus on the $u'v'$ chromaticity diagram) and the inaccuracy of all the predictions.

The difficulties with the CIELUV adaptation transform are reason enough to eliminate it from serious consideration as an appearance model. However, additional evidence is provided by its poor performance for predicting color differences. The current CIE recommendation for color-difference specification, CIE94 (CIE, 1995), is based on the CIELAB color space. Alman et al. (1989) provide experimental evidence for the poor performance of CIELUV as a color-difference equation. Additional comparisons between CIELUV and CIELAB have been made by Robertson (1990).

CHAPTER 11

The Nayatani et al. Model

11.1 Objectives and Approach
11.2 Input Data
11.3 Adaptation Model
11.4 Opponent-color Dimensions
11.5 Brightness
11.6 Lightness
11.7 Hue
11.8 Saturation
11.9 Chroma
11.10 Colorfulness
11.11 Inverse Model
11.12 Phenomena Predicted
11.13 Why Not Use Just the Nayatani Model?

THE NEXT FEW CHAPTERS describe details of the most widely discussed and used color appearance models. This chapter discusses the color appearance model developed by Nayatani and his coworkers that over the past two decades has evolved as one of the more important colorimetry-based models. This model, along with Hunt's model (described in Chapter 12), is a complete color appearance model capable of predicting the full array of color-appearance parameters defined in Chapter 4 for a fairly wide range of viewing conditions.

11.1   Objectives and Approach

The Nayatani et al. color appearance model evolved as a natural extension of their chromatic-adaptation model, described in Chapter 9 (Nayatani et al., 1981) in conjunction with the attributes of a color appearance model first outlined by Hunt (1982, 1985). The Nayatani model was first described in Nayatani et al. (1986, 1987) and most recently revised and summarized in Nayatani et al. (1995). The latter version of the model is described in this chapter.

As with any color appearance model, it is important to note the context in which the Nayatani model was formulated. The researchers who developed this model come from the field of illumination engineering. In that field, the critical application of color appearance models is the specification of the color rendering properties of light sources. This application provides significantly different challenges than those encountered in the field of image reproduction. Thus those interested in image reproduction might find some aspects of the Nayatani model inappropriate for their needs. The reverse is also true. Models derived strictly for imaging applications might not fulfill the requirements for illumination engineering applications. Despite the different pedigrees of the various models, it is worthwhile to stretch them to applications for which they were not designed. The best possible result will be that they work well (indicating good generality). The worst outcome is that something more is learned about the important differences in applications. Thus, while the Nayatani model was not designed for imaging applications, it is certainly worthy of evaluation in any application that might require a color appearance model.

The model attempts to predict a wide range of color-appearance phenomena, including the Stevens effect, the Hunt effect, and the Helson-Judd effect, in addition to the effects of chromatic adaptation. It is designed to predict the color appearance of simple patches on uniform mid-to-light-gray backgrounds. It is not designed for complex stimuli or changes in background or surround. The model includes output values designed to correlate with all of the important color-appearance attributes, including brightness, lightness, colorfulness, chroma, and hue. The model's design for simple stimuli on uniform backgrounds highlights the distinction between it and models such as Hunt's and RLAB that were designed with specific attributes for imaging applications. These apparent limitations of the model are not limitations at all for lighting and illumination color-rendering applications.

11.2   Input Data

The input data for the model include the colorimetric and photometric specification of the stimulus, the adapting illuminant, and the luminance factor of the background. Specifically, the required data include the following:

- The luminance factor of the achromatic background is expressed as a percentage, Y_o. Y_o is limited to values equal to or greater than 18%.
- The color of the illumination, x_o, y_o, is expressed in terms of its chromaticity coordinates for the CIE 1931 standard colorimetric observer.
- The test stimulus is specified in terms of its chromaticity coordinates, x, y, and its luminance factor, Y.
- The absolute luminance of the stimulus and adapting field is defined by the illuminance of the viewing field, E_o, expressed in lux.

In addition, two other parameters must be specified to define the model's required input:

- The normalizing illuminance, E_{or}, which is expressed in lux and is usually in the range of 1,000 to 3,000 lux
- The noise term, n, used in the nonlinear chromatic-adaptation model, which is usually taken to be 1

From these input data, a variety of intermediate and final output values are calculated according to the model. The equations necessary for determining these values are presented in the following sections. However, a few preliminary calculations are required before proceeding with the main features of the model. The first is the calculation of the adapting luminance and the normalizing luminance in cd/m² according to Equations 11-1 and 11-2:

$$L_o = \frac{Y_o E_o}{100\pi} \tag{11-1}$$

$$L_{or} = \frac{Y_o E_{or}}{100\pi} \tag{11-2}$$

Equations 11-1 and 11-2 are valid since the background is a Lambertian diffuser.

Second, in the Nayatani model, the transformation from CIE tristimulus values to cone responsivities for the adapting field is expressed in terms of chromaticity coordinates rather than tristimulus values. This necessitates the calculation of the intermediate values of ξ(xi), η(eta), and ζ(zeta), as illustrated in Equations 11-3 through 11-5:

$$\xi = (0.48105 x_o + 0.78841 y_o - 0.08081)/y_o \tag{11-3}$$

$$\eta = (-0.27200 x_o + 1.11962 y_o + 0.04570)/y_o \tag{11-4}$$

$$\zeta = 0.91822(1 - x_o - y_o)/y_o \tag{11-5}$$

11.3   Adaptation Model

As with all color appearance models, the first stage of the Nayatani color appearance model is a chromatic-adaptation transformation. The model used is a refined form of the nonlinear model of chromatic adaptation described in Chapter 9 (Nayatani et al., 1981; CIE, 1994). In the formulation of the color appearance model, the chromatic-adaptation model is embedded in other equations. Rather than separate the two, this book follows the treatment in Nayatani et al. (1995) and points out the important features of the chromatic-adaptation model that are embedded in other equations to clarify the formulation of the color appearance model.

First, the cone responses for the adapting field must be calculated in terms of the absolute luminance level. Doing this relies on the chromaticity transform described in Equations 11-3 through 11-5, the illuminance level, E_o, and the luminance factor of the adapting background, Y_o, as formulated in Equation 11-6:

$$\begin{vmatrix} R_o \\ G_o \\ B_o \end{vmatrix} = \frac{Y_o E_o}{100\pi} \begin{vmatrix} \xi \\ \eta \\ \zeta \end{vmatrix} \tag{11-6}$$

Next, the adapting-level cone responses from Equation 11-6 are used to calculate the exponents of the nonlinear model of chromatic adaptation, as described by Equations 11-7 through 11-9. Note that in this formulation, the exponent for the short-wavelength sensitive cones (B in Nayatani's notation) differs from the exponents for the middle- and long-wavelength sensitive cones (R and G in Nayatani's notation).

$$\beta_1(R_o) = \frac{6.469 + 6.362 R_o^{0.4495}}{6.469 + R_o^{0.4495}}$$

(11-7)

$$\beta_1(G_o) = \frac{6.469 + 6.362 G_o^{0.4495}}{6.469 + G_o^{0.4495}}$$

(11-8)

$$\beta_2(B_o) = \frac{8.414 + 8.091 B_o^{0.5128}}{8.414 + B_o^{0.5128}} \times 0.7844$$

(11-9)

Third, an additional exponential factor that depends on the normalizing luminance must be calculated using the same functional form as the exponents for the middle- and long-wavelength sensitive cones, as shown in Equation 11-10:

$$\beta_1(L_{or}) = \frac{6.469 + 6.362 L_{or}^{0.4495}}{6.469 + L_{or}^{0.4495}}$$

(11-10)

Fourth, the cone responses for the test stimulus are calculated from their tristimulus values using a more traditional linear transformation given in Equation 11-11:

$$\begin{vmatrix} R \\ G \\ B \end{vmatrix} = \begin{vmatrix} 0.40024 & 0.70760 & -0.08081 \\ -0.22630 & 1.16532 & 0.04570 \\ 0.0 & 0.0 & 0.91822 \end{vmatrix} \begin{vmatrix} X \\ Y \\ Z \end{vmatrix}$$

(11-11)

Finally, two scaling coefficients, $e(R)$ and $e(G)$, are calculated according to Equations 11-12 and 11-13:

$$e(R) = \begin{cases} 1.758 & R \geq 20\xi \\ 1.0 & R < 20\xi \end{cases}$$

(11-12)

$$e(G) = \begin{cases} 1.758 & G \geq 20\eta \\ 1.0 & G < 20\eta \end{cases}$$

(11-13)

All of these calculations provide the intermediate data necessary to implement the nonlinear chromatic-adaptation model within the color appearance model. The precise use of these values is described within the appearance equations as they come into play.

11.4   Opponent-color Dimensions

The cone responses are transformed directly into intermediate values representing classical opponent dimensions of visual response: an achromatic channel and two chromatic channels. These equations, used to model these opponent processes, also incorporate the nonlinear chromatic-adaptation model.

First, the achromatic response, Q, is calculated using Equation 11-14:

$$Q = \frac{41.69}{\beta_1(L_{or})}\left[\frac{2}{3}\beta_1(R_o)e(R)\log\frac{R+n}{20\xi+n} + \frac{1}{3}\beta_1(G_o)e(G)\log\frac{G+n}{20\eta+n}\right] \tag{11-14}$$

At first, Equation 11-14 looks fairly complex, but its components can be readily teased apart and understood. The general structure of the equation is such that the achromatic response is calculated as a weighted sum of the outputs of the long- and middle-wavelength cone responses (R and G), as is often postulated in color-vision theory. The outputs are summed with relative weights of 2/3 and 1/3, which correspond to their relative population in the human retina. They are first normalized after addition of noise, n, by the cone responses for the adapting stimulus represented by ξ and η according to a von Kries-type transformation. The value of n is typically taken to be 1.0, although it can vary.

A logarithmic transform is then taken to model the compressive nonlinearity that is known to occur in the human visual system.

Given the logarithmic transformation, the exponents (β terms) of the nonlinear chromatic-adaptation model become multiplicative factors along with the scaling factors [$e(R)$ and $e(G)$], as shown in Equation 11-14. All that remains to complete the equation is one more scaling factor, 41.69, and the luminance-dependent, exponential adjustment, $\beta_1(L_{or})$. Thus the achromatic response can be simply expressed as a weighted sum of the post-adaptation signals from the long- and middle-wavelength sensitive cones.

Next, preliminary chromatic channel responses, t (red-green) and p (yellow-blue), are calculated in a similar manner using Equations 11-15 and 11-16:

$$t = \beta_1(R_o)\log\frac{R+n}{20\xi+n} - \frac{12}{11}\beta_1(G_o)\log\frac{G+n}{20\eta+n} + \frac{1}{11}\beta_2(B_o)\log\frac{B+n}{20\zeta+n} \tag{11-15}$$

$$p = \frac{1}{9}\beta_1(R_o)\log\frac{R+n}{20\xi+n} + \frac{1}{9}\beta_1(G_o)\log\frac{G+n}{20\eta+n} - \frac{2}{9}\beta_2(B_o)\log\frac{B+n}{20\zeta+n} \tag{11-16}$$

The explanation of Equations 11-15 and 11-16 follow the same logic as that for the achromatic response, Equation 11-14. Begin with the t response. It is a weighted combination of the post-adaptation signals from each of the three cone types. The combination is the difference between the long- and middle-wavelength sensitive cones with a small input from the short-wavelength sensitive cones that adds with the long-wavelength response. This results in a red minus green response that also includes some reddish input from the short wavelength end of the spectrum that is often used to explain the violet (rather than blue) appearance of those wavelengths. It is also required for correct prediction of unique yellow. Next, the p response is calculated in a similar manner by adding the long- and middle-wavelength sensitive cone outputs to produce a yellow response and then subtracting the short-wavelength cone output to produce the opposing blue response. The weighting factors were those of the original Hunt model.

It is of interest to note that the t and p notations are derived from the terms "tritanopic" and "protanopic." A tritanope has only the red-green response (t), and a protanope has only the yellow-blue response (p). The Q, t, and p responses are used in further equations to calculate correlates of brightness, lightness, saturation, colorfulness, and hue.

One aspect of the hue correlate, hue angle, θ, is calculated directly from t and p as shown in Equation 11-17:

$$\theta = \tan^{-1}\left(\frac{p}{t}\right)$$

(11-17)

Hue angle is calculated as a positive angle from 0° to 360° beginning from the positive t axis, just as is done in the CIELAB color space (CIE, 1986). The hue angle is required to calculate some of the other appearance correlates, since a hue-dependent adjustment factor is required in some cases.

11.5   Brightness

The brightness, B_r, of the test sample is calculated using Equation 11-18:

$$B_r = Q + \frac{50}{\beta_1(L_{or})}\left[\frac{2}{3}\beta_1(R_o) + \frac{1}{3}\beta_1(G_o)\right]$$

(11-18)

Q is the achromatic response given by Equation 11-14, which is adjusted using the adaptation exponents in order to include the dependency upon absolute luminance level that is required for brightness, as opposed to lightness.

It is also necessary to calculate the brightness of an ideal white, B_{rw}, according to Equation 11-19. This equation is derived by substituting Equation 11-14 evaluated for a perfect reflector into Equation 11-18:

$$B_{rw} = \frac{41.69}{\beta_1(L_{or})}\left[\frac{2}{3}\beta_1(R_o)(1.758)\log\frac{100\xi+n}{20\xi+n} + \frac{1}{3}\beta_1(G_o)(1.758)\log\frac{100\eta+n}{20\eta+n}\right]$$
$$+ \frac{50}{\beta_1(L_{or})}\left[\frac{2}{3}\beta_1(R_o) + \frac{1}{3}\beta_1(G_o)\right]$$

(11-19)

11.6   Lightness

The achromatic lightness, L^*_P, of the test sample is calculated directly from the achromatic response, Q, by simply adding 50 as shown in Equation 11-20. This is the case since the achromatic response can take on both negative and positive values with a middle gray having $Q = 0.0$, while lightness is scaled from 0 for a black to 100 for a white.

$$L^*_P = Q + 50$$

(11-20)

A second lightness correlate—normalized achromatic lightness, L^*_N—is calculated according to the CIE definition that lightness is the brightness of the test sample relative to the brightness of a white, as shown in Equation 11-21:

$$L^*_N = 100\left(\frac{B_r}{B_{rw}}\right)$$

(11-21)

The differences between the two lightness correlates, L^*_P and L^*_N, are generally negligible. Neither of the lightness values correlate with the perceived lightness of chromatic object colors, since the model does not include the Helmholtz-Kohlrausch effect (e.g., Fairchild and Pirrotta, 1991; Nayatani et al., 1992). An additional model is necessary to include the Helmholtz-Kohlrausch effect, which is necessary for the comparison of the lightness or brightness of stimuli with differing hue and/or chroma.

11.7   Hue

Hue angle, θ, is calculated as shown in Equation 11-17, which is identical to the technique used in the CIELAB color space. More descriptive hue correlates can be obtained by determining the hue quadrature, H, and the hue composition, H_C.

Hue quadrature, H, is a 400-step hue scale on which the unique hues take on values of 0 (red), 100 (yellow), 200 (green), and 300 (blue). The hue quadrature is computed via linear interpolation using the hue angle, θ, of the test sample and the hue angles for the four unique hues, which are defined as 20.14° (red), 90.00° (yellow), 164.25° (green), and 231.00° (blue).

The hue composition, H_C, describes perceived hue in terms of percentages of two of the unique hues from which the test hue is composed. For example, an orange color might be expressed as 50Y 50R, thereby indicating that the hue is perceived to be halfway between unique red and unique yellow. Hue composition is computed by simply converting the hue quadrature into percent components of the unique hues falling on either side of the test color (again by a linear process). For example, a color stimulus with a hue angle of 44.99° will have a hue quadrature of 32.98 and a hue composition of 33Y 67R.

11.8   Saturation

In the Nayatani color appearance model, saturation is derived most directly. The measures of colorfulness and chroma are then derived from saturation. Saturation is expressed in terms of a red-green component, S_{RG}, derived from the t response, as shown in Equation 11-22, and a yellow-blue component, S_{YB}, derived from the p response, as shown in Equation 11-23:

$$S_{RG} = \frac{488.93}{\beta_1(L_{or})} E_s(\theta)t \tag{11-22}$$

$$S_{YB} = \frac{488.93}{\beta_1(L_{or})} E_s(\theta)p \tag{11-23}$$

The saturation predictors include a scaling factor, 488.93, for convenience; the luminance-dependent β terms required to predict the Hunt effect; and a chromatic strength function, $E_s(\theta)$, that was introduced to correct the saturation scale as a function of hue angle (Nayatani, 1995). It takes on the empirically derived form expressed in Equation 11-24:

$$E_s(\theta) = 0.9394 - 0.2478\sin\theta - 0.0743\sin2\theta + 0.0666\sin3\theta - 0.0186\sin4\theta$$
$$- 0.0055\cos\theta - 0.0521\cos2\theta - 0.0573\cos3\theta - 0.0061\cos4\theta \tag{11-24}$$

Finally, an overall saturation correlate, S, is calculated using Equation 11-25. This is precisely the same functional form as the chroma calculation in CIELAB (Euclidean distance from the origin):

$$S = \left(S_{RG}^2 + S_{YB}^2 \right)^{1/2}$$

$$(11\text{-}25)$$

11.9   Chroma

Using the correlates for saturation, one can easily derive correlates of chroma by considering their definitions. As was illustrated in Chapter 4, saturation can be expressed as chroma divided by lightness. Thus chroma can be described as saturation multiplied by lightness. This is almost exactly the functional form for chroma in the Nayatani model. The correlates for the red-green, yellow-blue, and overall chroma of the test sample are given in Equations 11-26 through 11-28:

$$C_{RG} = \left(\frac{L^*_P}{50} \right)^{0.7} S_{RG}$$

$$(11\text{-}26)$$

$$C_{YB} = \left(\frac{L^*_P}{50} \right)^{0.7} S_{YB}$$

$$(11\text{-}27)$$

$$C = \left(\frac{L^*_P}{50} \right)^{0.7} S$$

$$(11\text{-}28)$$

The only differences between the nominal definition of chroma and Equations 11-26 through 11-28 are the scaling factor of 50 and the slight nonlinearity introduced by the power function of lightness with an exponent of 0.7. This nonlinearity was introduced to better model constant chroma contours from the *Munsell Book of Color* (Nayatani et al., 1995).

11.10   Colorfulness

The predictors of colorfulness in the Nayatani model can also be derived directly from the CIE definitions of the appearance attributes. Recall that chroma is defined as colorfulness of the sample relative to the brightness of a white object under similar illumination. Thus colorfulness is simply the chroma of the sample multiplied by the brightness of an ideal white, as illustrated in Equations 11-29 through 11-31:

$$M_{RG} = C_{RG}\frac{B_{rw}}{100}$$

(11-29)

$$M_{YB} = C_{YB}\frac{B_{rw}}{100}$$

(11-30)

$$M = C\frac{B_{rw}}{100}$$

(11-31)

The normalizing value of 100 is derived as the brightness of an ideal white under illuminant D65 at the normalizing illuminance. It provides a convenient place to tie down the scale.

11.11   Inverse Model

In many applications, and particularly image reproduction, a color appearance model must be used in both forward and reverse directions. Thus it is important, or at least highly convenient, that the equations can be analytically inverted. Fortunately, the Nayatani color appearance model can be inverted analytically. Nayatani et al. (1990) published a paper that introduces the process for inverting the model for both brightness-colorfulness and lightness-chroma matches. While the model has changed slightly since, the same general procedure can be followed.

   In applying the model, it is often useful to consider its implementation as a simple step-by-step process. Here are the steps required to implement the model (and in reverse order, to invert it):

   1. Obtain physical data.
   2. Calculate Q, t, and p.
   3. Calculate θ, $E_s(\theta)$, H, and H_C.
   4. Calculate B_r, B_{rw}, L^*_p, L^*_N, and S.
   5. Calculate C.
   6. Calculate M.

11.12   Phenomena Predicted

The Nayatani color appearance model accounts for changes in color appearance due to chromatic adaptation and luminance level (the Stevens effect and the Hunt effect). It also predicts the Helson-Judd effect. The model can be used for different background luminance factors (greater than 18%); however, the model's authors caution against using it for comparisons between different luminance levels (Nayatani et al., 1990). It cannot be used to predict the effects of changes in background color (simultaneous contrast) or surround relative luminance (e.g., Bartleson-Breneman equations).

The Nayatani color appearance model also does not incorporate mechanisms for predicting incomplete chromatic adaptation or cognitive discounting-the-illuminant. Nayatani (1997) has outlined a procedure to estimate the level of chromatic adaptation from experimental data. This is useful to allow the model to be used to predict the results of visual experiments in which chromatic adaptation is incomplete. However, this technique is of little value in practical applications in which a prediction of the level of chromatic adaptation must be made for a set of viewing conditions for which prior visual data are not available.

Example calculations using the Nayatani color appearance model as described in this chapter are given for four samples in Table 11-1.

Table 11-1.   Example Nayatani color appearance model calculations.

Quantity	Case 1	Case 2	Case 3	Case 4
X	19.01	57.06	3.53	19.01
Y	20.00	43.06	6.56	20.00
Z	21.78	31.96	2.14	21.78
X_n	95.05	95.05	109.85	109.85
Y_n	100.00	100.00	100.00	100.00
Z_n	108.88	108.88	35.58	35.58
E_o	5,000	500	5,000	500
E_{or}	1,000	1,000	1,000	1,000
B_r	62.6	67.3	37.5	44.2
L^*_P	50.0	73.0	24.5	49.4
L^*_N	50.0	75.9	29.7	49.4
θ	268.0	7.4	183.2	249.4
H	324.8	391.5	228.4	312.3
H_C	75B 25R	92R 8B	72G 28B	88B 12R
S	0.0	108.1	272.6	39.6
C	0.0	140.9	165.4	39.3
M	0.0	125.0	208.5	35.2

11.13   Why Not Use Just the Nayatani Model?

Because of the extensive nature of the Nayatani color appearance model and its inclusion of correlates for all of the important color appearance attributes, it is reasonable to wonder why the CIE hasn't simply adopted this model as a recommended color appearance model. There are several reasons this has not happened.

First, it is worth reiterating the positive features of the model. It is a complete model in terms of output correlates. It is fairly straightforward (although the equations could be presented more simply) and it is analytically invertible.

However, there are some negative aspects of the model that prevent it from becoming the single best answer:

- It cannot account for changes in background, surround, or cognitive effects. Surround and cognitive factors are critical in image reproduction applications.
- It does not predict adaptation level, which is also important in cross-media reproduction applications.
- It has been derived and tested mainly for simple patches, which might limit its usefulness in more-complex viewing situations.
- It significantly over-predicts the Helson-Judd effect by predicting strong effects for illuminants that are not very chromatic (such as illuminant A). It is clear that the Helson-Judd effect does not occur under such conditions as were originally pointed out by Helson (1938) himself.
- In various tests of color appearance models described in later chapters, it has been generally shown to be not particularly accurate.
- It does not incorporate rod contributions to color appearance as can be found in the Hunt model.

These limitations indicate that this model cannot provide the ultimate answer for a single color appearance model. This does not lessen its significance and contributions to the development of such models. It is almost certain that some aspects of the Nayatani model will find their way into whatever model is agreed upon for general use in the future.

CHAPTER 12

The Hunt Model

12.1   Objectives and Approach
12.2   Input Data
12.3   Adaptation Model
12.4   Opponent-color Dimensions
12.5   Hue
12.6   Saturation
12.7   Brightness
12.8   Lightness
12.9   Chroma
12.10 Colorfulness
12.11 Inverse Model
12.12 Phenomena Predicted
12.13 Why Not Use Just the Hunt Model?

THIS CHAPTER CONTINUES the review of some of the most widely discussed and used color appearance models with a description of the model developed by Robert W. G. Hunt—the most extensive, complete, and complex model to date. Its roots can be traced to some of Hunt's early chromatic-adaptation studies (Hunt, 1952) up through its rigorous development in the 1980s and 1990s (Hunt, 1982, 1985, 1987, 1991b, 1994, 1995).

The Hunt color appearance model is not simple, but it is designed to predict a wide range of visual phenomena, and, as Hunt himself has stated, the human visual system is not simple, either. While there are applications in which simpler models such as those described in later chapters are adequate, it is certainly of great value to have a complete model that can be adapted to a wider range of viewing conditions for more well-defined or unusual circumstances. The Hunt model serves this purpose well, and many of the other color appearance models discussed in this book can trace many of their features back to ideas that originally appeared in Hunt's model.

12.1 Objectives and Approach

Hunt spent 36 years of his career in the Kodak Research Laboratories. Thus the Hunt model has been developed in the context of the requirements for color-image reproduction. This is a significantly different point of view than found in the field of illumination engineering from which the Nayatani model discussed in Chapter 11 was developed. The imaging science influence on the Hunt model can be easily witnessed by examining its input parameters. For example, the surround relative luminance is an important factor that is not present in the Nayatani model. Other examples can be found in parameters that are set to certain values for "transparencies projected in a dark room" or "television displays in a dim surround" or "normal scenes." Such capabilities clearly indicate that the model was intended to be applied to imaging situations. However, this is not the limit of the model's applicability. For example, it has also been extended for unrelated colors such as those found in traditional vision-science experiments.

The Hunt model is designed to predict a wide range of visual phenomena, including the appearance of related and unrelated colors in various backgrounds, surrounds, illumination colors, and luminance levels ranging from low scotopic to bleaching levels. In this sense, it is a complete model of color appearance for static stimuli. Hunt's model, like the others described in this book, does not attempt to incorporate complex spatial or temporal characteristics of appearance.

To make reasonable predictions of appearance over such a wide range of conditions, the Hunt model requires a more rigorous definition of the viewing field. Thus Hunt (1991b) defined the components of the viewing field as described in Chapter 7. These components include the stimulus, the prox-imal field, the background, and the surround. The Hunt model is the only available model that treats each of these components of the viewing field separately.

While Hunt's model has been continuously evolving over the past two decades (see Hunt 1982, 1985, 1987, 1991b, and 1994 for major milestones), a comprehensive review of the model's current formulation can be found in Chapter 31 of the fifth edition of Hunt's book, *The Reproduction of Colour* (Hunt, 1995). The treatment that follows is adapted from that chapter. Those desiring more detail on Hunt's model should refer to Chapter 31 of Hunt's book.

12.2   Input Data

The Hunt model requires an extensive list of input data. All colorimetric coordinates are typically calculated using the CIE 1931 standard colorimetric observer (2°). The chromaticity coordinates (x, y) of the illuminant and the adapting field are required. Typically, the adapting field is taken to be the integrated chromaticity of the scene, which is assumed to be identical to that of the illuminant (or source). Next, the chromaticities (x, y) and luminance factors (Y) of the background, proximal field, reference white, and test sample are required. If separate data are not available for the proximal field, the field is generally assumed to be identical to the background. Also, the reference white is often taken to have the same chromaticities as the illuminant with a luminance factor of 100 if specific data are not available.

All of these data are relative colorimetric values. Absolute luminance levels are required to predict several luminance-dependent appearance phenomena. Thus the absolute luminance levels, in cd/m², are required for the reference white and the adapting field. If the specific luminance of the adapting field is not available, it is taken to be 20% of the luminance of the reference white under the assumption that scenes integrate to a gray with a reflectance factor of 0.2. Additionally, scotopic luminance data are required in order to incorporate rod responses into the model (another feature unique to the Hunt model). Thus the scotopic luminance of the adapting field in scotopic cd/m² is required. Since scotopic data are rarely available, the scotopic luminance of the illuminant, L_{AS}, can be approximated from its photopic luminance, L_A, and correlated color temperature, T, by using Equation 12-1:

Table 12-1.   Values of the chromatic and brightness surround
induction factors.

Situation	N_c	N_b
Small areas in uniform backgrounds and surrounds	1.0	300
Normal scenes	1.0	75
Television and CRT displays in dim surrounds	1.0	25
Large transparencies on light boxes	0.7	25
Projected transparencies in dark surrounds	0.7	10

$$L_{AS} = 2.26L_A\big[(T/4000) - 0.4\big]^{1/3}$$

(12-1)

The scotopic luminance of the test stimulus relative to the scotopic luminance of the reference white is also required. Again, since such data are rarely available, an approximation is often used by substituting the photopic luminance of the sample relative to the reference white for the scotopic values.

Lastly, there are several input variables that are decided based on the viewing configuration. Two of these are the chromatic (N_c) and brightness (N_b) surround induction factors. Hunt (1995) suggests using values optimized for the particular viewing situation. Since this is often not possible, the nominal values listed in Table 12-1 are recommended.

The last two input parameters are the chromatic (N_{cb}) and brightness (N_{bb}) background induction factors. Again, Hunt recommends optimized values. Assuming these are not available, one can calculate the background induction factors from the luminances of the reference white, Y_W, and background, Y_b, by using Equations 12-2 and 12-3:

$$N_{cb} = 0.725\big(Y_W/Y_b\big)^{0.2}$$

(12-2)

$$N_{bb} = 0.725\big(Y_W/Y_b\big)^{0.2}$$

(12-3)

A final decision must be made regarding discounting-the-illuminant. Certain parameters in the model are assigned different values for situations in which discounting-the-illuminant occurs. Given all of the previous data, one can then continue with the calculations of the Hunt model parameters.

12.3   Adaptation Model

As with all of the models described in this book, the first step is a transformation from CIE tristimulus values to cone responses. In Hunt's model, the cone responses are denoted ργβ rather than *LMS*. The transformation used (referred to as the *Hunt, Pointer, Estevez transformation*, also used in the Nayatani et al. and RLAB models) is given in Equation 12-4. For the Hunt model, this transformation is normalized such that the equal-energy illuminant has equal ργβ values.

$$
\begin{vmatrix} \rho \\ \gamma \\ \beta \end{vmatrix} = \begin{vmatrix} 0.38971 & 0.68898 & -0.07868 \\ -0.22981 & 1.18340 & 0.04641 \\ 0.0 & 0.0 & 1.0 \end{vmatrix} \begin{vmatrix} X \\ Y \\ Z \end{vmatrix}
$$

(12-4)

The transformation from *XYZ* to ργβ values must be completed for the reference white, background, proximal field, and test stimulus.

The chromatic-adaptation model embedded in Hunt's color appearance model is a significantly modified form of the von Kries hypothesis. The adapted cone signals, $\rho_a\gamma_a\beta_a$, are determined from the cone responses for the stimulus, ργβ, and those for the reference white, $\rho_W\gamma_W\beta_W$, by using Equations 12-5 through 12-7:

$$
\rho_a = B_\rho \left[f_n \left(F_L F_\rho \rho / \rho_W \right) + \rho_D \right] + 1
$$

(12-5)

$$
\gamma_a = B_\gamma \left[f_n \left(F_L F_\gamma \gamma / \gamma_W \right) + \gamma_D \right] + 1
$$

(12-6)

$$
\beta_a = B_\beta \left[f_n \left(F_L F_\beta \beta / \beta_W \right) + \beta_D \right] + 1
$$

(12-7)

The von Kries hypothesis can be recognized in Equations 12-5 through 12-7 by noting the ratios ρ/ρ_W, γ/γ_W, β/β_W at the heart of the equations. Clearly, there are many other parameters in Equations 12-5 through 12-7 that require definition and explanation; these are given next. First, $f_n()$ is a general hyperbolic function given in Equation 12-8 that is used to model the nonlinear behavior of various visual responses:

$$
f_n [I] = 40 \left[I^{0.75} / (I^{0.75} + 2) \right]
$$

(12-8)

Figure 12-1 illustrates the form of Hunt's nonlinear function on log-log axes. In the central operating range, the function is linear and therefore equivalent to a simple power function (in this case, with an exponent of about 1/2). However, this function has the advantage that it models threshold behavior at low levels (the gradual increase in slope) and saturation behavior at high levels (the decrease in slope back to zero). Such a nonlinearity is required to model the visual system over the large range in luminance levels that the Hunt model addresses.

F_L is a luminance-level adaptation factor incorporated into the adaptation model to predict the general behavior of light adaptation over a wide

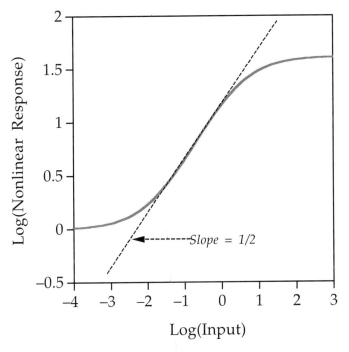

Figure 12-1. The nonlinear response function, $f_n()$, of the Hunt color appearance model.

range of luminance levels. It also reintroduces the absolute luminance level prior to the nonlinearity, thus allowing appearance phenomena such as the Stevens effect and Hunt effect to be predicted. F_L is calculated by using Equations 12-9 and 12-10:

$$F_L = 0.2k^4\left(5L_A\right)+0.1\left(1-k^4\right)^2\left(5L_A\right)^{1/3}$$

(12-9)

$$k = 1/\left(5L_A+1\right)$$

(12-10)

F_p, F_γ, and F_β are chromatic-adaptation factors that are introduced to model the fact that chromatic adaptation is often incomplete. These factors are designed such that chromatic adaptation is always complete for the equal-energy illuminant (sometimes referred to as illuminant E). This means that the chromaticity of illuminant E always appears achromatic according to the model; thus F_p, F_γ, and F_β are all equal to one. Such a prediction is supported by experimental results of Hurvich and Jameson (1951), Hunt and Winter (1975), and Fairchild (1991b). As F_p, F_γ, and F_β depart from 1 (in either direction depending on the adapting field color), chromatic adaptation is predicted to be less complete. The formulation of F_p, F_γ, and F_β is given in Equations

12-11 through 12-16, and the behavior of these functions is illustrated in Figure 12-2 for F_ρ as an example:

$$F_\rho = \left(1 + L_A^{1/3} + h_\rho\right)/\left(1 + L_A^{1/3} + 1/h_\rho\right) \tag{12-11}$$

$$F_\gamma = \left(1 + L_A^{1/3} + h_\gamma\right)/\left(1 + L_A^{1/3} + 1/h_\gamma\right) \tag{12-12}$$

$$F_\beta = \left(1 + L_A^{1/3} + h_\beta\right)/\left(1 + L_A^{1/3} + 1/h_\beta\right) \tag{12-13}$$

$$h_\rho = 3\rho_W /\left(\rho_W + \gamma_W + \beta_W\right) \tag{12-14}$$

$$h_\gamma = 3\gamma_W /\left(\rho_W + \gamma_W + \beta_W\right) \tag{12-15}$$

$$h_\beta = 3\beta_W /\left(\rho_W + \gamma_W + \beta_W\right) \tag{12-16}$$

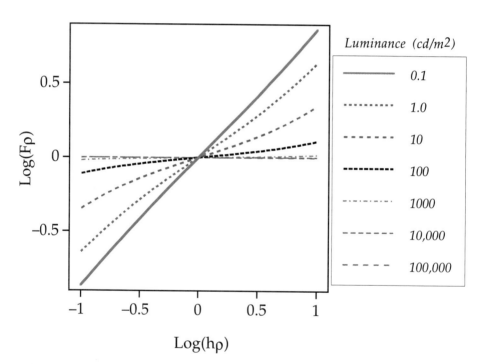

Figure 12-2.   One of the chromatic-adaptation factors, F_ρ, plotted as a function of the adapting chromaticity for a variety of adapting luminance levels. This function illustrates how adaptation becomes less complete (F_ρ departs from 1.0) as the purity of the adapting stimulus increases (h_ρ departs from 1.0) and the luminance level decreases.

The parameters h_ρ, h_γ, and h_β can be thought of as chromaticity coordinates scaled relative to illuminant E (since $\rho\gamma\beta$ themselves are normalized to illuminant E). They take on values of 1.0 for illuminant E and depart further from 1.0 as the reference white becomes more saturated. These parameters, taken together with the luminance-level dependency (L_A) in Equations 12-11 through 12-13, produce values that depart from 1.0 by increasing amounts as the color of the reference white moves away from illuminant E (becoming more saturated) and the adapting luminance increases. The feature that chromatic adaptation becomes more complete with increasing adapting luminance is also consistent with the visual experiments cited previously. This general behavior is illustrated in Figure 12-2 with a family of curves for various adapting luminance levels.

If discounting-the-illuminant occurs, then chromatic adaptation is taken to be complete and F_ρ, F_γ, and F_β are set equal to values of 1.0.

The parameters ρ_D, γ_D, and β_D are included to allow prediction of the Helson-Judd effect. This is accomplished by additive adjustments to the cone signals that depend on the relationship between the luminance of the background, Y_b, the reference white, Y_W, and the test stimulus as given in Equations 12-17 through 12-19:

$$\rho_D + f_n\left[\left(Y_b / Y_W\right)F_L F_\gamma\right] - f_n\left[\left(Y_b / Y_W\right)F_L F_\rho\right] \tag{12-17}$$

$$\gamma_D = 0.0 \tag{12-18}$$

$$\beta_D = f_n\left[\left(Y_b / Y_W\right)F_L F_\gamma\right] - f_n\left[\left(Y_b / Y_W\right)F_L F_\beta\right] \tag{12-19}$$

The Helson-Judd effect does not occur in most typical viewing situations (Helson, 1938). In such cases ρ_D, γ_D, and β_D should be set equal to 0.0. In cases for which discounting-the-illuminant occurs, there is no Helson-Judd effect and ρ_D, γ_D, and β_D are forced to 0.0, since F_ρ, F_γ, and F_β are set equal to 1.0. There are some situations in which it is desirable to have F_ρ, F_γ, and F_β take on their normal values while ρ_D, γ_D, and β_D are set to 0.0. These include the viewing of images projected in a darkened surround or viewed on CRT displays.

The last factors in the chromatic-adaptation formulas (Equations 12-5 through 12-7) are the cone bleach factors B_ρ, B_γ, and B_β. Once again, these factors are necessary only to model visual responses over extremely large ranges in luminance level. They are formulated to model photopigment depletion (i.e., bleaching) that occurs at high luminance levels and results in decreased photoreceptor output, as shown in Equations 12-20 through 12-22:

$$B_\rho = 10^7 / \left[10^7 + 5L_A\left(\rho_W / 100\right)\right] \tag{12-20}$$

$$B_\gamma = 10^7 / \left[10^7 + 5L_A\left(\gamma_W / 100\right)\right] \tag{12-21}$$

$$B_\beta = 10^7 / \left[10^7 + 5L_A\left(\beta_W/100\right)\right]$$

$$(12\text{-}22)$$

The cone bleaching factors are essentially 1.0 for most normal luminance levels. As the adapting luminance, L_A, reaches extremely high levels, the bleaching factors begin to decrease, thereby resulting in decreased adapted cone output. In the limit, the bleaching factors will approach zero as the adapting luminance approaches infinity. This would result in no cone output when the receptors are fully bleached (sometimes referred to as a *retinal burn*). Such adapting levels are truly dangerous to the observer and would cause permanent damage. However, the influence of the cone bleaching factors does begin to take effect at high luminance levels that are below the threshold for retinal damage, such as those found outdoors on a sunny day. In such situations, one can observe the decreased range of visual response due to "too much light"; a typical response is to put on sunglasses. These high-luminance levels are not found in typical image reproduction applications (except, perhaps, in some original scenes).

The adaptation formulas (Equations 12-5 through 12-7) are completed with the addition of 1.0 designed to represent noise in the visual system.

If the proximal field and background differ from a gray, chromatic induction is modeled by adjusting the cone signals for the reference white used in the adaptation equations. This suggests that the state of adaptation is being influenced by the local color of the proximal field and background in addition to the color of the reference white. This type of modeling is completely consistent with observed visual phenomena. Hunt (1991b) has suggested one algorithm for calculating adjusted reference white signals ρ'_W, γ'_W, and $\beta'w_W$ from the cone responses for the background, ρ_b, γ_b, and β_b, and proximal field, ρ_p, γ_p, and β_p, given in Equations 12-23 through 12-28:

$$\rho'_W = \frac{\rho_W\left[(1-p)p_\rho + (1+p)/p_\rho\right]^{1/2}}{\left[(1+p)p_\rho + (1-p)/p_\rho\right]^{1/2}}$$

$$(12\text{-}23)$$

$$\gamma'_W = \frac{\gamma_W\left[(1-p)p_\gamma + (1+p)/p_\gamma\right]^{1/2}}{\left[(1+p)p_\gamma + (1-p)/p_\gamma\right]^{1/2}}$$

$$(12\text{-}24)$$

$$\beta'_W = \frac{\beta_W\left[(1-p)p_\beta + (1+p)/p_\beta\right]^{1/2}}{\left[(1+p)p_\beta + (1-p)/p_\beta\right]^{1/2}}$$

$$(12\text{-}25)$$

$$p_\rho = \left(\rho_p/\rho_b\right)$$

$$(12\text{-}26)$$

$$p_\gamma = \left(\gamma_p / \gamma_b \right) \tag{12-27}$$

$$p_\beta = \left(\beta_p / \beta_b \right) \tag{12-28}$$

Values of p in Equations 12-23 through 12-25 are taken to be between 0 and -1 when simultaneous contrast occurs and between 0 and +1 when assimilation occurs. In most practical applications, the background and proximal field are assumed to be achromatic and adjustments such as those given in Equations 12-23 through 12-25 are not used.

Now that the adapted cone signals ρ_a, γ_a, and β_a are available, it is possible to move onward to the opponent responses and color-appearance correlates. The rod signals and their adaptation will be treated at the point at which they are incorporated in the achromatic response.

12.4   Opponent-color Dimensions

Given the adapted cone signals, ρ_a, γ_a, and β_a, one can calculate opponent-type visual responses very simply, as shown in Equations 12-29 through 12-32:

$$A_a = 2\rho_a + \gamma_a + \left(1/20 \right)\beta_a - 3.05 + 1 \tag{12-29}$$

$$C_1 = \rho_a - \gamma_a \tag{12-30}$$

$$C_2 = \gamma_a - \beta_a \tag{12-31}$$

$$C_3 = \beta_a - \rho_a \tag{12-32}$$

The achromatic post-adaptation signal, A_a, is calculated by summing the cone responses with weights that represent their relative population in the retina. The subtraction of 3.05 and the addition of 1.0 represents removal of the earlier noise components followed by the addition of new noise. The three color-difference signals, C_1, C_2, and C_3, represent all of the possible chromatic opponent signals that could be produced in the retina. These may or may not have direct physiological correlates, but they are convenient formulations and used to construct more traditional opponent responses, as described in the next sections.

12.5   Hue

Hue angle in the Hunt color appearance model is calculated just as it is in other models, once red-green and yellow-blue opponent dimensions are specified as appropriate combinations of the color-difference signals described in Equations 12-30 through 12-32. Hue angle, h_s, is calculated using Equation 12-33:

$$h_s = \tan^{-1}\left[\frac{(1/2)(C_2 - C_3)/4.5}{C_1 - (C_2/11)}\right]$$

(12-33)

Given the hue angle, h_s, one can calculate a hue quadrature value, H, by interpolation between specified hue angles for the unique hues with adjustment of an eccentricity factor, e_s. The interpolating function is given by Equation 12-34:

$$H = H_1 + \frac{100[(h_s - h_1)/e_1]}{[(h_s - h_1)/e_1 + (h_2 - h_s)/e_2]}$$

(12-34)

H_1 is defined as 0, 100, 200, or 300 based on whether red, yellow, green, or blue, respectively, is the unique hue with the hue angle nearest to and less than that of the test sample. The values of h_1 and e_1 are taken from Table 12-2 as the values for the unique hue having the nearest lower value of h_s. h_2 and e_2 are taken as the values of the unique hue with the nearest higher value of h_s.

Hue composition, H_C, is calculated directly from the hue quadrature just as it was in the Nayatani et al. model described in Chapter 11. Hue composition is expressed as the percentages of two unique hues that describe the composition of the test stimulus hue.

Finally, an eccentricity factor, e_s, must be calculated for the test stimulus to be used in further calculations of appearance correlates. This is accomplished through linear interpolation using the hue angle, h_s, of the test stimulus and the data in Table 12-2.

Table 12-2.   Hue angles, h_s, and eccentricity factors, e_s, for the unique hues.

Hue	h_s	e_s
Red	20.14	0.8
Yellow	90.00	0.7
Green	164.25	1.0
Blue	237.53	1.2

12.6   Saturation

As a step toward calculating a correlate of saturation, yellowness-blueness and redness-greenness responses must be calculated from the color difference signals according to Equations 12-35 and 12-36:

$$M_{YB} = 100\left[(1/2)(C_2 - C_3)/4.5\right]\left[e_s(10/13)N_c N_{cb} F_t\right]$$
(12-35)

$$M_{RG} = 100\left[C_1 - (C_2/11)\right]\left[e_s(10/13)N_c N_{cb}\right]$$
(12-36)

The constant values in Equations 12-34 and 12-35 are simply scaling factors. N_c and N_{cb} are the chromatic surround and background induction factors determined at the outset. F_t is a low-luminance tritanopia factor calculated by using Equation 12-37:

$$F_t = L_A/(L_A + 0.1)$$
(12-37)

Low-luminance tritanopia is a phenomenon whereby observers with normal color vision become more and more tritanopic (yellow-blue deficient) as luminance decreases. This is because the luminance threshold for short-wavelength sensitive cones is higher than that for the other two cone types. As can be seen in Equation 12-37, F_t is essentially 1.0 for all typical luminance levels. It approaches zero as the adapting luminance, L_A, approaches zero, thus forcing the yellowness-blueness response to decrease at low luminance levels. This factor is of little importance in most practical situations, but it is necessary in order to model appearance over an extremely wide range of luminance levels. For most applications, it is better to avoid this situation by viewing samples at sufficiently high luminance levels.

Given the yellowness-blueness and redness-greenness responses defined above, one can calculate an overall chromatic response, M, as their quadrature sum, as shown in Equation 12-38:

$$M = \left(M_{YB}^2 + M_{RG}^2\right)^{1/2}$$
(12-38)

Finally, saturation, s, is calculated from M and the adapted cone signals by using Equation 12-39. This calculation follows the definition of saturation (colorfulness of stimulus relative to its own brightness) if one takes the overall chromatic response, M, to approximate colorfulness and the sum of the adapted cone signals in the denominator of Equation 12-39 to approximate the stimulus brightness.

$$s = 50M/(\rho_a + \gamma_a + \beta_a)$$
(12-39)

12.7   Brightness

Further development of the achromatic signals is required to derive correlates of brightness and lightness. Recall that the Hunt color appearance model is designed to function over the full range of luminance levels. In so doing, it must also incorporate the response of the rod photoreceptors, which are active at low luminance levels. The rod response is incorporated into the achromatic signal (which in turn impacts the predictors of chroma and colorfulness). The rod response after adaptation, A_S, is given by Equation 12-40. The S subscript is derived from the word "scotopic," which is used to describe vision at the low luminance levels for which the rod response dominates.

$$A_S = 3.05B_S\left[f_n\left(F_{LS}S/S_W\right)\right] + 0.3$$

$$(12\text{-}40)$$

The formulation of the adapted rod signal is analogous to the formulation of the adapted cone signals described in Equations 12-5 through 12-7. At its heart is a von Kries-type scaling of the scotopic response for the stimulus, S, by that for the reference white, S_W. F_{LS} is a scotopic luminance level adaptation factor given by Equations 12-41 and 12-42 that is similar to the cone luminance adaptation factor:

$$F_{LS} = 3800j^2\left(5L_{AS}/2.26\right) + 0.2\left(1-j^2\right)^4\left(5L_{AS}/2.26\right)^{1/6}$$

$$(12\text{-}41)$$

$$j = 0.00001/\left[\left(5L_{AS}/2.26\right) + 0.00001\right]$$

$$(12\text{-}42)$$

The same nonlinear photoreceptor response function, $f_n()$, is used for the scotopic response (see Equation 12-8). The value 3.05 is simply a scaling factor, and the noise level of 0.3 (rather than 1.0) is chosen, since the rods are more sensitive than cones. Finally, a rod bleaching factor, B_S, is added to reduce the rod contribution to the overall color appearance as luminance level increases and the rods become less active. This factor is calculated using Equation 12-43. Examination of the rod bleaching factor in comparison with the cone bleaching factors given in Equation 12-20 through 12-22 shows that the rods will become saturated and their response will become significantly decreased at much lower luminance levels.

$$B_S = 0.5/\left\{1 + 0.3\left[\left(5L_{AS}/2.26\right)\left(S/S_W\right)\right]^{0.3}\right\} + 0.5/\left\{1 + 5\left[5L_{AS}/2.26\right]\right\}$$

$$(12\text{-}43)$$

Given the achromatic cone signal, A_a (Equation 12-29), the adapted scotopic signal, A_S (Equation 12-40), and the brightness background induction factor determined at the outset, one can calculate an overall achromatic signal, A, by using Equation 12-44:

$$A = N_{bb} \left[A_a - 1 + A_S - 0.3 + \left(1^2 + 0.3^2 \right)^{1/2} \right]$$

(12-44)

All of the constant values in Equation 12-44 represent removal of the earlier noise terms and then their reintroduction through quadrature summation.

The achromatic signal, A, is then combined with the overall chromatic signal, M, to calculate a correlate of brightness, Q, by using Equation 12-45:

$$Q = \left\{ 7 \left[A + (M/100) \right] \right\}^{0.6} N_1 - N_2$$

(12-45)

The correlate of brightness, Q, depends on both the achromatic, A, and chromatic, M, responses in order to appropriately model the Helmholtz-Kohlrausch effect. Equation 12-45 also includes two terms, N_1 and N_2, that account for the effects of surround on perceived brightness (e.g, Stevens effect and the Bartleson-Breneman results discussed in Chapter 6). These terms are calculated from the achromatic signal for the reference white, A_W, and the brightness surround induction factor, N_b (also determined at the outset), through Equations 12-46 and 12-47:

$$N_1 = \left(7 A_W \right)^{0.5} / \left(5.33 N_b^{0.13} \right)$$

(12-46)

$$N_2 = 7 A_W N_b^{0.362} / 200$$

(12-47)

Note that since the achromatic signal for the reference white, A_W, is required, it is necessary to carry through all of the model calculations described previously for the reference white in addition to the test stimulus. The brightness of the reference white, Q_W, must also be calculated for use in later equations.

Another form of brightness, referred to as whiteness-blackness, Q_{WB}, can be calculated in the Hunt model. This is a bipolar value similar to the Q value in the Nayatani et al. model that illustrates that black objects look darker and white objects look brighter as the adapting luminance level increases (another way to state the Stevens effect). Q_{WB} is calculated according to Equation 12-48 using the brightness of the background (which also must be calculated through the model):

$$Q_{WB} = 20 \left(Q^{0.7} - Q_b^{0.7} \right)$$

(12-48)

12.8   Lightness

Given the brightness of the test stimulus, Q, and the brightness of the reference white, Q_W, the Hunt color appearance model correlate of lightness, J, is calculated as shown in Equation 12-49:

$$J = 100(Q/Q_W)^z \tag{12-49}$$

This formulation for lightness follows the CIE definition that lightness is the brightness of the test stimulus relative to the brightness of a white. This ratio is raised to a power, z, that models the influence of the background relative luminance on perceived lightness according to Equation 12-50. The exponent, z, increases as the background becomes lighter, thus indicating that dark test stimuli will appear relatively more dark on a light background than they would on a dark background. This follows the commonly observed phenomenon of simultaneous lightness contrast.

$$z = 1 + (Y_b/Y_W)^{1/2} \tag{12-50}$$

12.9   Chroma

The Hunt color appearance model correlate of chroma, C_{94}, is determined from saturation, s, and the relative brightness (approximately lightness) following the general definitions given in Chapter 4 that indicate chroma can be represented as saturation multiplied by lightness. The precise formulation is given in Equation 12-51:

$$C_{94} = 2.44s^{0.69}(Q/Q_W)^{Y_b/Y_W}(1.64 - 0.29^{Y_b/Y_W}) \tag{12-51}$$

Equation 12-51 illustrates that chroma depends on the relative brightness of the stimulus, Q/Q_W, and on the relative luminance of the background, Y_b/Y_W. The formulation for chroma given by Equation 12-51 was derived empirically, based upon the results of a series of appearance scaling experiments (Hunt, 1994; Hunt and Luo, 1994).

12.10   Colorfulness

Given chroma, one can determine colorfulness by factoring in the brightness (or at least the luminance level). This is accomplished in the Hunt color appearance model by multiplying chroma, C_{94}, by the luminance level adaptation factor F_L (Equation 12-9) raised to a power of 0.15, as shown in Equation 12-52:

$$M_{94} = F_L^{0.15} C_{94} \tag{12-52}$$

Thus M_{94} is the colorfulness correlate for the Hunt color appearance model. It was also derived empirically through the analysis of visual scaling results.

12.11   Inverse Model

Unfortunately, the complete Hunt color appearance model cannot be analytically inverted because of its complexity. It is an even more severe problem if one has only lightness, chroma, and hue correlates to start from, which is often the case. Many applications, particularly image reproduction, require a color appearance model to be used in both forward and reverse directions. Thus the lack of an analytical inverse introduces some difficulty in using the Hunt model for these applications. Hunt (1995) provides some suggestions for how to deal with this difficulty.

In all cases, it is easier to reverse the model if all, rather than just three, of the appearance correlates are available. One alternative is to use the model without the scotopic response. This simplifies the inversion process, since the introduction of the scotopic terms into higher-level equations is one feature that prevents analytical inversion. The predictions of the model are slightly changed when the scotopic response is ignored, but this difference might be negligible for many applications. Hunt (1995) suggests that this technique is appropriate for reference white luminances greater than 10 cd/m². Most situations in which careful judgments of color reproduction are made are at luminance levels above 10 cd/m².

Other techniques suggested by Hunt (1995) require successive approximation for some parts of the reverse model. In most applications, it is simpler to use successive approximation for the whole model and iterate until appropriate output tristimulus values are obtained that produce the appearance correlates that are available at the outset. This technique can be accomplished with a method such as a Newton-Raphson optimization, and it can be applied when only lightness, chroma, and hue are available (in fact, it is the only option). While a successive-approximation technique can be very time-consuming for large data sets, such as images, this drawback is overcome by using the forward and reverse models to build three-dimensional lookup tables, which are then used with an interpolation technique to convert image data. Then, the time-consuming model inversion process need be performed only once for each viewing condition in order to build the lookup table. This approach helps. But if users want to vary the viewing conditions, they must wait a long time for the lookup table to be recalculated before an image can be processed. The delay might be significant enough to render the full Hunt model impractical for some applications.

In applying the model, it is often useful to consider its implementation as a simple step-by-step process. Here are the steps required to implement the model (and in reverse, as possible, to invert it):

1. Obtain physical data and decide on other parameters.
2. Calculate the cone excitations, $\rho\gamma\beta$, for the various stimulus fields.
3. Calculate the relative cone excitations.
4. Calculate the luminance-level adaptation factor, F_L.
5. Calculate the chromatic-adaptation factors, F_ρ, F_γ, F_β.
6. Calculate the Helson-Judd effect parameters, ρ_D, γ_D, β_D.
7. Calculate the adapted cone signals, ρ_a, γ_a, β_a.
8. Calculate the achromatic, A_a, and color difference, C_1, C_2, C_3, signals.
9. Calculate the hue angle, h_S.
10. Calculate the hue quadrature, H.
11. Calculate the hue composition, H_C.
12. Calculate the eccentricity factor, e_S.
13. Calculate the low-luminance tritanopia factor, F_t.
14. Calculate the chromatic responses, M, and saturation, s.
15. Calculate the scotopic luminance adaptation factor, F_{LS}.
16. Calculate the scotopic response, A_S.
17. Calculate the complete achromatic response, A.
18. Calculate the brightness, Q.
19. Calculate the lightness, J.
20. Calculate the chroma, C_{94}.
21. Calculate the colorfulness, M_{94}.
22. Calculate the whiteness-blackness, Q_{WB}.

The Hunt color appearance model is the most complex to implement of the models described in this book. There are a couple of implementation techniques that can simplify program development and greatly improve computational speed. One is to go through the model and calculate first, in a precalculation routine, all of the parameters that are constant for a given set of viewing conditions. This can be applied to all of the adaptation factors, correlates for the reference white, and so on. Then these precalculated data can be used for the remaining data calculations rather than the values for each stimulus (or pixel in an image) having to be recalculated. This can be particularly useful when transforming image data (or lookup tables) that might require the model calculations to be completed millions of times. Secondly, many of the equations, for example the achromatic response, A (Equation 12-44), include operations on constants. The number of computations can be reduced by combining all of these constants into a single number first (some compilers will do this for you). For example, the $-1-0.3+(1^2+0.3^2)^{1/2}$ can be converted into simply 1.044.

12.12   Phenomena Predicted

As stated previously, the Hunt color appearance model is the most extensive and complete model available. It has the following features:

- It is designed to predict the appearance of stimuli in a variety of backgrounds and surrounds at luminance levels ranging from the absolute threshold of human vision to cone bleaching.
- It can be used for related or unrelated stimuli. (See Hunt, 1991b, for an explanation of how the model is applied to unrelated colors.)
- It predicts a wide range of color-appearance phenomena, including the Bezold-Brücke hue shift, Abney effect, Helmholtz-Kohlrausch effect, Hunt effect, simultaneous contrast, Helson-Judd effect, Stevens effect, and Bartleson-Breneman observations.
- It predicts changes in color appearance due to light and chromatic adaptation and cognitive discounting-the-illuminant.
- It is unique in that it includes the contributions of rod photoreceptors.

While the list of appearance phenomena that the Hunt model addresses is extensive, it comes at the price of complexity (with no apologies required—the visual system is complex) that makes the model difficult to use in some applications.

Example calculations using the Hunt color appearance model as described in this chapter are given for four samples in Table 12-3. These results were calculated with the assumptions that the proximal field and background were both achromatic (same chromaticity as the source) with luminance factors of 20% and that the reference white also had the same chromaticity as the source with a luminance factor of 100%. Scotopic input was calculated using the approximate equations described in this chapter. The Helson-Judd parameters were always set to 0.0.

Table 12-3.   Example Hunt color appearance model calculations.

Quantity	Case 1	Case 2	Case 3	Case 4
X	19.01	57.06	3.53	19.01
Y	20.00	43.06	6.56	20.00
Z	21.78	31.96	2.14	21.78
X_W	95.05	95.05	109.85	109.85
Y_W	100.00	100.00	100.00	100.00
Z_W	108.88	108.88	35.58	35.58
L_A	318.31	31.83	318.31	31.83
N_c	1.0	1.0	1.0	1.0
N_b	75	75	75	75
Discounting?	Yes	Yes	Yes	Yes
h_S	268.8	18.5	178.3	263.2
H	316.9	398.8	222.2	313.6
H_C	83B 17R	99R 1B	78G 22B	86B 14R
s	0.03	157.98	249.72	214.66
Q	33.26	30.49	21.21	27.53
J	44.62	65.50	23.08	55.48
C_{94}	0.16	65.04	77.23	78.54
M_{94}	0.16	59.33	79.05	71.64

12.13   Why Not Use Just the Hunt Model?

The Hunt color appearance model seems to be able to do everything that anyone could ever want from a color appearance model. So why isn't it adopted as
the single, standard color appearance model for all applications? The main reason could be the very fact that it is so complete. In its completeness also lies its
complexity. And its complexity makes its application to practical situations
range from difficult to impossible. However, its complexity also allows it to be
extremely flexible. As is shown in Chapter 15, the Hunt model is generally capable of making accurate predictions for a wide range of visual experiments.
This is because the model is flexible enough to be adjusted to the required situations. This flexibility and general accuracy are great features of the Hunt
model. However, often it is not possible to know just how to apply the Hunt
model (i.e., decide on the appropriate parameter values) until after the visual
data have been obtained. In other cases, the parameters actually need to be optimized, not just chosen, for the particular viewing situation. This is not a
problem if the resources are available to derive the optimized parameters.
However, when such resources are not available and the Hunt model must be
used "as is" with the recommended parameters, the model can perform extremely poorly (see Chapter 15). This is because the nominal parameters used

for a given viewing condition are being used to make specific predictions of phenomena that may or may not be important in that situation. After the fact, adjustments can be made, but that might be too late. Thus, if it is not possible to optimize (or optimally choose) the implementation of the Hunt model, its precision might result in predictions that are worse than those of much simpler models for some applications.

Other negative aspects that counteract the positive features of the Hunt color appearance model include the following:

- It cannot be easily inverted.
- It is computationally expensive, difficult to implement, and requires significant user knowledge to use consistently.
- It uses functions with additive offsets to predict contrast changes due to variation in surround relative luminance. These functions can result in predicted corresponding colors with negative tristimulus values for some changes in surround.

CHAPTER 13

The RLAB Model

13.1   Objectives and Approach
13.2   Input Data
13.3   Adaptation Model
13.4   Opponent-color Dimensions
13.5   Lightness
13.6   Hue
13.7   Chroma
13.8   Saturation
13.9   Inverse Model
13.10 Phenomena Predicted
13.11 Why Not Use Just the RLAB Model?

THIS CHAPTER COMPLETES the discussion of some of the most widely used color appearance models with a description of the RLAB model (RLAB). The Hunt and Nayatani et al. models discussed in the preceding chapters are designed to predict all of the perceptual attributes of color appearance for a wide range of viewing conditions. The RLAB model, however, was designed with other considerations in mind. It was developed with the intent of producing a simple color appearance model capable of predicting the most significant appearance phenomena in practical applications. It also was developed with cross-media image reproduction as its target application, and it has been effectively applied to such situations.

13.1 Objectives and Approach

The RLAB color appearance model evolved from studies of chromatic adaptation (Fairchild, 1990), chromatic-adaptation modeling (Fairchild, 1991a, b), fundamental CIE colorimetry (CIE, 1986), and practical implications in cross-media image reproduction (Fairchild and Berns, 1993; Fairchild, 1994). The starting point for RLAB is the CIELAB color space. While CIELAB can be used as an approximate color appearance model, it does have significant limitations. The most important for many practical applications are an inaccurate chromatic-adaptation transform, no luminance-level dependency, no surround dependency, and no distinction for when discounting-the-illuminant occurs. Thus RLAB was designed to build on the positive aspects of CIELAB, and, through additions, to address its limitations.

CIELAB provides good perceptual uniformity with respect to color appearance for average daylight illuminants. This is illustrated by the spacing of constant hue and chroma contours from the *Munsell Book of Color* shown in Figure 13-1. The contours plotted in the figure are as good as, and in some cases better than, those produced using any other color appearance model. However, due to CIELAB's "wrong von Kries" chromatic-adaptation transform, the good perceptual spacing of CIELAB quickly degrades as the illuminant moves away from average daylight. The concept of RLAB is to take advantage of the good spacing under daylight and familiarity of the CIELAB space, while improving its applicability to nondaylight illuminants. This was accomplished by defining a reference set of viewing conditions (illuminant D65, 318 cd/m², average surround, discounting-the-illuminant) for which the

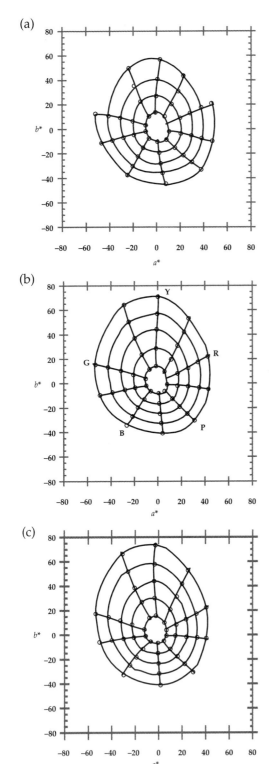

Figure 13-1. Contours of constant Munsell hue and chroma plotted in the CIELAB/RLAB color space at (a) Value 3, (b) Value 5, and (c) Value 7.

CIELAB space is used. Then a more accurate chromatic-adaptation transform (Fairchild, 1991b, 1994, 1996) is used to determine corresponding colors between the test viewing conditions and the reference viewing conditions. Thus test tristimulus values are first transformed into corresponding colors under the reference viewing condition. Then a modified CIELAB space is used to describe appearance correlates.

In addition, the compressive nonlinearity (cube root) of the CIELAB space is adapted to become a function of the surround relative luminance. This allows the prediction of decreases in perceived image contrast as the surround becomes darker, as suggested in the work of Bartleson (1975). The improved chromatic-adaptation transform and the surround dependence enhance CIELAB in the two areas that are most critical in image reproduction.

RLAB was designed to include predictors of only relative color-appearance attributes. Thus it can be used to calculate correlates of lightness, chroma, saturation, and hue, but it cannot be used to predict brightness or colorfulness. This limitation was imposed to keep the model as simple as possible and because the predictions of brightness and colorfulness have little importance in most image reproduction applications.

The fact that RLAB is based on the CIELAB space has benefit in addition to familiarity. Since the RLAB spacing is essentially identical to CIELAB spacing, color-difference formulas such as CIELAB ΔE^*_{ab} (CIE, 1986), CMC (Clark et al., 1984), and CIE94 (CIE, 1995) can be used with results that are similar to those obtained when using CIELAB alone under average daylight illuminants.

A more detailed description of the RLAB model, as presented in this chapter, can be found in Fairchild (1996).

13.2   Input Data

Input data for the RLAB model include the relative tristimulus values of the test stimulus (XYZ) and the white point ($X_n Y_n Z_n$), the absolute luminance of a white object in the scene in cd/m^2, the relative luminance of the surround (dark, dim, average), and a decision on whether discounting-the-illuminant is taking place. The surround relative luminance is generally taken to be average for reflection prints, dim for CRT displays or televisions, and dark for projected transparencies—in all cases, it is assumed that these media are being viewed in their typical environments. The surround is not directly tied to the medium. Thus it is certainly possible to have reflection prints viewed in a dark surround and projected transparencies viewed in an average surround. Discounting-the-illuminant is assumed to occur for object color stimuli such as prints and not to occur for emissive displays such as CRTs. Intermediate levels of discounting-the-illuminant are likely to occur in some situations, such as the viewing of projected transparencies.

13.3   Adaptation Model

The following equations describe the chromatic-adaptation model built into
RLAB. The model is based on the model of incomplete chromatic adaptation
described by Fairchild (1991b) and later modified (Fairchild, 1994, 1996).
This transformation is also discussed in Chapter 9.

One begins with a conversion from CIE tristimulus values ($Y = 100$ for
white) to fundamental tristimulus values as illustrated in Equations 13-1 and
13-2. All CIE tristimulus values are normally calculated using the CIE 1931
Standard Colorimetric Observer (2°). The transformation must also be completed for the tristimulus values of the adapting stimulus.

$$\begin{vmatrix} L \\ M \\ S \end{vmatrix} = \mathbf{M} \begin{vmatrix} X \\ Y \\ Z \end{vmatrix}$$

(13-1)

$$\mathbf{M} = \begin{vmatrix} 0.3897 & 0.6890 & -0.0787 \\ -0.2298 & 1.1834 & 0.0464 \\ 0.0 & 0.0 & 1.0000 \end{vmatrix}$$

(13-2)

The transformation to cone responses is the same as that used in the
Hunt model. Matrix $\mathbf{M}$ is normalized such that the tristimulus values for the
equal-energy illuminant ($X = Y = Z = 100$) produce equal cone responses ($L = M = S = 100$). The next step is the calculation of the $\mathbf{A}$ matrix that is used to
model the chromatic-adaptation transformation:

$$\mathbf{A} = \begin{vmatrix} a_L & 0.0 & 0.0 \\ 0.0 & a_M & 0.0 \\ 0.0 & 0.0 & a_S \end{vmatrix}$$

(13-3)

The $\mathbf{A}$ matrix represents von Kries adaptation coefficients that are applied to the cone responses for the test stimulus (LMS). The von Kries–type coefficients are calculated using Equations 13-4 through 13-12:

$$a_L = \frac{p_L + D(1.0 - p_L)}{L_n}$$

(13-4)

$$a_M = \frac{p_M + D(1.0 - p_M)}{M_n}$$

(13-5)

$$a_S = \frac{p_S + D(1.0 - p_S)}{S_n}$$

(13-6)

The p terms describe the proportion of complete von Kries adaptation that is occurring. They are calculated using formulas that predict chromatic adaptation to be more complete as the luminance level increases and less complete as the color of the adapting stimulus departs from that of the equal-energy illuminant. These terms are equivalent to the chromatic-adaptation factors in the Hunt model and are calculated using the same equations given in Equations 13-7 through 13-12:

$$p_L = \frac{\left(1.0 + Y_n^{1/3} + l_E\right)}{\left(1.0 + Y_n^{1/3} + 1.0/l_E\right)} \tag{13-7}$$

$$p_M = \frac{\left(1.0 + Y_n^{1/3} + m_E\right)}{\left(1.0 + Y_n^{1/3} + 1.0/m_E\right)} \tag{13-8}$$

$$p_S = \frac{\left(1.0 + Y_n^{1/3} + s_E\right)}{\left(1.0 + Y_n^{1/3} + 1.0/s_E\right)} \tag{13-9}$$

$$l_E = \frac{3.0L_n}{L_n + M_n + S_n} \tag{13-10}$$

$$m_E = \frac{3.0M_n}{L_n + M_n + S_n} \tag{13-11}$$

$$s_E = \frac{3.0S_n}{L_n + M_n + S_n} \tag{13-12}$$

Y_n in Equations 13-7 through 13-9 is the absolute adapting luminance in cd/m^2. The cone response terms with n subscripts (L_n, M_n, S_n) refer to values for the adapting stimulus derived from relative tristimulus values. The D factor in Equations 13-4 through 13-6 allows various proportions of cognitive discounting-the-illuminant. D should be set equal to 1.0 for hard-copy images, 0.0 for soft-copy displays, and an intermediate value for situations such as projected transparencies in completely darkened rooms. The exact value of the D factor can be used to account for the various levels of chromatic adaptation found in the infinite variety of practical viewing situations. The exact choice of intermediate values will depend upon the specific viewing conditions. Katoh (1994) has illustrated an example of intermediate adaptation in direct comparison between soft- and hard-copy displays, and Fairchild (1992a) has reported a case of intermediate discounting-the-illuminant for a soft-copy display. When no visual data are available and an intermediate value is necessary, a value of 0.5 should be chosen and refined with experience.

Note that if discounting-the-illuminant occurs and D is set equal to 1.0, then the adaptation coefficients described in Equations 13-4 through 13-6 reduce to the reciprocals of the adapting cone excitations exactly as would be implemented in a simple von Kries model.

After the **A** matrix is calculated, the tristimulus values for a stimulus color are converted to corresponding tristimulus values under the reference viewing conditions by using Equations 13-13 and 13-14:

$$
\begin{vmatrix} X_{ref} \\ Y_{ref} \\ Z_{ref} \end{vmatrix} = \mathbf{RAM} \begin{vmatrix} X \\ Y \\ Z \end{vmatrix}
$$

(13-13)

$$
\mathbf{R} = \begin{vmatrix} 1.9569 & -1.1882 & 0.2313 \\ 0.3612 & 0.6388 & 0.0 \\ 0.0 & 0.0 & 1.0000 \end{vmatrix}
$$

(13-14)

The R matrix represents the inverse of the **M** and **A** matrices for the reference viewing conditions ($A^{-1}M^{-1}$), plus a normalization, discussed next, that are always constant and can therefore be precalculated. Thus Equation 13-13 represents a modified von Kries chromatic-adaptation transform that converts test-stimulus tristimulus values to corresponding colors under the RLAB reference viewing conditions (illuminant D65, 318 cd/m², discounting-the-illuminant). The next step is to use these reference tristimulus values to calculate modified CIELAB appearance correlates, as shown in the following sections.

13.4   Opponent-color Dimensions

Opponent-type responses in RLAB are calculated using Equations 13-15 through 13-17:

$$
L^R = 100\left(Y_{ref}\right)^{\sigma}
$$

(13-15)

$$
a^R = 430\left[\left(X_{ref}\right)^{\sigma} - \left(Y_{ref}\right)^{\sigma}\right]
$$

(13-16)

$$
b^R = 170\left[\left(Y_{ref}\right)^{\sigma} - \left(Z_{ref}\right)^{\sigma}\right]
$$

(13-17)

L^R represents an achromatic response analogous to CIELAB L^*. The red-green chromatic response is given by a^R (analogous to CIELAB a^*), and the yellow-blue chromatic response is given by b^R (analogous to CIELAB b^*). Recall that for the reference viewing conditions, the RLAB coordinates are nearly identical to CIELAB coordinates. They are not fully identical because Equations 13-15 through 13-17 have been simplified from the CIELAB equations for computational efficiency. The conditional compressive nonlinearities (i.e., different functions for low tristimulus values) of CIELAB have been replaced with simple power functions. This results in the exponents and the scaling factors being slightly different than the CIELAB equations. Fairchild (1996) provides more details on these differences.

It is also worth noting that the divisions by the tristimulus values of the white point that are incorporated in the CIELAB equations are missing from Equations 13-15 through 13-17. This is because these normalizations are constant in the RLAB model and they have been built into the **R** matrix given in Equation 13-14.

The exponents in Equations 13-15 through 13-17 vary depending on the relative luminance of the surround. For an average surround, $\sigma = 1/2.3$; for a dim surround, $\sigma = 1/2.9$; and for a dark surround, $\sigma = 1/3.5$. The ratios of these exponents are precisely in line with the contrast changes suggested by Bartleson (1975) and Hunt (1995) for image reproduction. More detail on the exponents can be found in Fairchild (1995b).

As nominal definitions, a dark surround is considered essentially zero luminance, a dim surround is considered a relative luminance less than 20% of white in the image, and an average surround is considered a relative luminance equal to or greater than 20% of the image white. The precise nature and magnitude of the contrast changes required for various changes in image-viewing conditions are still a topic of research and debate. Thus it is best to use these parameters with some flexibility.

In some applications, it might be desired to use intermediate values for the exponents in order to model less severe changes in surround relative luminance. This requires no more than a substitution in the RLAB equations, since they do not include the conditional functions that are found in the CIELAB equations. In addition, it might be desirable to use different exponents on the lightness, L^R, dimension than on the chromatic, a^R and b^R, dimensions. This can also be easily accommodated. The equations have been formulated as simple power functions to encourage the use of different exponents, which might be more appropriate than the nominal exponents for particular, practical viewing conditions.

Figure 13-2 illustrates the effect of changing the exponents. The left image (a) is a typical printed reproduction. The other two images (b and c) show the change in contrast necessary to reproduce the same appearance if the image were viewed in a dark surround. Note the increase in contrast required for viewing in a dark surround. In the middle image (b), the adjustment has been made on lightness contrast only. In the right image (c), the surround compensation has been applied to the chromatic dimensions, as well as to lightness. The right image (c) is similar to the reproduction that would be produced in a

Print (Original)   Slide (*Y* Adjusted)   Slide (*XYZ* Adjusted)

Figure 13-2.   Images illustrating the change in image contrast necessary to account for appearance differences due to change in surround relative luminance: (a) original print image; (b) image for viewing in a dark surround, adjusted only in lightness contrast; (c) image for viewing in a dark surround, adjusted in both lightness and chromatic contrast.

simple photographic system (since the contrast of all three film layers must be changed together) and is generally preferable. Note, however, that interimage effects in real film could be used to compensate for some of the increase in saturation with contrast. The images in Figure 13-2 should be used only to judge the relative impact of the adjustments, since they are not being viewed in the appropriate viewing conditions.

13.5   Lightness

The RLAB correlate of lightness is L^R, given in Equation 13-15. No further calculations are required.

13.6   Hue

Hue angle, h^R, is calculated in the RLAB space using the same procedure as CIELAB. As in CIELAB, h^R is expressed in degrees ranging from 0 to 360 measured from the positive a^R axis calculated according to Equation 13-18:

$$h^R = \tan^{-1}(b^R / a^R) \tag{13-18}$$

Table 13-1.   Data for conversion from hue angle to hue composition.

h^R	R	B	G	Y	H^R
24	100	0	0	0	R
90	0	0	0	100	Y
162	0	0	100	0	G
180	0	21.4	78.6	0	B79G
246	0	100	0	0	B
270	17.4	82.6	0	0	R83B
0	82.6	17.4	0	0	R17B
24	100	0	0	0	R

Hue composition can be determined in RLAB using a procedure similar to that of the Hunt model and the Nayatani et al. model. This is useful when testing a color appearance model against magnitude estimation data and when it is desired to reproduce a named hue. Hue composition, H^R, can be calculated via linear interpolation of the values in Table 13-1. These were derived based on the notation of the Swedish Natural Color System and are illustrated in Figure 13-3. This figure is a useful visualization of the loci of the unique hues, since they do not correspond to the principal axes of the color space. The unique hue locations are the same as those in the CIELAB space under only the reference conditions. Example hue composition values are listed in Table 13-1 in italics.

13.7   Chroma

RLAB chroma, C^R, is calculated in the same way as CIELAB chroma, as shown in Equation 13-19:

$$C^R = \sqrt{(a^R)^2 + (b^R)^2}$$

(13-19)

13.8   Saturation

In some applications, such as the image-color manipulation required for gamut mapping, it might be desirable to change colors along the lines of constant saturation rather than constant chroma. Wolski, Allebach, and Bouman (1994) have proposed such a technique; Montag and Fairchild (1996, 1997) also describe such situations. *Saturation* is colorfulness relative to brightness, *chroma* is colorfulness relative to the brightness of a white, and *lightness* is

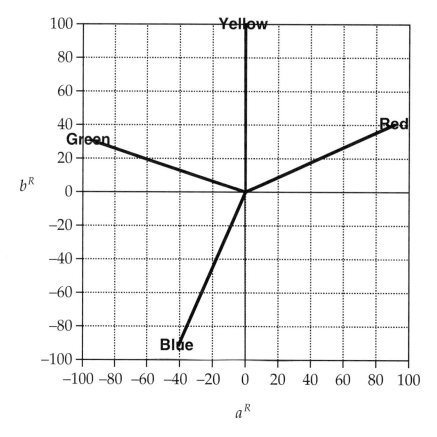

Figure 13-3. Illustration of the hue angles of the perceptually unique hues in the RLAB color space.

brightness relative to the brightness of a white. Therefore saturation can be defined as chroma relative to lightness. Chroma, C^R, and lightness, L^R, are already defined in RLAB; thus saturation, s^R, is defined as shown in Equation 13-20:

$$s^R = C^R / L^R \qquad (13\text{-}20)$$

It is of interest to note that a progression along a line of constant saturation is the series of colors that can be observed when an object is viewed in ever deepening shadows. This could be why transformations along lines of constant saturation, rather than chroma, are sometimes useful in gamut-mapping applications.

13.9 Inverse Model

Since the RLAB model was designed with image-reproduction applications in mind, computational efficiency and simple inversion were considered of significant importance. Thus the RLAB model is very easy to invert and requires a minimum of calculations. A step-by-step procedure for implementing the RLAB model is given next:

1. Obtain the colorimetric data for the test and adapting stimuli and the absolute luminance of the adapting stimulus. Decide on the discounting-the-illuminant factor and the exponent (based on surround relative luminance).
2. Calculate the chromatic-adaptation matrix, **A**.
3. Calculate the reference tristimulus values.
4. Calculate the RLAB parameters, L^R, a^R, and b^R.
5. Use a^R and b^R to calculate C^R and h^R.
6. Use h^R to determine H^R.
7. Calculate s^R using C^R and L^R.

In typical color-reproduction applications, it is not enough to know the appearance of image elements; one must reproduce those appearances in a second set of viewing conditions as illustrated in Figure 13-4. To accomplish this, one must be able to calculate CIE tristimulus values, XYZ, from the appearance parameters, $L^R a^R b^R$, and the definition of the new viewing conditions.

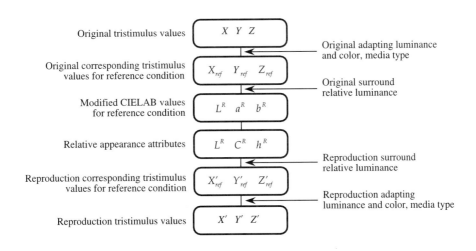

Figure 13-4. A flow chart of the application of the RLAB model to image reproduction applications.

These tristimulus values are then used, along with the imaging-device characterization, to determine device color signals such as RGB or CMYK. Equations
13-21 through 13-24 outline how to calculate CIE tristimulus values from
RLAB $L^R a^R b^R$. If starting with $L^R C^R h^R$, one must first transform back
to $L^R a^R b^R$ using the usual transformation from cylindrical coordinates to
rectangular coordinates.

The reference tristimulus values are calculated from the RLAB parameters using Equations 13-21 through 13-23 with an exponent, σ, appropriate
for the second viewing condition:

$$Y_{ref} = \left(\frac{L^R}{100}\right)^{1/\sigma}$$

(13-21)

$$X_{ref} = \left[\left(\frac{a^R}{430}\right) + \left(Y_{ref}\right)^\sigma\right]^{1/\sigma}$$

(13-22)

$$Z_{ref} = \left[\left(Y_{ref}\right)^\sigma - \left(\frac{b^R}{170}\right)\right]^{1/\sigma}$$

(13-23)

The reference tristimulus values are then transformed to tristimulus values for the second viewing condition by using Equation 13-24 with an **A** matrix calculated for the second viewing condition:

$$\begin{vmatrix} X \\ Y \\ Z \end{vmatrix} = (\mathbf{RAM})^{-1} \begin{vmatrix} X_{ref} \\ Y_{ref} \\ Z_{ref} \end{vmatrix}$$

(13-24)

13.10   Phenomena Predicted

The RLAB model provides correlates for relative color-appearance attributes
only (lightness, chroma, saturation, and hue). It cannot be used to predict
brightness and colorfulness. This limitation is of little practical importance for
image reproduction, since there are few applications in which brightness and
colorfulness are required. RLAB includes a chromatic-adaptation transform,
with a parameter for discounting-the-illuminant, and predicts incomplete chromatic adaptation to certain stimuli (e.g., RLAB correctly predicts that a CRT
display with a D50 white point will retain a yellowish appearance). It also includes variable exponents, σ, that modulate image contrast as a function of the

Table 13-2.   Example RLAB color appearance model calculations.

Quantity	Case 1	Case 2	Case 3	Case 4
X	19.01	57.06	3.53	19.01
Y	20.00	43.06	6.56	20.00
Z	21.78	31.96	2.14	21.78
X_n	95.05	95.05	109.85	109.85
Y_n	100.00	100.00	100.00	100.00
Z_n	108.88	108.88	35.58	35.58
Y_n (cd/m^2)	318.31	31.83	318.31	31.83
σ	0.43	0.43	0.43	0.43
D	1.0	1.0	1.0	1.0
L^R	49.67	69.33	30.78	49.83
a^R	0.00	46.33	-40.96	15.57
b^R	-0.01	18.09	2.25	-52.61
h^R	270.0	21.3	176.9	286.5
H^R	R83B	R2B	B74G	R71B
C^R	0.01	49.74	41.02	54.87
s^R	0.00	0.72	1.33	1.10

surround relative luminance. These are the most important color-appearance phenomena for cross-media image reproduction applications. If it is necessary to predict absolute appearance attributes (i.e., brightness and colorfulness) or more unusual appearance phenomena, then more extensive appearance models such as the Nayatani et al. and Hunt models described in preceding chapters should be considered.

Example calculations for the RLAB color appearance model are given in Table 13-2.

13.11   Why Not Use Just the RLAB Model?

The RLAB model is simple, straightforward, and easily invertible. It also has been found to be as accurate as, or better than, more complicated color appearance models for many practical applications. Because of these features, why wouldn't RLAB be considered as a recommendation for a single, universal color appearance model?

RLAB's weaknesses are the same as its strengths:

- Since it is such a simple model, it is not exhaustive in its prediction of color-appearance phenomena.
- It does not include correlates of brightness and colorfulness.

- It cannot be applied over a wide range of luminance levels.
- It does not predict some color-appearance phenomena such as the Hunt effect, the Helson-Judd effect, and the Stevens effect (although the latter could be simulated by making σ change with adapting luminance rather than with surround relative luminance).

In practical imaging applications, these limitations are of little consequence due to device-gamut limitations. In other applications, where these phenomena might be important, a different color appearance model is required. In summary, RLAB performs well for the image-reproduction applications for which it was designed, but it is not comprehensive enough for all color-appearance situations that might be encountered.

CHAPTER 14

Other Models

14.1 Overview
14.2 ATD Model
14.3 LLAB Model

THE PRECEDING CHAPTERS have reviewed four of the most widely used and most general color appearance models. These and other models are continually being modified and introduced. This chapter provides overviews of two more color appearance models—the ATD model, which has been evolving for many years, and the LLAB model, which is very new. For various reasons discussed herein, these models are not as well suited for general application as those described in earlier chapters. However, various aspects of their formulation are of interest and are likely to find their way into future models or applications. Thus they have been included to appropriately cover their potential impact on the field.

14.1 Overview

The formulation of color appearance models is an area of active, ongoing research activity. That explains why this book can provide only an overview of the associated problems and several approaches to solving them. It is not possible to present a single model of color appearance that will solve all problems if the model is followed in a "cookbook" manner. This is the "holy grail" for researchers in the field, but it is not likely to be achieved in short order. Chapters 10 through 13 presented some of the best available approaches to problems of color-appearance specification and reproduction. One of those models is likely to be the most appropriate solution for any given application. However, one model is not likely to be the best solution for all applications. Since the development of color appearance models is ongoing, it is important to describe other models that have had, or are likely to have, significant impact on the field.

Two such models are described in this chapter:

1. The ATD model formulated by Guth (1995), which has a history of development dating back over 20 years (Guth, 1994a).
2. A very recent model derived by Luo et al. (1996) called LLAB.

A third model is under development by the CIE TC1-34 as an attempt to promote uniformity of practice in color-appearance specification for industry. The current status of this CIE color appearance model is described in Appendix A.

14.2   ATD Model

The ATD model, developed by Guth, is a different type of model than those described in earlier chapters. In fact, according to the CIE TC1-34 definition of what makes a color appearance model, it cannot be considered a true color appearance model. This is because the model was developed with different aims. It is described as a model of color vision, or, more appropriately, the first few stages of color vision.

Guth (1994a, b) has given a good overview of the development and performance of the model. The model's history dates back to 1972. It was developed to predict a wide range of vision-science data on phenomena such as chromatic discrimination, absolute thresholds, the Bezold-Brücke hue shift, the Abney effect, heterochromatic brightness matching, light adaptation, and chromatic adaptation. The ATD model is capable of making remarkable predictions of vision data for these phenomena. However, most such experiments have been performed using unrelated stimuli. Thus the model is designed for unrelated stimuli, and only by somewhat arbitrary modification of the model can it be applied to related colors. This background explains why the model incorporates measures of color discrimination, brightness, saturation, and hue but does not distinguish other important appearance attributes such as lightness, colorfulness, and chroma. This limits the model's applicability somewhat, but its general structure and chromatic-adaptation transformation are certainly worthy of further study and attention.

The ATD model has been modified and used in practical imaging applications. Such applications have been described by Granger (1994, 1995). Granger took advantage of the opponent-colors encoding of the ATD model to develop a space that is useful for describing color appearance and doing color manipulations in desktop publishing. However, he did not incorporate any chromatic-adaptation transformation in his modified ATD model. Thus the utility of the model is limited to a single illumination white point, unless a user is willing to make the erroneous assumption that printed color images represent color-constant stimuli.

Objectives and Approach

Guth (1995) refers to the ATD model as a "model for color perception and visual adaptation." With its long history of development aimed at the prediction of various color-vision data, that is certainly an appropriate description. Regarding what the model is intended for, Guth states that it "should now be seriously considered by the vision community as a replacement for all models that are currently used to make predictions (or to establish standards) that concern human color perception." This is an ambitious goal that clearly overlaps the objectives of some of the other models described in this book. As discussed later in the chapter, it is also an extreme overstatement of the model's capabilities.

The latest revision of the ATD model, called "ATD95," is described in the following sections. Guth's treatment has been followed. Earlier papers, such as Guth (1991), should also be referred to in order to obtain a more global understanding of the model's derivation and capabilities.

The model begins with nonlinear cone responses followed by a nonlinear von Kries-type receptor gain control and two stages of opponent responses necessary for predicting various discrimination and appearance phenomena. Finally, the model includes a neural compression of the opponent signals. The letters A, T, and D are abbreviations for achromatic, tritanopic, and deuteranopic mechanisms, respectively. The A system signals brightness, the T system redness-greenness, and the D system yellowness-blueness.

Input Data

The input data for the ATD model operating on unrelated colors are the $X'Y'Z'$ tristimulus values (Judd's modified tristimulus values, not CIE XYZ tristimulus values) expressed in absolute luminance units. The $X'Y'Z'$ tristimulus values are scaled such that Y' is set equal to Y (photopic luminance) expressed in trolands. A *troland* is a measure of retinal illuminance that factors in the eye's pupil diameter. Since the pupil diameter is controlled by the scene luminance (to some degree), Guth suggests converting from luminance in cd/m^2 to retinal illuminance in trolands by raising the luminance to the power of 0.8 and multiplying by 18 as a reasonable approximation. Strictly speaking, the ATD model is incompatible with CIE colorimetry, since it is based on Judd-modified $X'Y'Z'$ tristimulus values rather than CIE XYZ tristimulus values. However, Guth states that "it is probably true that XYZs rather than $X'Y'Z'$s can be used in most situations." For the remainder of this chapter, CIE XYZ tristimulus values are assumed to be used.

For predictions involving related colors, the absolute tristimulus values expressed in trolands must also be available for the adapting stimulus, $X_0Y_0Z_0$. Guth is equivocal on how to obtain these values, so it will be assumed that they are the tristimulus values for a perfect white under illumination similar to the test stimulus.

No other input data are required, or used, in the ATD model. Thus it is clear that it cannot be used to account for background, surround, or cognitive effects.

Adaptation Model

As with all of the models presented in this book, the first step of the ATD model is a transformation from CIE (or Judd) tristimulus values to cone responses. However, a significant difference in the ATD model is that the cone responses are nonlinear and additive noise signals are incorporated at this stage. The transformations are given in Equations 14-1 through 14-3:

$$L = \left[0.66\big(0.2435X + 0.8524Y - 0.0516Z\big)\right]^{0.70} + 0.024 \tag{14-1}$$

$$M = \left[1.0\big(-0.3954X + 1.1642Y + 0.0837Z\big)\right]^{0.70} + 0.036 \tag{14-2}$$

$$S = \left[0.43\big(0.04Y + 0.6225Z\big)\right]^{0.70} + 0.31 \tag{14-3}$$

Chromatic adaptation is then modeled using a modified form of the von Kries transformation, as illustrated in Equations 14-4 through 14-6:

$$L_g = L\big[\sigma/(\sigma + L_a)\big] \tag{14-4}$$

$$M_g = M\big[\sigma/(\sigma + M_a)\big] \tag{14-5}$$

$$S_g = S\big[\sigma/(\sigma + S_a)\big] \tag{14-6}$$

L_g, M_g, and S_g are the post-adaptation cone signals. The constant, σ, is varied to predict different types of data, but it is nominally set equal to 300. The cone signals for the adapting light, L_a, M_a, and S_a, are determined from a weighted sum of the tristimulus values for the stimulus itself and for a perfect white (or other adapting stimulus), as shown in Equations 14-7 through 14-9. These are then transformed to cone signals using Equations 14-1 through 14-3:

$$X_a = k_1 X + k_2 X_0 \tag{14-7}$$

$$Y_a = k_1 Y + k_2 Y_0 \tag{14-8}$$

$$Z_a = k_1 Z + k_2 Z_0 \tag{14-9}$$

For unrelated colors, there is only self-adaptation and k_1 is set to 1.0, while k_2 is set to 0.0. For related colors such as typical colorimetric applications, k_1 is set to 0.0 and k_2 is set to a value between 15 and 50 (Guth, 1995). In some cases, the observer might adapt to both the test stimulus and the white point to some degree, and some other combination of values, such as $k_1 = 1.0$ and $k_2 = 5.0$, would be used (Guth, 1995). Guth makes no specific recommendation on how to calculate cone signals for the adapting light. It is left to interpretation. However, it is worth noting that as the value of k_2 increases, the adaptation transform of Equations 14-4 through 14-6 becomes more and more like the nominal von Kries transformation. This is also true as σ is decreased.

Thus it is not difficult to make the ATD model perform almost the same as a simple von Kries model for color-reproduction applications. It is therefore recommended that Guth's maximum value of k_2 equal to 50 be used with k_1 set to 0.0.

Opponent-color Dimensions

The adapted cone signals are then transformed into two sets of initial opponent signals. The first-stage initial signals are denoted A_{1i}, T_{1i}, and D_{1i} and calculated according to Equations 14-10 through 14-12. The second-stage initial signals, calculated using Equations 14-13 through 14-15, are denoted A_{2i}, T_{2i}, and D_{2i}:

$$A_{1i} = 3.57L_g + 2.64M_g \tag{14-10}$$

$$T_{1i} = 7.18L_g - 6.21M_g \tag{14-11}$$

$$D_{1i} = -0.70L_g + 0.085M_g + 1.00S_g \tag{14-12}$$

$$A_{2i} = 0.09A_{1i} \tag{14-13}$$

$$T_{2i} = 0.43T_{1i} + 0.76D_{1i} \tag{14-14}$$

$$D_{2i} = D_{1i} \tag{14-15}$$

The final *ATD* responses after compression are calculated for both the first and second stages according to Equations 14-16 through 14-21:

$$A_1 = A_{1i} / \left(200 + |A_{1i}|\right) \tag{14-16}$$

$$T_1 = T_{1i} / \left(200 + |T_{1i}|\right) \tag{14-17}$$

$$D_1 = D_{1i} / \left(200 + |D_{1i}|\right) \tag{14-18}$$

$$A_2 = A_{2i} / \left(200 + |A_{2i}|\right) \tag{14-19}$$

$$T_2 = T_{2i} / \left(200 + |T_{2i}|\right) \tag{14-20}$$

$$D_2 = D_{2i} / \left(200 + |D_{2i}|\right) \tag{14-21}$$

The first-stage opponent responses are used to model apparent brightness and discriminations (absolute and difference thresholds). Discriminations are modeled using Euclidean distance in the $A_1T_1D_1$ three-space with a visual threshold set to approximately 0.005 unit. The second-stage mechanisms are used to model large color differences, hue, and saturation.

Perceptual Correlates

The ATD model incorporates measures to predict brightness, saturation, and hue. The brightness correlate is the quadrature summation of the A_1, T_1, and D_1 responses, as illustrated in Equation 14-22:

$$Br = \left(A_1^2 + T_1^2 + D_1^2\right)^{1/2}$$

$$(14\text{-}22)$$

Saturation is calculated as the quadrature sum of the second-stage chromatic responses, T_2 and D_2, divided by the achromatic response, A_2, as shown in Equation 14-23:

$$C = \left(T_2^2 + D_2^2\right)^{0.5} / A_2$$

$$(14\text{-}23)$$

Guth (1995) incorrectly uses the terms "saturation" and "chroma" interchangeably. However, it is clear that the formulation of Equation 14-23 is a measure of saturation rather than chroma, since it is measured relative to the achromatic response for the stimulus rather than that of a similarly illuminated white.

Guth (1995) indicates that hue is directly related to H as defined in Equation 14-24. However, the ratio in Equation 14-24 is equivocal (giving equal values for complementary hues, infinite values for some, and undefined values for others). So it is necessary to add an inverse tangent function, as is typical practice.

$$H = T_2 / D_2$$

$$(14\text{-}24)$$

There are no correlates of lightness, colorfulness, chroma, or hue composition in the ATD model.

Phenomena Predicted

The ATD model accounts for chromatic adaptation, heterochromatic brightness matching (Helmholtz-Kohlrausch effect), the Bezold-Brücke hue shift, the Abney effect, and various color-discrimination experiments. It includes correlates for brightness and saturation. A correlate for hue can also be easily calculated, although hue composition has not been specifically defined. The model is inadequately defined for related stimuli, since it does not include correlates for lightness and chroma. There is no way to distinguish between brightness-colorfulness matching and lightness-chroma matching using the ATD model. It is unclear which type of matching is predicted by the adaptation transform, but it is likely to be more similar to brightness-colorfulness matching due to the way absolute units are used in the ATD model. This model can-

Table 14-1.   Example ATD color-vision model calculations.

Quantity	Case 1	Case 2	Case 3	Case 4
X	19.01	57.06	3.53	19.01
Y	20.00	43.06	6.56	20.00
Z	21.78	31.96	2.14	21.78
X_0	95.05	95.05	109.85	109.85
Y_0	100.00	100.00	100.00	100.00
Z_0	108.88	108.88	35.58	35.58
Y_0 (cd/m^2)	318.31	31.83	318.31	31.83
σ	300	300	300	300
k_1	0.0	0.0	0.0	0.0
k_2	50.0	50.0	50.0	50.0
A_1	0.1788	0.2031	0.1068	0.1460
T_1	0.0287	0.0680	−0.0110	0.0007
D_1	0.0108	0.0005	0.0044	0.0130
A_2	0.0192	0.0224	0.0106	0.0152
T_2	0.0205	0.0308	−0.0014	0.0102
D_2	0.0108	0.0005	0.0044	0.0130
Br	0.1814	0.2142	0.1075	0.1466
C	1.206	1.371	0.436	1.091
H	1.91	63.96	−0.31	0.79

not be used (without modification) to predict background or surround effects or effects based on medium changes, such as discounting-the-illuminant. Examples of calculated values using the ATD model as described in this chapter are given in Table 14-1.

Why Not Use Just the ATD Model?

The ATD model provides a simple, elegant framework for the early stages of signal processing in the human color-vision system. While the framework is clearly sound, because of the wide range of data the model can be used to predict, it is not well defined for particular applications. There are several aspects of the model that require further definition or specification for practical application. Thus the flexibility of the model that allows it to predict a wide range of data also precludes it from being practically useful. It can be applied to practical applications with some modification, as has been done by Granger (1994, 1995). However, even this formulation is incomplete as a color appearance model because it neglects chromatic adaptation.

The ATD model has the advantages of being fairly simple and generally easily invertible (for $k_1 = 0.0$). Its disadvantages include the lack of a strict definition of its implementation, inadequate treatment of related colors (necessary for most applications), and lack of cognitive factors. Strictly speaking, the model also is not directly relatable to CIE tristimulus values. Since it does not incorporate distinct predictors for lightness and chroma, it actually cannot be considered a color appearance model. However, as a framework for visual processing and discrimination, it certainly warrants some attention.

14.3   LLAB Model

The LLAB model (LLAB) is a very recent entry into the field of color appearance models. It is similar in structure to the RLAB model described in Chapter 13. However, LLAB incorporates a different range of effects than RLAB. LLAB was originally developed by Luo, Lo, and Kuo (1996). However, prior to the article's publication, LLAB was revised by Luo and Morovic (1996) in a conference proceeding. The treatment in this chapter follows the revised model but includes some comments on the original.

LLAB is designed as a colorimetric model and is clearly an extension of CIE colorimetry (as opposed to a vision model). It was synthesized from the results of a series of experiments on color appearance and on color-difference scaling.

LLAB is designed to be a universal model for color matching, color-appearance specification, and color-difference measurement. It therefore incorporates features from previous work in both areas of study. Like RLAB, it is designed to be relatively simple and not inclusive of all visual phenomena. It is not as simple as RLAB. It does, however, predict some effects that RLAB cannot (the converse is also true).

Objectives and Approach

LLAB, as described by Luo et al. (1996), is derived from an extensive series of data derived by Luo and his coworkers on color-appearance scaling and color discrimination. This work has resulted in tests of color appearance models and the development of color-difference equations, as summarized by Luo et al. LLAB is an attempt to synthesize this work into a single, coherent model.

The formulation of LLAB is similar to RLAB in concept, but it differs markedly in detail. It begins with a chromatic-adaptation transform called the BFD transform (developed at Bradford University and previously unpublished) from the test-viewing conditions to defined reference conditions. Then, modified CIELAB coordinates are calculated under the reference conditions and appearance correlates are specified. The surround relative luminance is accounted for using variable exponents as in RLAB. The colorfulness scale is adjusted based on the nonlinear chroma functions incorporated into the CMC color-difference equation (Clarke et al., 1984).

LLAB also incorporates a factor for lightness contrast due to the relative luminance of the background. Hue angle is defined in the same way as in CIELAB, and hue composition is specified according to techniques similar to those used by Nayatani et al., Hunt, and RLAB.

Lastly, lightness and chroma weighting factors can be applied for the calculation of color differences in a manner identical to that used in the CMC and CIE94 color-difference equations. Thus the full designation of the model is LLAB(l:c).

Input Data

The LLAB model requires the following input data:

- Relative tristimulus values of the stimulus, XYZ
- The reference white, $X_0 Y_0 Z_0$
- The luminance (in cd/m^2) of the reference white, L
- The luminance factor of the background, Y_b

It also requires choices regarding:

- The discounting-the-illuminant factor, D
- The surround induction factor, F_S
- The lightness induction factor, F_L
- The chroma induction factor, F_C

Values for specified viewing conditions are given in Table 14-2.

Table 14-2. Values of induction factors for the LLAB model.

	D	F_S	F_L	F_C
Reflection samples and images in average surround				
Subtending >4°	1.0	3.0	0.0	1.00
Subtending <4°	1.0	3.0	1.0	1.00
Television and VDU displays in dim surround	0.7	3.5	1.0	1.00
Cut-sheet transparency in dim surround	1.0	5.0	1.0	1.10
35-mm projection transparency in dark surround	0.7	4.0	1.0	1.00

Adaptation Model

In the LLAB model, the BFD adaptation transform is used to calculate corresponding colors under a reference viewing condition. The BFD transform is a modified von Kries transform in which the short-wavelength sensitive cone signals are subjected to an adaptation-dependent nonlinearity, while the middle- and long-wavelength sensitive cone signals are subject to a simple von Kries transform. The first step is a transform from CIE XYZ tristimulus values to normalized cone responses, denoted RGB, as shown in Equations 14-25 and 14-26:

$$\begin{vmatrix} R \\ G \\ B \end{vmatrix} = \mathbf{M} \begin{vmatrix} X/Y \\ Y/Y \\ Z/Y \end{vmatrix}$$

(14-25)

$$\mathbf{M} = \begin{vmatrix} 0.8951 & 0.2664 & -0.1614 \\ -0.7502 & 1.7135 & 0.0367 \\ 0.0389 & -0.0685 & 1.0296 \end{vmatrix}$$

(14-26)

Note that the transform in Equations 14-25 and 14-26 is atypical in two ways. First, the CIE tristimulus values are always normalized to Y prior to the transform. This results in all stimuli with identical chromaticity coordinates having the same cone signals (a luminance normalization). This normalization is required to preserve achromatic scales through the nonlinear chromatic-adaptation transform described later in the chapter.

Second, the transform itself does not represent plausible cone responses, but rather "spectrally sharpened" cone responses with negative responsivity at some wavelengths. These responsivities tend to preserve saturation across changes in adaptation and impact predicted hue changes across adaptation. Despite the fact that the BFD transformation results in RGB signals that cannot be considered physiologically plausible cone responses, they are referred to here as cone responses for simplicity. The cone responses are then transformed to the corresponding cone responses for adaptation to the reference illuminant. The reference illuminant is defined to be CIE illuminant D65 using the 1931 standard colorimetric observer (X_{0r} = 95.05, Y_{0r} = 100.0, Z_{0r}=108.88). The transformation is performed using Equations 14-27 through 14-30:

$$R_r = \left[D\left(R_{0r} / R_0 \right) + 1 - D \right] R$$

(14-27)

$$G_r = \left[D\left(G_{0r} / G_0 \right) + 1 - D \right] G$$

(14-28)

$$B_r = \left[D\left(B_{0r} / B_0^{\beta} \right) + 1 - D \right] B^{\beta}$$

(14-29)

$$\beta = \left(B_0 / B_{0r}\right)^{0.0834}$$

$$(14\text{-}30)$$

In the event that the B response is negative, Equation 14-29 is replaced with Equation 14-31 to avoid taking a root of a negative number:

$$B_r = -\left[D\left(B_{0r} / B_0{}^{\beta}\right) + 1 - D\right]|B|^{\beta}$$

$$(14\text{-}31)$$

The D factors in Equations 14-27 through 14-31 allow for discounting-the-illuminant. When discounting occurs, $D = 1.0$ and observers completely adapt to the color of the light source. If there is no adaptation, $D = 0.0$ and observers are always adapted to the reference illuminant. For intermediate values of D, observers are adapted to chromaticities intermediate to the light source and the reference illuminant (with D specifying the proportional level of adaption to the source). This is different than the D value in RLAB, which allows for various levels of incomplete adaptation that depend on the color and luminance of the source. ($D = 0.0$ in RLAB does not mean no adaptation. Rather, it means incomplete adaptation to that particular source.)

The final step of the chromatic-adaptation transformation is the conversion from the cone signals for the reference viewing condition to CIE tristimulus values, $X_r Y_r Z_r$, by using Equation 14-32:

$$\begin{vmatrix} X_r \\ Y_r \\ Z_r \end{vmatrix} = \mathbf{M}^{-1} \begin{vmatrix} R_r Y \\ G_r Y \\ B_r Y \end{vmatrix}$$

$$(14\text{-}32)$$

Opponent-color Dimensions

The corresponding tristimulus values under the reference illuminant (D65) are then transformed to preliminary opponent dimensions using modified CIELAB formulas, as illustrated in Equations 14-33 through 14-36:

$$L_L = 116 f\left(Y_r / 100\right)^Z - 16$$

$$(14\text{-}33)$$

$$z = 1 + F_L\left(Y_b / 100\right)^{1/2}$$

$$(14\text{-}34)$$

$$A = 500\left[f\left(X_r / 95.05\right) - f\left(Y_r / 100\right)\right]$$

$$(14\text{-}35)$$

$$B = 200\left[f\left(Y_r / 100\right) - f\left(Z_r / 108.88\right)\right]$$

$$(14\text{-}36)$$

The z exponent is incorporated to account for lightness contrast from the background. It is similar to the form used in the Hunt model. Since the L_I, A, and B dimensions follow the definitions of the CIELAB equations, the non-linearity is dependent on the relative tristimulus values as shown in Equations 14-37 and 14-38. For values of $\omega > 0.008856$, Equation 14-37 is used:

$$f(\omega) = (\omega)^{1/F_s} \tag{14-37}$$

For values of $\omega \leq 0.008856$, Equation 14-38 is used:

$$f(\omega) = \left[\left(0.008856^{1/F_s} - 16/116\right)/0.008856\right]\omega + 16/116 \tag{14-38}$$

The value of F_S depends on the surround relative luminance, as specified in Table 14-2. This is similar to the surround dependency incorporated in RLAB.

Perceptual Correlates

The LLAB model includes predictors for lightness, chroma, colorfulness, saturation, hue angle, and hue composition. The lightness predictor, L_I, is defined in Equation 14-33. The chroma predictor, Ch_L, and colorfulness predictor, C_L, are derived using a nonlinear function similar to that incorporated in the CMC color-difference equation as a chroma-weighting function. This incorporates the behavior of the CMC color-difference equation into the LLAB color space, as shown in Equations 14-39 through 14-43:

$$C = \left(A^2 + B^2\right)^{1/2} \tag{14-39}$$

$$Ch_1 = 25\ln(1 + 0.05C) \tag{14-40}$$

$$C_L = Ch_L S_M S_C F_C \tag{14-41}$$

$$S_C = 1.0 + 0.47\log(L) - 0.057\left[\log(L)\right]^2 \tag{14-42}$$

$$S_M = 0.7 + 0.02L_L - 0.0002L_L^{\,2} \tag{14-43}$$

F_C is the chroma induction factor defined in Table 14-2. S_C provides the luminance dependency necessary to predict an increase in colorfulness with luminance. Thus C_L is truly a colorfulness predictor. S_M provides a similar lightness dependency.

Saturation is defined in LLAB as the ratio of chroma to lightness, as shown in Equation 14-44:

$$s_L = Cb_L / L_L \qquad (14\text{-}44)$$

LLAB hue angle, h_L, is calculated in the usual way according to Equation 14-45:

$$h_L = \tan^{-1}(B/A) \qquad (14\text{-}45)$$

Hue composition in NCS notation is calculated via linear interpolation between the hue angles for the unique hues, which are defined as 25° (red), 93° (yellow), 165° (green), and 254° (blue).

The final opponent signals are calculated using the colorfulness scale, C_L, and the hue angle, h_L, as shown in Equations 14-46 and 14-47:

$$A_L = C_L \cos(h_L) \qquad (14\text{-}46)$$

$$B_L = C_L \sin(h_L) \qquad (14\text{-}47)$$

There were no predictors of brightness, chroma, or saturation defined in LLAB as initially published by Luo, Lo, and Kuo (1996). In the revised version (Luo and Morovic, 1996), the predictors of chroma and saturation were added.

Color Differences

LLAB incorporates the chroma-weighting function from the CMC color-difference equation. This chroma dependency is the most important factor that produces the improved performance of the CMC color-difference equation over the simple CIELAB ΔE^*_{ab} equation. Thus LLAB(*l:c*) color differences are defined by Equation 14-48:

$$\Delta E_L = \left[\left(\Delta L_L / l \right)^2 + \Delta A_L^2 + \Delta B_L^2 \right] \qquad (14\text{-}48)$$

The lightness weight, *l*, is defined to be 1.0, 1.5, and 0.67 for perceptibility, acceptability, and large color differences, respectively. The chroma weight, *c* (not present in this formulation), is always set equal to 1.0.

Phenomena Predicted

The revised LLAB model accounts for chromatic adaptation, lightness induction, surround relative luminance, discounting-the-illuminant, and the Hunt effect. It cannot predict the Stevens effect, incomplete chromatic adaptation, or the Helmholtz-Kohlrausch effect. It also cannot predict the Helson-Judd effect.

LLAB includes predictors for lightness, chroma, saturation, colorfulness, and hue in its latest revision (Luo and Morovic, 1996). This revision corrected the unnatural combination of appearance attributes (lightness, colorfulness, and hue) in the original version (Luo, Lo, and Kuo, 1996). The natural sets are lightness, chroma, and hue or brightness, colorfulness, and hue. The original LLAB formulation could not be used to calculate either brightness-colorfulness matches or lightness-chroma matches. It could be used to predict lightness-colorfulness matches, which probably have little practical utility. If corresponding colors are predicted at constant luminance, then the lightness-colorfulness matches become equivalent to lightness-chroma matches. Interestingly, the original LLAB formulation did not meet the CIE TC1-34 requirement of a color appearance model, that is, it must include predictors of at least lightness, chroma, and hue. These limitations were corrected in the Luo and Morovic (1996) formulation presented in this chapter.

Table 14-3 includes example calculations of LLAB appearance attributes for a few stimuli.

Table 14-3. Example LLAB color appearance model calculations.

Quantity	Case 1	Case 2	Case 3	Case 4
X	19.01	57.06	3.53	19.01
Y	20.00	43.06	6.56	20.00
Z	21.78	31.96	2.14	21.78
X_0	95.05	95.05	109.85	109.85
Y_0	100.00	100.00	100.00	100.00
Z_0	108.88	108.88	35.58	35.58
L (cd/m^2)	318.31	31.83	318.31	31.83
Y_b	20.0	20.0	20.0	20.0
F_S	3.0	3.0	3.0	3.0
F_L	1.0	1.0	1.0	1.0
F_C	1.0	1.0	1.0	1.0
L_L	37.73	61.92	15.97	39.43
Ch_L	0.01	30.15	29.71	29.71
C_L	0.02	55.74	52.34	55.20
s_L	0.00	0.49	1.86	0.75
h_L	230.4	24.2	174.2	269.0
H_L	73B 27G	99R 1B	90G 10B	89B 11R
A_L	−0.01	50.85	−52.07	−0.97
B_L	−0.01	22.82	5.31	−55.19

Why Not Use Just the LLAB Model?

LLAB has the advantages of being fairly simple, incorporates a potentially accurate adaptation model, includes surround effects, and has a reliable built-in measure of color differences. However, its original formulation could not be considered a complete appearance model, since it had predictors for the unusual combination of lightness, colorfulness, and hue with no predictor of chroma. The revised formulation includes both chroma and saturation predictors.

LLAB has a serious drawback in that it is not analytically invertible. It also is not capable of predicting incomplete chromatic adaptation. The revised formulation does include the D factor that can be used to model cognitive effects that occur upon changes in media.

Lastly, since the model is very new, it has not been tested with data independent of those from which it was derived. It is possible that some of the parameters in the model have been too tightly fit to one collection of visual data and might not generalize well to other data. However, its good performance on the LUTCHI data from which it was derived suggests that it has the potential to be quite useful for some applications.

CHAPTER 15

Testing Color Appearance Models

15.1 Overview
15.2 Qualitative Tests
15.3 Corresponding-colors Data
15.4 Magnitude-estimation Experiments
15.5 Direct Model Tests
15.6 CIE Activities
15.7 A Pictorial Review of Color Appearance Models

CHAPTERS 10 THROUGH 14 described several color appearance models with little reference to data on the visual phenomena they are intended to predict. In contemplating the existence of this variety of color appearance models, one may logically wonder just how well they work. Quantitative tests of the models are certainly required (as with any scientific theory, they must be supported by data). Unfortunately, far more has been published on the formulation of the models than on their actual performance. There are two reasons for this. The first is the paucity of reliable data measuring observers' perceptions of color appearance. And second, the models themselves are evolving at a rate that outpaces researchers' abilities to evaluate their performances. Fortunately, both of these situations are changing.

This chapter reviews some of the recent experimental work to test color appearance models and collect additional color-appearance data for future model testing. This is an active area of research, and additional tests (and model revisions) are expected to be published in the near future.

15.1 Overview

One might expect that the derivation of color appearance models of the sophistication presented in Chapters 10 through 14 would require extensive data. This is true. However, the data used to derive the models come from a long history of vision experiments, each aimed at one particular aspect of color appearance. These include the experiments described in Chapter 6 on color appearance phenomena. The models were then formulated to simultaneously predict a wide variety of these phenomena. As a result, quite sophisticated models can be derived with little or no data that test the wide variety of predictions. A true test of the models requires visual data on scaling the appearance attributes of brightness, lightness, colorfulness, chroma, and hue. Alternatively, visual evaluations of performance can be made for existent models.

A variety of tests of color appearance models have been performed. Unfortunately, none of these tests are completely satisfactory and the question of which model is best cannot be unequivocally answered. The various tests can be classified into four general groups, described in the following sections:

- Qualitative tests of various phenomena
- Predictions of corresponding-colors data
- Magnitude estimation of appearance attributes
- Direct psychophysical comparison of the predictions of various models

Each of these types of tests contributes to the evaluation of models in a unique way, and none is adequate, on its own, to completely specify the best color appearance model.

There are also organized activities within the CIE technical committee structure aimed at the evaluation of color appearance models. These activities draw upon evaluations in each of the four groups with the ultimate aim to recommend procedures for color-appearance specification.

15.2   Qualitative Tests

There are a variety of model tests that can be considered qualitative for one reason or another. A test is qualitative if the results show general trends in the predictions rather than providing numerical comparisons of various models. Such tests include

- Calculations showing that a particular appearance phenomenon can be predicted (e.g., colorfulness increases with luminance, and so on)
- Prediction of trends from historical data
- Comparisons with color-order systems
- Abridged visual experiments on model features

Nayatani et al. (1988) published an early example of qualitative evaluation of their color appearance model. They looked at two sets of data. The first was the predicted color appearance of Munsell samples under CIE illuminant A. The second was the evaluation of results from a color-rendering experiment of Mori and Fuchida (1982). Their calculations for the Munsell samples showed that their model predicted the Helson-Judd effect under illuminant A, while a von Kries transformation would not. Nayatani followed this up with a brief visual experiment in which three observers (including the authors) observed a small Helson-Judd effect for samples viewed under an incandescent lamp. They also examined some color-discrimination observations that correlated better with results from their model than did predictions made using a von Kries transformation. Further, they showed that the Nayatani color appearance model made reasonable predictions of the Mori and Fuchida corresponding-colors data. However, they did not compare these results with other models.

As illustrated in the Nayatani study, color-order systems are often used to evaluate the performance of appearance models. The systems used are those based on color appearance (Munsell and NCS) as described in Chapter 5. In many cases, when formulating a model, its authors will plot contours of Munsell hue, Value, and chroma in order to evaluate the perceptual uniformity of the model. Alternatively, contours of constant NCS hue, whiteness-blackness, and chromaticness can be examined. The assumption is that the color-order systems have been constructed with accurate appearance scales. Thus, for example, a model should be able to predict that samples with constant Munsell hue have the same predicted hue and make analogous predictions for Value (lightness) and chroma. Many authors have included plots of Munsell or NCS contours along with the formulation of their models. Examples include Nayatani et al. (1987, 1990), Guth (1991), Hunt (1995), and Fairchild and Berns (1993). In fact, the latest revision of the Nayatani et al. (1995) model was formulated to correct discrepancies in plots of Munsell hue and chroma contours at various value levels. Seim and Valberg (1986) provide a more quantitative analysis of the Munsell system in the CIELAB color space and propose alternative equations similar to those found in the Hunt model.

Nayatani et al. (1990) published an interesting comparison of the Hunt and Nayatani et al. models. This included plots of Munsell contours in the color spaces of each model as well as flow charts comparing the computational procedure for each model. While this work is interesting, at this point it is largely historical, since both models have been revised significantly since it was published. Plots of Munsell and NCS contours do provide some insight into the performance and properties of various color appearance models. However, none of the published results include a quantitative comparison between the color-order systems and color appearance models. For example, constant hue contours should plot as straight radial lines and constant chroma contours should plot as concentric, evenly spaced circles. It would be possible to derive measures of how close the models come to producing these results. Perhaps this has not been done because the results would not be terribly impressive. Another reason is that the perceptual uncertainty of the color-order systems is not well defined, thus making it difficult to know how good the predictions should be. Examination of the published results suggests that the models perform about equally well in these qualitative comparisons. This conclusion is confirmed by a study of constant-hue contours completed by Hung and Berns (1995) in which extensive visual evaluations were made. Quantitative analysis by Hung and Berns (1995) showed that the observed constant hue contours were not adequately predicted by any color appearance model and that no single model was clearly superior.

Hunt (1991b) provides an excellent example of qualitative evaluation of his appearance model. For example, he shows how the model predicts cone and rod saturation, the Stevens effect, the Hunt effect, and the effect of surround relative luminance on image contrast, as well as other effects. One fascinating demonstration that Hunt predicts is the appearance of objects in a filtered slide. In a classic demonstration (Hunt, 1995), a cyan filter is superimposed over a yellow cushion in a slide, resulting in the cushion's appearing green. However, if the same filter is placed over the entire slide, the cushion retains much of its yellow appearance due to chromatic adaptation. This effect is strongest for projected slides, but it also can be observed in printed images (Hunt, 1995). The effect is also simulated in Figure 15-1 with the filter's being added over the banana. Hunt (1991b) shows that his model is capable of predicting this effect. It is worth noting that simpler models, including CIELAB and RLAB, can also predict this effect (Fairchild and Berns, 1993).

Perhaps the most useful result of qualitative analysis of color appearance models is a summary of the various effects that can be predicted by each model. Table 15-1 provides such a summary. While this table is useful to gauge the capabilities of each model, it is important to remember that it includes no information on how accurately each model can predict the various phenomena. This lack of information on accuracy is the most significant drawback of qualitative model tests and necessitates the additional tests described in the following sections.

Table 15-1   Color-appearance phenomena predicted by various color appearance models. Check marks indicate that the model is capable of directly making the prediction.

	ATD	CIELAB	LLAB	RLAB	Nayatani	Hunt
Lightness		√	√	√	√	√
Brightness	√				√	√
Chroma		√	√	√	√	√
Saturation	√		√	√	√	√
Colorfulness			√		√	√
Hue angle	√	√	√	√	√	√
Hue			√	√	√	√
Helson-Judd effect					√	√
Stevens effect					√	√
Hunt effect	√		√		√	√
Helmholtz-Kohlrausch effect	√				√	√
Bartleson-Breneman results			√	√		√
Discounting-the-illuminant			√	√		√
Incomplete adaptation				√		√
Color difference		√	√	√		
Others	√				√	√

(a)

(b)

(c)

Figure 15-1. An illustration of chromatic adaptation: (a) original image; (b) simulation of a cyan filter placed over the yellow banana, thereby resulting in the appearance of a green banana; (c) simulation of the same cyan filter placed over the entire image. Note how the banana in (c) returns to a yellowish appearance despite its being physically identical to the banana in (b). Original image part of the ISO SCID set.

15.3   Corresponding-colors Data

Corresponding-colors data were described in Chapter 8 with respect to the study of chromatic adaptation. In addition, corresponding-colors data can be collected and analyzed for a wide range of color appearance phenomena in addition to simple chromatic adaptation. Corresponding colors are defined by two sets of tristimulus values that specify stimuli that match in color appearance for two disparate sets of viewing conditions. Recall that if the change in viewing conditions has an impact on color appearance, then the corresponding tristimulus values of the stimuli will differ in absolute values.

Corresponding-colors data are used to test color appearance models by taking the tristimulus values for the first viewing condition and transforming them to the matching tristimulus values under the second viewing condition. The predicted corresponding colors can then be compared with visually observed corresponding colors to determine how well the model performs. The results are often analyzed in terms of RMS deviations between predicted and observed colors in either a uniform chromaticity space (e.g., $u'v'$) or a uniform color space (e.g., CIELAB). Recall that Nayatani et al. (1990) illustrated the important distinction between lightness-chroma and brightness-colorfulness matches. Complete color appearance models can be used to predict either type of match. The two types of matches will differ if there is a change in luminance level between the two viewing conditions in question.

One of the most extensive series of experiments measuring corresponding-colors data for color appearance analysis was completed by the Color Science Association of Japan (CSAJ) and reported by Mori et al. (1991). Mori summarized data from four experiments performed by CSAJ:

1. An experiment on chromatic adaptation from illuminant D65 to illuminant A simulators at an illuminance of 1,000 lux. Judgments were made by 104 observers on 87 samples using a modified haploscopic matching technique.
2. An experiment that collected data consisting of measurements of the Hunt effect using five colored samples judged under illuminant D65 simulators at five different illuminance levels by 40 observers.
3. An experiment that collected data that represent measurements of the Stevens effect using five neutral samples viewed under five different illuminance levels by 31 observers.
4. An experiment that examined the Helson-Judd effect for achromatic samples viewed under highly chromatic fluorescent light sources. These data represent one of the most extensive studies, with the largest numbers of observers, completed in the area of color appearance to date.

Unfortunately, Mori et al. (1991) reported only qualitative analyses of the experimental results. They showed plots of predicted and observed corresponding colors for the chromatic-adaptation experiment and the Nayatani, von Kries, and Hunt models. Mori concluded that the Nayatani model made the best predictions. However, examination of their plots suggests that Hunt's model provides similar performance to it, and the von Kries transform perhaps works better than both of them. They illustrated that the Nayatani et al. model could predict the Hunt effect data well, but they did not compare the results with predictions of the Hunt model. Similar analyses were performed for the Stevens effect and the Helson-Judd effect data. The results showed a fairly small Stevens effect that was over-predicted by the Nayatani model. The Helson-Judd effect, while observed in this experiment, was also over-predicted by the Nayatani model. Further, quantitative analyses of these data have been carried out through CIE TC1-34 and are described later in the chapter.

Breneman (1987) collected a fairly extensive set of corresponding-colors data for changes in chromatic adaptation and luminance level. These data were used to evaluate various chromatic-adaptation transforms and color appearance models by Fairchild (1991a, b). The models were compared in terms of RMS deviations between observed and predicted results in the CIE 1976 $u'v'$ chromaticity diagram. The chromatic-adaptation data were best predicted by the Hunt and RLAB models followed by the Nayatani and von Kries models. The CIELAB and CIELUV models performed the worst for these data. Breneman's data showed a small Hunt effect that was over-predicted by the Hunt and Nayatani et al. models and not predicted at all by the other models. The RMS deviations produced by both sets of models are similar in magnitude, suggesting that making no prediction is as accurate as an over-prediction for these particular data.

Luo et al. (1991b) converted some of their magnitude-scaling data (described in Section 15.4) to sets of corresponding colors for various changes in viewing conditions. They generated three sets of corresponding-colors data for changes in chromatic adaptation from CIE illuminant D65 to D50, D65 to A, and D65 to white fluorescent. They then evaluated six different chromatic-adaptation transforms using mean and RMS color differences in the CIELAB space. The results showed that the Bradford model, the basis of LLAB, performed best. The Hunt, Nayatani et al., and CIELAB models performed similarly and almost as well. These were followed by the simple von Kries transformation and a transformation proposed by Bartleson. Additional results can be found in Kuo et al. (1995).

Braun and Fairchild (1997b) performed an experiment in which observers were asked to adjust CRT-displayed images to match printed images viewed under a different white point. Matching images were obtained for five observers using two different images for white-point changes from 3000 K to 6500 K and 9300 K to 6500 K. The data were analyzed by segmenting the images into meaningful object regions to avoid overly weighting large image areas. The corresponding colors were analyzed in terms of average and RMS CIELAB color differences. The results showed that the RLAB, LLAB, and

CIELAB models best predicted the observed corresponding colors. The Hunt and Nayatani et al. models did not perform as well.

The above-mentioned studies illustrate the variety of corresponding-colors experiments that have been completed. Unfortunately, a clear picture of relative model performance does not emerge from the analysis of these results. This is partly because the models have changed and new ones have emerged since some of the results were published.

15.4   Magnitude-estimation Experiments

Magnitude-estimation experiments involve asking observers to directly assign numerical values to the magnitude of their perceptions. The utility of such experimental techniques was brought into focus by the classic study of Stevens (1961). Magnitude-estimation experiments allow the rather direct measurement of the magnitudes of color-appearance attributes such as lightness, chroma, and hue for various stimuli and viewing conditions. These data can then be used to evaluate various color appearance models and derive new models.

The most extensive series of experiments on the magnitude scaling of color appearance has been carried out through the Loughborough University of Technology Computer Human Interface (LUTCHI) Research Centre as published in a series of papers by Luo et al. (1991a, b; 1993a, b; 1995), the results of which have been summarized by Hunt and Luo (1994). The LUTCHI studies consisted of five parts:

1. Luo et al. (1991a)

    Six or seven observers were asked to scale the lightness, color-fulness, and hue of between 61 and 105 stimuli presented in a series of 21 different viewing conditions. The viewing conditions varied in white point, medium, luminance level, and background. The results showed that the most significant influences on color appearance were background and white point. Other effects were not clearly present in the data. One reason for this is the intrinsically high uncertainty in magnitude-estimation experiments. The uncertainty in the data was expressed in terms of coefficients of variation, CV, which can be thought of as percent standard deviations. The overall CV values for intraobserver variability were about 13 for lightness, 18 for colorfulness, and 9 for hue. No color appearance models were evaluated in part one.

    One problem with the LUTCHI studies is the rather unusual choice of appearance attributes that were scaled—lightness, colorfulness, and hue. The authors claim that colorfulness is more natural than chroma. However, chroma is the more appropriate attribute to scale with lightness. It is the attribute that observers normally associate with objects, and it does not require that the observers be taught its definition. At a single luminance level, it is probably reasonable to assume that chroma and colorfulness are related by a simple scaling factor. However, there is no reason to believe that chroma and colorfulness are linearly related across changes in luminance.

2. Luo et al. (1991b)

   The part one data were used to evaluate various color appearance and chromatic-adaptation models. The models were analyzed by calculating CV values between the observed results and the model predictions. To summarize: The Hunt model performed best for lightness, followed by CIELAB and then Nayatani. For colorfulness, no model performed particularly well, but Hunt's model (and a version of Hunt's modified with respect to these data) performed slightly better than the others. For hue, the Hunt model performed significantly better than the Nayatani et al. model. Other models were not tested for hue. These data were also used to formulate later versions of the Hunt model.

3. Luo et al. (1993a)

   Additional data were collected to check previous results, extend the range of conditions, and include the scaling of brightness. They were then used to test various models. Four observers took part in the scaling for collection of new data for a CIE illuminant D50 simulator at six different luminance levels. Analyses of the results showed that the Hunt model performed best overall. For lightness scaling, CIELAB performed nearly as well when the lowest luminance level was ignored. For colorfulness and hue scaling, the Hunt model performed substantially better than the Nayatani et al. model.

4. Luo et al. (1993b)

   In this part, they extended their experimental technique to the evaluation of transmissive media, including both projected transparencies and transparencies viewed on light boxes. Between five and eight observers took part in these experiments. They scaled lightness, colorfulness, and hue for a total of 16 different sets of viewing conditions. They found that the Hunt model did not perform as well as in previous experiments and proposed some changes that have been incorporated in the latest version of the model. CIELAB performed very well for these data— better, in fact, than the unmodified Hunt model. The Nayatani model performed worse than both CIELAB and the unmodified Hunt model. The Hunt model with modifications based on the experimental data performed best overall.

5. Luo et al. (1995)

The phenomenon of simultaneous contrast was specifically examined. Five or six observers scaled lightness, colorfulness, and hue of samples in systematically varied proximal fields presented on a CRT display. The results showed that all three dimensions of color appearance are influenced by induction (as expected). Evaluation of the Hunt model (the only model capable of directly accounting for simultaneous contrast) showed that it did not perform well and required further modification.

Hunt and Luo (1994) summarize the first four parts of the LUTCHI experiments and how the results were used to refine the Hunt color appearance model. Overall, they show that the Hunt model predicts the hue results with CVs between 7 and 8, while the inter-observer variability CV is 8. For lightness, the model CVs range from 10 to 14 with the inter-observer variability of 13. For colorfulness, the model CVs are around 18 with inter-observer variability of 17. Thus they concluded that the Hunt model is capable of predicting the experimental results as well as the results from one observer would predict the mean. This is impressive, and certainly a good result. However, it should be kept in mind that these data are not independent of the model formulation.

Many of the results of the LUTCHI experiments have also been contributed to the efforts of CIE TC1-34 in order to allow the evaluation of more recent color appearance models. The results of these analyses are described later in the chapter. Unfortunately, the data themselves have been deemed proprietary by the research sponsors and have not been released. It is expected that these data will be made publicly available in the near future.

15.5   Direct Model Tests

One way to overcome the limited precision of magnitude-estimation experiments is to take advantage of more refined psychophysical techniques to evaluate model performance. One such technique involves paired-comparison experiments in which observers view two stimuli at a time and simply choose which is better. The results are then analyzed using the law of comparative judgments to derive an interval scale and associated uncertainties. To evaluate color appearance models in this fashion, one must begin with an original stimulus (or image) in one set of viewing conditions and then calculate the corresponding stimulus (or image) for the second set of viewing conditions using each model to be tested. The observers then look at each possible pair of stimuli and choose which is a better reproduction of the original in its viewing conditions. The interval-scale results are then used to measure the relative performance of each model.

The significant drawback of this approach is that the results cannot be used to derive new models and are limited to the models available and included at the time of the experiment. An extensive series of these experiments has been conducted at the Munsell Color Science Laboratory at the Rochester Institute of Technology and summarized by Fairchild (1996). The results of these and other experiments are described in the following paragraphs.

Fairchild and Berns (1993) described an early and simplified form of these experiments to confirm the utility of color appearance models in cross-media image reproduction applications. They examined the transformation from prints viewed under either illuminant A or D50 simulators to CRT displays with a D65 white point and various backgrounds using a simple successive binocular viewing technique. Six different images were used, and 14 observers took part in the experiment. Comparisons were made among no model (CIE XYZ reproduction), CIELAB, and RLAB. The results indicated that observers chose the RLAB reproduction as the best nearly 70% of the time, the CIELAB image about 30% of the time, and the XYZ image almost never. This result showed that a color-appearance transformation was indeed required for these viewing conditions and that the RLAB model outperformed CIELAB.

Kim et al. (1993) examined the performance of eight color-appearance transformations for printed images viewed under different viewing conditions. Original prints were viewed under a CIE illuminant A simulator, and reproductions that were calculated using the various appearance transformations were viewed under CIE illuminant D65 simulators at three different luminance levels. A paired-comparison experiment was completed, and an interval scale was derived using the law of comparative judgments. A successive-*Ganzfeld* haploscopic viewing technique (Fairchild, Pirrotta, and Kim, 1994) was used with 30 observers.

The results showed that the Hunt, RLAB, CIELAB, and von Kries models performed similarly to one another and significantly better than the other models. The Nayatani et al. model performed worse than each of the previous models. Three models performed significantly worse and were not included in further experiments: CIELUV, LABHNU2, and a proprietary model. The Nayatani et al. model performed poorly due to its prediction that the Helson-Judd effect would result in yellowish highlights and bluish shadows in the reproductions. The Helson-Judd effect cannot be observed for complex stimuli under these viewing conditions. CIELUV and the others performed poorly due to their intrinsically flawed chromatic-adaptation transformations.

Pirrotta et al. (1995) performed a similar experiment using simple stimuli on gray backgrounds rather than images. The first phase of this study was a computational comparison of the various models in order to find the stimuli and viewing conditions for which the models differed the most such that the visual experiments could concentrate on these differences. It is useful to examine a few of these results.

Figure 15-2 shows the CIELAB coordinates of corresponding colors under CIE illuminant A at 1,000 lux for neutral Munsell samples of Values 3, 5, and 7 viewed under CIE illuminant D65 at either 1,000 or 10,000 lux. The points labeled *F* illustrate the incomplete adaptation predicted by the Fairchild (1991b) model used in RLAB. The points labeled *N* show the prediction of the Helson-Judd and Stevens effects incorporated in the Nayatani et al. model. The points labeled *H* show the prediction of the Stevens effect for the condition with a luminance change according to the Hunt model. The figure illustrates the extreme prediction of the Helson-Judd effect for illuminant *A* according to the Nayatani et al. model.

Figure 15-3 illustrates the wide range of corresponding-colors predictions for a 5PB 5/12 Munsell sample under the same viewing conditions. Note the extreme differences in the predictions of the various models as the scales of the plots in Figure 15-3 encompass 50 CIELAB units. The Pirrotta et al. (1995) visual experiment used a paired-comparison technique with 26 observers, ten stimulus colors, and a change in viewing conditions from an illuminant A simulator at 76 cd/m² to an illuminant D65 simulator at 763 cd/m². The results showed that the Hunt model performed best. The von Kries, CIELAB, and Nayatani et al. models performed similarly to one another but not as well as the Hunt model. CIELUV and the Fairchild (1991b) model performed significantly worse. These results led to the revision of the adaptation model incorporated in RLAB (Fairchild, 1996).

Braun et al. (1996) investigated viewing techniques and the performance of color appearance models for changes in image medium and viewing conditions. They examined the reproduction of printed images viewed under CIE illuminant D50 and A simulators on a CRT display with a CIE illuminant D65 white point. Fifteen observers took part in this experiment using five different viewing techniques. It was concluded that a successive binocular viewing technique with a 60-second adaptation period provided the most reliable results. Interestingly, a simultaneous binocular viewing technique, common in many practical situations, produced completely unreliable results. The experiment utilized five different pictorial images with a variety of content. The result showed significant differences in the performance of each model tested. The order of performance from best to worst was RLAB, CIELAB, von Kries, Hunt, and Nayatani et al.

Fairchild et al. (1996) performed a similar experiment on the reproduction of CRT-displayed images as projected 35-mm slides. The CRT display was set up with both illuminant D65 and D93 white points at 60 cd/m² and viewed in dim surround. The projected image had a 3900-K white point at 109 cd/m² with a dark surround. Fifteen observers completed the experiment with three different pictorial images. The RLAB model performed best, followed by CIELAB and von Kries with similar performances and then the Hunt model. The Nayatani et al. model was not included in the final visual experiments,

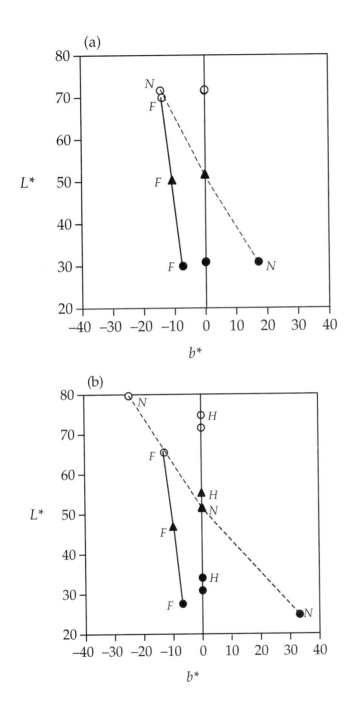

Figure 15-2. Illustration of differences among predictions of various appearance models represented in the CIELAB L*-b* plane. Neutrals at (a) 1,000 lux and (b) 10,000 lux.

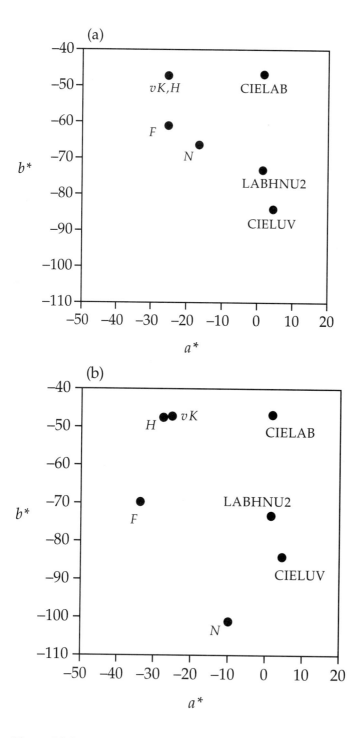

Figure 15-3. Illustration of differences among predictions of various appearance models represented in the CIELAB a*-b* plane. 5PB 5/12 at (a) 1,000 lux and (b) 10,000 lux.

since it produced clearly inferior images due to its prediction of the Helson-Judd effect and the limited number of images that could be included in the experiment.

Braun and Fairchild (1997a) extended the experiments of Braun et al. (1996) to a wide variety of viewing conditions. Ten different sets of viewing conditions were investigated using between 14 and 24 observers. The viewing conditions varied in white point, luminance level, background, and surround. Overall, RLAB performed best. For changes in white point, CIELAB and a von Kries transformation also performed well. The Hunt and Nayatani et al. models did not perform as well as those three.

Similar results were obtained for changes in luminance level. For changes in background, the Hunt model performed poorly, apparently because it over-predicted the effect for complex images. The models that did not account for changes in background performed better. This result was as expected, since the background of an image does not coincide with what is normally considered the background of a stimulus (i.e., an image element). For changes in surround, RLAB performed worst, with the Hunt model performing almost as poorly. These were the only two models that accounted for the surround change and, while they were predicting the correct trend, both models over-predicted the effect for these viewing conditions.

Braun and Fairchild (1997b) extended the corresponding-colors experiment described in Section 15.3 by including the results of the image-matching technique in a paired-comparison experiment with other color appearance models. In this research, a paired-comparison experiment was carried out that included model predictions as well as the matching images generated by the observers and statistical linear transformations between white points derived from the match data. Five observers took part in the image-matching experiment, and 32 were used in the paired-comparison experiment. The results showed that RLAB produced a matching image that was as good as the image produced by observers and the statistical models. The CIELAB, von Kries, and Hunt models did not perform as well as RLAB.

Lo et al. (1996) and Luo et al. (1996) presented the results of paired-comparison experiments for the reproduction of printed images on CRT displays. They evaluated five different changes in white point at constant luminance using 9 to 18 observers. The results show that the CIELUV model performs significantly worse than all of the other models tested. The other models performed similarly to one another, with LLAB performing slightly better for adaptation from illuminant A to illuminant D65.

Other similar experiments are underway in a variety of laboratories. The results of these ongoing experiments and those described previously are being collected by CIE technical committees as described in the following sections. It is likely that this type of research will continue for a while as existing models are refined and new ones introduced.

15.6   CIE Activities

It is certainly difficult to digest all of the previously given results and come up with a single answer to the question of which color appearance model is best or which should be used in a particular application. There are too many experimental variables in addition to ongoing refinement of the models to draw conclusions from the literature alone. These are some of the reasons why there are a large number of published appearance models and no international consensus on a single model for various applications. The CIE is actively addressing these issues through the activities of three of its technical committees, as described next.

TC1-34, Testing Colour-Appearance Models

CIE Technical Committee 1-34 (TC1-34), Testing Colour-Appearance Models, was established to evaluate the performance of various models for the prediction of color appearance of object colors. TC1-34 has published guidelines for coordinated research on testing color appearance models (Fairchild, 1995a) that outline its plan of work with the intention of motivating researchers to perform additional model tests. In addition, TC1-34 has collected various sets of data and test results and completed additional tests. These results will be published in a CIE technical report on the committee's progress. This report will include additional analyses on the CSAJ, LUTCHI, and RIT experiments discussed previously.

The TC1-34 analyses of the CSAJ data included calculations of RMS deviations in CIELAB space for the data on chromatic adaptation, the Stevens effect, and the Hunt effect. For the chromatic-adaptation data, the Hunt, RLAB, and CIELAB models perform similarly to one another and better than the others. They are followed, in order of performance, by the Nayatani et al., LABHNU, and CIELUV models. For the Stevens effect data, the Hunt model performs best, followed by RLAB, CIELAB, LABHNU, and CIELUV, all of which perform identically, since they predict no effect. The Nayatani model performs worst because it over-predicts the effect. For the Hunt effect data, the Hunt model performs best, followed by Nayatani and then the models that do not predict any effect (RLAB, CIELAB, LABHNU, and CIELUV).

Additional analyses of the LUTCHI data contributed to TC1-34 show that the Hunt model performs best, followed by the RLAB, the CIELAB, and then finally the Nayatani models. The TC1-34 summary of the RIT direct model tests shows differing results for images and simple stimuli. For images, the Hunt, CIELAB, and RLAB models perform similarly and best, followed by Nayatani and LABHNU in a tie, and then by CIELUV, with the worst performance. For simple stimuli, the Hunt model performed best and the CIELUV model worst. The others performed similarly in between those two. An overall ranking of the TC1-34 analyses results in the following ordering of model performance: Hunt, RLAB, CIELAB, Nayatani, LABHNU, and CIELUV. Analyses of LLAB for all of the data have not been completed, but it performs better than the Hunt model for the LUTCHI data and is likely to also do well on the other data.

At this time, TC1-34 has concluded that one or two of the published color appearance models cannot be recommended for general use. There are various reasons for this. One of the most significant is that the models are still evolving and more tests are required to make strong conclusions. To this end, TC1-34 is formulating a CIE color appearance model that incorporates the best features of the published models, while avoiding their various pitfalls. It is expected that this model will be recommended by the CIE for general use to promote uniformity of practice and further evaluation to promote the future development of an even better model. The status of this CIE model is summarized in Appendix A.

TC1-27, Specification of Colour Appearance for Reflective Media and Self-luminous Display Comparisons

CIE Technical Committee 1-27 (TC1-27) was established to evaluate the performance of various color appearance models in CRT-to-print image reproduction. TC1-27 has also published a set of guidelines for coordinated research (Alessi, 1994). The various experiments of Braun et al. and Lo et al. described in Section 15.5 represent contributions to the activities of TC1-27. Additional experiments are being carried out in three or four other laboratories that will be contributed to TC1-27. It is expected that TC1-27 will collect the various results, summarize them, and prepare a progress report within the next few years. This committee will work in conjunction with TC1-34 with respect to the evaluation of a new CIE color appearance model.

TC1-33, Color Rendering

As described in Chapter 16, the CIE procedure for calculating a color-rendering index for light sources is based on an obsolete color space. CIE Technical Committee 1-33 (TC1-33) was established to formulate a new procedure for calculating a color-rendering index for light sources. There are two aspects to this problem. The first is the specification of a calculation procedure, and the second is the selection of a color space in which to do the calculations. A color appearance model is necessary, since color-rendering indices must be compared for light sources of various colors. TC1-33 has developed a new procedure and will publish a report. Originally TC1-33 was waiting for the results of TC1-34 before choosing a model, but it has apparently decided to go on without an agreed-upon appearance model.

15.7   A Pictorial Review of Color Appearance Models

No sets of equations or lists of RMS deviations or coefficients of variation can truly communicate the differences among the various color appearance models. To appreciate these differences, it is useful to view images that have been calculated using the various models. Figures 15-4 through 15-6 illustrate the predictions of various models (Fairchild and Reniff, 1996). These figures cannot be generally used to indicate which models are best. They should just be considered a display of the relative performance of the various models for these types of predictions. With such a consideration in mind, one will find that the viewing conditions for these figures are not critical (although a high-luminance, D65 simulator would be ideal).

Figure 15-4 illustrates the images viewed under illuminant D65 that would be predicted as matches to an original image viewed under CIE illuminant A for 14 different color models. The models include CIE *XYZ* to illustrate the original image, CIELAB, CIELUV, LABHNU2 (Richter, 1985), von Kries, spectrally sharpened von Kries (Drew et al., 1994), ATD, Nayatani et al., Hunt (discounting), Hunt (no discounting), Hunt (incomplete adaptation, no Helson-Judd effect), RLAB (discounting), RLAB (partial discounting), and RLAB (no discounting).

CIE *XYZ* (original image date)	CIELAB	CIELUV
LLAB	Spectrally Sharpened von Kries	ATD 1994
Hunt 1994 (Hard copy)	Hunt 1994 (Soft copy)	Hunt 1994 (Slides)
RLAB (Hard copy)	RLAB (Soft copy)	RLAB (Slides)
von Kries	Nayatani et al. 1995	LABHNU

Figure 15-4.   Key.

Figure 15-4. (See key opposite.) Comparison of the predictions of various appearance models for change in chromatic adaptation from illuminant A to illuminant D65. Original image data represent the reproduction of tristimulus values with no adjustment for adaptation. Original images: *Portland Head Light*, Kodak Photo Sampler PhotoCD, ©1991, Eastman Kodak; *Picnic*, Courtesy Eastman Kodak; Macbeth ColorChecker® Color Rendition Chart.

There are several features to note in this set of images:

- The *XYZ* image shows a rendering of the original illuminant A image data with no adaptation model.
- The hard-copy versions of RLAB and Hunt are very similar to the von Kries model in this situation. These produce what have

been found to be generally the best results in experiments completed to date.

- LLAB produces more saturated reddish hues due to the characteristics of its "cone responses" and a bluish hue shift due to its nonlinear adaptation model for the blue channel.
- CIELAB produces hue shifts (in comparison with von Kries et al.), particularly noticeable in the sky and grass colors, due to its "wrong von Kries" adaptation model.
- Incomplete adaptation can be noted by the yellowness of the RLAB soft-copy image, along with the intermediate level of adaptation in the RLAB slide image.
- The Hunt soft-copy image includes the Helson-Judd effect (yellow highlights and blue shadows), which can be seen even more strongly in the Nayatani et al. image.
- The Hunt slide image is more similar to the RLAB soft-copy image.
- The ATD model also predicts incomplete levels of adaptation due to the nature of its formulation that treats stimuli in a more absolute, rather than relative, sense.
- The spectrally sharpened von Kries transform produces highly saturated reddish hues. This is to be expected from the "color-constancy preserving" nature of sharpened responsivities.
- The CIELUV and LABHNU2 models produce unusual hue shifts due to their subtractive adaptation models. In fact, they produce predictions outside the gamut of physically realizable colors if a D65-to-A transformation is performed rather than the A-to-D65 transformation illustrated.

Figure 15-5 illustrates predictions for changes in adapting luminance from 100 cd/m^2 to 10,000 cd/m^2 with a constant D65 white point. These predictions are presented for the models with luminance dependencies (ATD, LLAB, Nayatani et al., Hunt, and RLAB) in addition to a single image representing all of the other models. The original image was "gamut compressed" to allow all of the model predictions to remain within gamut. The RLAB model has very little luminance dependency and therefore produces an image very similar to the original. The Hunt and Nayatani models produce images of lower contrast. This is to be expected according to the Hunt and Stevens effects. These low-contrast images would appear to be of higher contrast when viewed at a high luminance level. The Nayatani model predicts a larger luminance-dependent effect than does the Hunt model. The ATD model makes the opposite prediction. Since it is based on absolute rather than relative signals, the ATD model predicts that a brighter, higher contrast image will be required at the higher luminance levels. This prediction is incorrect.

Figure 15-6 shows predictions for a change in surround from average to dark at a constant D65 white point for the surround-sensitive models (LLAB,

Hunt, and RLAB) in addition to a single image representing all of the other models. All three models illustrate the increase in contrast required for image viewing in a dark surround. LLAB and RLAB have similar predictions, with RLAB predicting a bit stronger effect than LLAB does. The Hunt model uses

CIE XYZ (and all other models)	
RLAB	Hunt 1994
Nayatani et al. 1995	ATD 1994

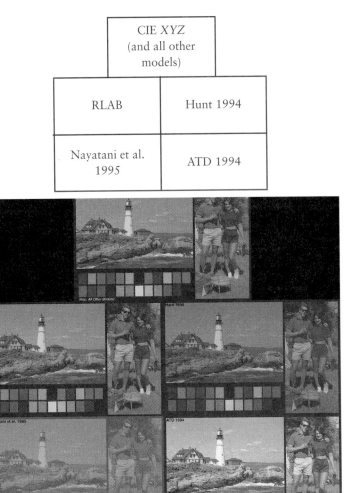

Figure 15-5. Comparison of the predictions of various appearance models for change in luminance from illuminant D65 at 100 cd/m² to illuminant D65 at 10,000 cd/m². Original image data represent the reproduction of tristimulus values with no adjustment for luminance and therefore the predictions of all models that do not account for luminance-level changes. See Figure 15-4 caption for image credits.

CIE XYZ (and all other models)	RLAB
LLAB	Hunt 1994

Figure 15-6. Comparison of the predictions of various appearance models for change in surround relative luminance from illuminant D65 with an average surround to illuminant D65 with a dark surround. Original image data represent the reproduction of tristimulus values with no adjustment for surround and therefore the predictions of all models that do not account for surround changes. See Figure 15-4 caption for image credits.

functions with additive offsets to predict the surround-dependent contrast changes. These offsets force some dark colors to have predicted corresponding colors with negative tristimulus values. Since this is physically impossible, pixels with such colors have been mapped to black. This illustrates one practical limitation of using the Hunt model for changes in surround.

CHAPTER 16

Traditional Colorimetric Applications

16.1 Color Rendering
16.2 Color Differences
16.3 Indices of Metamerism
16.4 A General System of Colorimetry?

GIVEN ALL OF THE EFFORT put into the formulation, evaluation, and refinement of color appearance models, one could naturally wonder if the models have practical application beyond the natural scientific curiosity that has driven much of the historical research on color-appearance phenomena. In recent years, it has been the development of technology, and thus applications, that has really pushed the scientific investigation of color appearance and the development of models. These applications can be divided into two general categories:

1. Image reproduction, which is the subject of Chapter 17
2. The area of color measurement and specification, the subject of this chapter

Colorimetry has steadily evolved over the past century. For many applications, simple tristimulus (XYZ) colorimetry or CIELAB-type color-difference specifications are sufficient. However, there are some applications in the traditional fields of colorimetry that require further evolution. A few of these are discussed in this chapter.

16.1   Color Rendering

Color rendering is the way in which various light sources influence, or "render," the color appearance of objects. For example, it is possible for two light sources to be of the same color even though one is, for example, natural daylight and the other is a fluorescent source made up of two narrow-band phosphors that happen to add together to make the same white. While the color of the two sources will match, the appearances of objects illuminated by these two sources will differ tremendously. This phenomenon is obviously important in the engineering of artificial illumination and the choice of light sources for various installations. If illumination was specified strictly based on its efficiency, or its efficacy, the appearance of objects in our environment would be quite disturbing. To aid in this application, the CIE has defined a color-rendering index as a measure of the quality with which a light source renders the colors of objects.

Current Techniques and Recommendations

The current CIE-recommended techniques for calculating color-rendering in-dices are described in CIE Publication 13.3 (CIE, 1995). Light sources are eval-uated relative to reference illuminants. These illuminants are defined to be the CIE D-series of illuminants for correlated color temperatures greater than or equal to 5000 K and Planckian radiators for correlated color temperatures less than 5000 K. The reference illuminant is chosen such that it has the same cor-related color temperature as the test source. Differences in color between the test source and reference illuminant are accounted for by using a von Kries-type chromatic-adaptation transform. The CIE technique defines a special color-rendering index according to Equation 16-1:

$$R_i = 100 - \Delta E_i \qquad (16\text{-}1)$$

The color-difference calculation, ΔE_i, is based on the Euclidean distance between the color of the sample under the test and reference sources in the now obsolete CIE $U^*V^*W^*$ space. A general color-rendering index, R_a, is defined as the average of the special color-rendering indices for eight specified Munsell samples. Examples of general color-rendering indices for various light sources are given in Table 16-1.

Application of Color Appearance Models

The fundamental question in the specification of color-rendering properties of light sources is the specification of the appearance of colored objects under differ-ent light sources. In some cases, one must compare the appearance of objects under sources of different colors. It might also be of interest to compare the

Table 16-1.   Example values
of color-rendering indices.

Source	R_a
Tungsten halogen	100
Illuminant D65	100
Xenon	93
Daylight fluorescent	92
Cool white fluorescent	58
Tri-band fluorescent	85
High-pressure sodium	25
Mercury	45
Metal halide	80

rendering properties of sources at different luminance levels. To make comparisons of color appearance across changes in illumination color and luminance level, one needs an accurate color appearance model. Thus, given appropriate standard references, one would be able to compare the quality of color rendering of a tungsten source with that of a daylight source and have meaningful results.

Future Directions

While there is certainly room for improved measures of color rendering based on an accurate color appearance model, the current status of color appearance models precludes a large change in the capabilities of an index. Currently, CIE TC1-33, Colour Rendering, is in the process of revising the procedures for the specification of color rendering. The first step has been to define a new procedure for calculation that is independent of the color space used.

This procedure uses the CIELAB color space combined with an improved chromatic-adaptation transform (currently the Nayatani et al. nonlinear transform is being considered) and will be recommended as an improvement over the current color-rendering index. When a CIE color appearance model becomes available, the recommended technique for calculating color rendering should take advantage of it.

In the current type of color-rendering index, sources are compared only with a standard illuminant of the same correlated color temperature. The color appearance model is not being used for a large change in chromatic adaptation. Thus it is doubtful whether any color appearance model would be significantly better than any other, including CIELAB, for this calculation. Only when a more sophisticated technique for specifying color rendering across large changes in light source color and/or luminance level is formulated will a more complicated color appearance model be required.

16.2   Color Differences

The measurement of color differences has wide application in a variety of industries. Such measurements are required to set and maintain color tolerances for the production and sale of all colored materials. Ideally, one would be able to take a metric of color difference, such as CIELAB ΔE^*_{ab}, and consider it as a ratio scale of perceived color differences. This would require that color differences of the same perceived magnitude have the same ΔE^*_{ab} for all areas of color space. Another requirement would be for the perceptions of color differences to scale linearly with measured ΔE^*_{ab}. A desirable feature would be for ΔE^*_{ab} values measured under one illuminant to be perceptually equal to ΔE^*_{ab} values measured under any other illuminant such that color differences could be directly compared across different light sources. It has been well established that the CIELAB color space is not uniform for the measurement of color differences and does not meet any of these requirements. In fact, this might not be possible in any Euclidean color space.

Current Techniques and Recommendations

The weaknesses of the simple CIELAB ΔE^*_{ab} formula have been addressed both within and outside CIE activities. For example, the CMC color-difference formula is designed to address some of the nonuniformities in the CIELAB space in order to make color-difference measurements in one region of color space equivalent to measurements in other regions of color space. CIE TC1-29 investigated the CMC and other equations as possible refinements of the CIELAB ΔE^*_{ab} formula. It concluded that the CMC color-difference equation was more complex than warranted by the available visual data, and it created a simplified formula called CIE ΔE^*_{94} (CIE, 1995). The ΔE^*_{94} equations are specified in Equations 16-2 through 16-5:

$$\Delta E^*_{94} = \left[\left(\frac{\Delta L^*}{k_L S_L}\right)^2 + \left(\frac{\Delta C^*_{ab}}{k_C S_C}\right)^2 + \left(\frac{\Delta H^*_{ab}}{k_H S_H}\right)^2\right]^{1/2}$$

(16-2)

$$S_L = 1$$

(16-3)

$$S_C = 1 + 0.045 C^*_{ab}$$

(16-4)

$$S_H = 1 + 0.015 C^*_{ab}$$

(16-5)

C^*_{ab} in Equations 16-4 and 16-5 refers to the CIELAB chroma of the standard sample of the color-difference pair or, alternatively, to the geometric mean of the two chroma values. Parametric factors, k_L, k_C, and k_H, are introduced to correct for variation in perceived color difference caused by certain experimental variables such as sample size, texture, separation, and so on. Under reference conditions, the parametric factors are all set to 1.0. Reference conditions are defined as follows:

Illumination: CIE illuminant D65 simulator
Illuminance: 1,000 lux
Observer: Normal color vision
Viewing Mode: Object
Sample Size: Greater than 4° visual angle
Sample Separation: Minimal, direct edge contact
Color-difference Magnitude: 0 to 5 CIELAB units
Structure: Visually homogeneous

The reference conditions of the ΔE^*_{94} equations illustrate the limitations of the CIELAB space for the specification of color appearance. These are the areas in which it might be possible for a color appearance model to make a contribution to the specification of color differences.

Application of Color Appearance Models

In the area of color-difference specification, color appearance models could be used to incorporate some of the parametric effects directly into the equation. For example, an accurate color appearance model could incorporate the effects of the background and luminance level on color-difference perception. A color appearance model would also make it possible to directly compare color differences measured for different viewing conditions. This has applications in the calculation of indices of metamerism, as described in following sections.

Color appearance models would also enable the calculation of color differences between a sample viewed in one condition and a second sample viewed in other, different conditions. This could be useful for critical colors, such as those on warning signs or for corporate trademarks.

Future Directions

Currently, there is little activity aimed at incorporating color appearance models beyond CIELAB into a practical color-difference specification. Perhaps this is because of the effort already invested in fine-tuning CIELAB within various industries. Instead, research activity (which is not abundant) is aimed at further refining equations, such as ΔE^*_{94}, within CIELAB and defining the influence of parametric effects such as gloss, texture, sample separation, sample size, and so on. Also, the majority of effort in the formulation of color appearance models has been in the area of chromatic-adaptation transforms. Little attention has been paid to color-difference specification within the color-appearance spaces. The notable exception is the formulation of the LLAB space (Luo et al., 1996) in which color appearance and color difference were treated simultaneously. Also, the RLAB space (Fairchild, 1996) has been formulated to preserve the CIELAB spacing so that CIE color-difference formulas such as ΔE^*_{94} could still be used.

Another interesting future direction for color-difference specification is the incorporation of the spatial characteristics of human visual performance into the difference metric such that the relative sensitivity to color variations of various spatial frequencies is appropriately treated. Examples of this type of metric can be found in the work of Maximus et al. (1994) and Zhang and Wandell (1996).

16.3   Indices of Metamerism

Metamerism, the fact that two stimuli can match in color while having disparate spectral power distributions, is both a great benefit and a severe detriment to a variety of industries. Techniques to quantify the degree of metamerism for various stimuli are of significant value. There are two types of metamerism to be measured:

1. Illuminant
2. Observer

Measures of the degree of metamerism for specific stimuli are called *indices of metamerism.*

Illuminant metamerism is generally of most concern. It occurs when two objects match in color for one illuminant but mismatch for a second illuminant. This happens when the spectral reflectance functions of the two stimuli differ, but those differences are unimportant with respect to the visual response functions (color-matching functions) when integrated with the spectral power distribution of the first illuminant. When the illuminant is changed, these differences might become apparent to an observer. Illuminant metamerism is often a problem in industries that produce colored materials. If they produce two materials that are a metameric match to one another, they might mismatch under some practical viewing conditions. If the two materials are an identical match, meaning their spectral reflectance functions are identical, then they are not metameric and will match for any illuminant.

Observer metamerism is more difficult to quantify, but it is perhaps equally important. It is caused by the normal variations in the color responsivities of various observers. Observer metamerism is defined by two stimuli that have differing spectral power distributions that match for a given observer. When these stimuli are examined by a second observer, they might no longer match. Again, stimuli that are identical spectral matches will match for any observer. Thus illuminant metamerism becomes apparent when the illuminant is changed and observer metamerism becomes apparent when the observer is changed.

Current Techniques and Recommendations

CIE Publication 15.2 (CIE, 1986) describes a technique to calculate an index of metamerism for change in illuminant. Essentially, the recommendation is to calculate a CIELAB ΔE^*_{ab}, or any other color-difference metric, for the illuminant under which the two stimuli do not match. This could be any illuminant of interest as long as the two stimuli match under the illuminant of primary interest. There is no clear recommendation on how to calculate this index of metamerism when the two stimuli are not perfect matches under the primary illuminant. In such a case, technically there is no metamerism, but rather simply a pair of stimuli with an unstable color difference. Techniques for overcoming this limitation have been discussed by Fairman (1987).

CIE Publication 80 (CIE, 1989) describes a technique for calculating an index of metamerism for change in observer. Essentially, the standard colorimetric observer is replaced with a standard deviate observer and the color difference between the two stimuli is calculated for this new observer. The concept is sound, but the data on which the standard deviate observer were based had been normalized, thus resulting in an under-prediction of the degree of observer metamerism (Alfvin and Fairchild, 1997). Nimeroff et al. (1961) described a technique whereby a complete standard observer system, including mean and covariance color-matching functions, could be specified. This concept is similar to the idea behind the CIE (1989) technique, but it has never been fully implemented due to a lack of data.

Application of Color Appearance Models

Color appearance models could be of some utility in the quantification of illuminant metamerism, since it involves the comparison of stimuli across changes in illumination. Essentially, the contribution to an index of illuminant metamerism would be the availability of a color-difference metric that is consistent across a variety of illuminants. Also, an accurate color appearance model would allow the creation of a new type of metric for single stimuli. A single sample cannot be considered metameric, since it does not match anything. However, it is common to talk of a single sample as being metameric when its apparent color changes significantly with a change in illuminant. This is really a lack of color constancy. A good color appearance model would allow one to calculate a color-difference metric between a sample under one illuminant and the same sample under a second illuminant, thus allowing the creation of an index of color constancy. This could be useful for objects that are intended to appear to be the same color under various illumination conditions, such as those containing safety colors.

There really is little use for a color appearance model in the specification of observer metamerism beyond the potential for a better color-difference metric. The measurement of observer metamerism is an excellent example of a situation in which the problem needs to be first completely addressed at the level of basic colorimetry. In other words, the observer variability in tristimulus values must first be adequately specified before one needs to be concerned about the improvements that a color appearance model could make. This path should be taken as a model for all potential applications of color appearance models.

Future Directions

There is little activity aimed at improving indices of metamerism. For illuminant metamerism, effort is concentrated on improving color-difference metrics. For observer metamerism, there seems to be little call for a better metric, despite the flaws in the current metric (which is not widely used). This could be because it is difficult enough to address problems of illuminant metamerism to cause the difficulties associated with observer metamerism to be considered of second order at this time.

16.4   A General System of Colorimetry?

Consideration of some of the problems of traditional colorimetry described in this chapter leads one to wonder whether it might be possible to create a general system of colorimetry that could be used to address all of the problems of interest. Currently, colorimetry has a very evolutionary form of development, moving from CIE XYZ tristimulus values to the CIELAB color space, to enhancements of CIELAB for measuring color difference and color appearance. This development is useful to ensure compatibility with industrial practices that are based on previous standard procedures. However, the level of complexity is getting to the point where there might be multiple color models, each more appropriate for a different application. CIELAB and CIELUV were recommended by the CIE in 1976 to limit to two the number of color-difference formulae being used internationally, rather than the ever-increasing number that had been used. That recommendation was quite successful and has resulted in CIELAB becoming essentially the only color space in use (along with a few color-difference equations based on it).

Perhaps a similar state of affairs is developing in the area of color appearance models and forthcoming recommendations from the CIE will bring about some order. However, it is still likely that systems for color appearance and for color difference will be separate in practice. The LLAB model (Luo et al., 1996) represents one interesting attempt to bring colorimetry together with one general model. Perhaps this approach should be pursued.

An alternative approach is to start over from scratch, taking advantage of the progress in visual science and colorimetry over the last century, to create a new system of colorimetry that is superior for all steps in the process and that can find a wide range of applications in science, technology, and industry. Color appearance models take a step in this direction by first transforming from CIE tristimulus values to cone responses and then building up color-appearance correlates from there. There is also activity within the CIE (e.g., TC1-36, Fundamental Chromaticity Diagram with Physiologically Significant Axes) to develop a system of colorimetry based on more-accurate cone responsivities that are not necessarily tied to a CIE Standard Colorimetric Observer. It is thought that such a system would find wide use in color-vision research. Boynton (1996) has reviewed the history and status of such work. Perhaps some convergence between the two activities is needed to develop a better, general system of colorimetry that could be used by everyone.

CHAPTER 17

Device-independent
Color Imaging

17.1 The Problem
17.2 Levels of Color Reproduction
17.3 General Solution
17.4 Device Calibration and Characterization
17.5 The Need for Color Appearance Models
17.6 Definition of Viewing Conditions
17.7 Viewing-conditions-independent Color Space
17.8 Gamut Mapping
17.9 Color Preferences
17.10 Inverse Process
17.11 Example System
17.12 ICC Implementation

A COMPUTER USER TAKES a photograph, has it processed and printed, scans the print into the system, displays it on the monitor, and finally prints it on a digital printer. In so doing, the user has completed at least three input-process-display cycles on this image. And despite the user's having spent large sums of money on various imaging hardware, it is extremely unlikely that the colors in the final print look anything like the original object or that they are even satisfactory. There are also intermediate images that might be compared to one another and the original object.

In this scenario, the user chose each of the components and put them together in what is called an "open system." Each of the imaging devices has its own intrinsic physical process for image input, processing, and/or display, and none is necessarily designed to function with any of the others. If each component functions in its native mode, then the results produced with an open system are nearly unpredictable. One reason for this is the open nature of the system; that is, there are too many possible combinations of devices so that making them all work well with one another is difficult, if not impossible.

The devices are allowed to function in their own intrinsic color dimensions in what is called *device-dependent color imaging*. The difficulty with this is that the RGB coordinates from a scanner might not mean the same thing as the RGB signals used to drive a monitor or printer. To solve these problems and produce reliable results with open systems, one must use device-independent color-imaging processes. Device-independent color imaging is intended to provide enough information along with the image color data such that the image data can, if necessary, be described in coordinates that are not necessarily related to any particular device. Transformations are then performed to represent those colors on any particular device.

The strong technological push for reliable device-independent color imaging has stressed the scientific capabilities in the area of color appearance modeling, since the various images are typically viewed in a wide variety of viewing conditions. While it has been recognized for some time that a color appearance model is necessary for successful device-independent color imaging, there has not been a simple solution to that problem available. The first 15 chapters of this book present some of the issues and problems that must be addressed. This chapter provides an overview of the basic concepts required to put the pieces together and build systems.

Device-independent color imaging has become the focus of many scientists and engineers over recent years. It is impossible to cover in a single chapter the scope of issues involved. Entire books dedicated to this topic are available (Giorgianni and Madden, 1997; Kang, 1997). Also, Hunt's text on color reproduction (Hunt, 1995) provides much necessary insight into the fundamentals of traditional and digital color reproduction. Sharma and Trussel (1997) have published a review paper on the field of digital color imaging that includes hundreds of references. The treatment in this chapter is culled from a series of previous works (Fairchild, 1994, 1995b, 1996).

17.1   The Problem

The application of basic colorimetry produces significant improvement in the construction of open color-imaging systems by defining the relationships between device coordinates (e.g., RGB, CMYK) and the colors detected or produced by the imaging systems. However, it is important to recall that matching CIE tristimulus values across various imaging devices is only part of the story. If an image is reproduced such that it has CIE tristimulus values identical to those of the original image, then the reproduction will match the original in appearance as long as the two are viewed under identical viewing conditions (those conditions that match those for which the tristimulus values were calculated). Since originals, reproductions, and intermediate images are rarely viewed under identical conditions, it becomes necessary to introduce color appearance models to the system in order to represent the appearance of the image at each stage of the process.

Issues in device-independent color imaging that color appearance models can be used to address include changes in white point, luminance level, surround, medium (viewing mode), and so on. Since these parameters normally vary for different imaging modalities, the need for color appearance models is clear. The introduction of such models allows systems to be set up and used to preserve, or purposefully manipulate, the appearances of image elements in a controlled manner at each step. Thus users can view an image on a CRT display, manipulate it as they choose, and then make prints that accurately reproduce the appearance of the image on the CRT, all with the aid of color appearance models.

Of course, it is not always possible, or desirable, to exactly reproduce the appearance of an original image. Color appearance models can be useful in these situations as well. One reason is that different imaging devices can produce different ranges of colors; this is their *color gamut*. A given stimulus on a CRT display produces a certain appearance. It might not be possible to produce a stimulus on a given printer that can replicate that appearance. In such cases, a color appearance model can be used to adjust the image in a perceptually meaningful way to produce the best possible result. In other cases, the viewing conditions might limit the gamut of a reproduction. For example, photographic prints of outdoor scenes are often viewed under artificial illumination at significantly lower luminance levels than that of the original scene. At the lower luminance level, producing the range of luminance and chromatic contrast of the original scene is impossible. Thus it is common for consumer photographic prints to be produced with increased physical contrast to overcome this change in viewing conditions. Color appearance models can be used to predict such effects and guide the design of systems to address them.

Another advantage of color appearance models in device-independent color imaging is in the area of image editing. It is more intuitive for untrained users to manipulate the colors in images along perceptual dimensions such as lightness, hue, and chroma, rather than through device coordinates such as CMYK. A good color appearance model can improve the correlation between tools intended to manipulate these dimensions and the changes that users implement on their images.

17.2 Levels of Color Reproduction

Hunt (1970, 1995) has defined six different objectives for color reproduction:

1. Spectral color reproduction

   Spectral color reproduction involves identical reproduction of the spectral reflectance curves of the original image or objects. Two techniques that are so impractical as to be of only historical interest, the Lippman and microdispersion methods (see Hunt, 1995), managed to fulfill this difficult objective. Modern color-reproduction techniques take advantage of metamerism by using RGB additive primaries or CMY subtractive primaries, thus eliminating the possibility of spectral reproduction except in cases in which the original is comprised of the same imaging materials. Recently developed, and currently developing, printing techniques that utilize six or more inks provide an opportunity for better approximations to spectral color reproduction that might be useful in applications such as mail-order catalogs or fine-art reproductions (in addition to expanding the output gamut).

2. Colorimetric color reproduction

   Colorimetric color reproduction is defined via metameric matches between the original and the reproduction such that they both have the same CIE *XYZ* tristimulus values. This will result in the reproduction of color appearances in cases for which the original and reproduction of the same size are viewed under illuminants that have the same relative spectral power distribution, luminance, and surround. Hunt, however, does not make equality of luminance level a requirement for colorimetric color reproduction.

3. Exact color reproduction

Exact color reproduction is colorimetric color reproduction with the additional constraint that the luminance levels be equal for the original and the reproduction.

4. Equivalent color reproduction

Equivalent color reproduction applies in situations in which the color of illumination for the original and the reproduction differs. In such cases, precise reproduction of CIE tristimulus values would result in images that were clearly incorrect, since nothing has been done to account for chromatic adaptation. Equivalent color reproduction thus requires the tristimulus values and the luminances of the reproduction to be adjusted such that they produce the same appearances as found in the original. This requires the differences between the original and the reproduction viewing conditions to be incorporated using some form of color appearance or chromatic-adaptation model. When there are large changes in luminance level between the original and the reproduction, it might be impossible to produce appearance matches, especially if the objective is brightness-colorfulness matching rather than lightness-chroma matching.

5. Corresponding color reproduction

Corresponding color reproduction addresses the luminance issue by neglecting it to a degree. A corresponding color reproduction is one that is adjusted such that its tristimulus values are those required to produce appearance matches if the original and the reproduction were viewed at equal luminance levels. This eliminates the problems that arise when trying to reproduce brightly illuminated originals in dim viewing conditions and vice versa. It can be thought of as an approximation to lightness-chroma matching, if one were willing to assume (incorrectly) that lightness and chroma are constant across changes in luminance level. Since lightness and chroma are far more constant across luminance changes than are brightness and colorfulness, this assumption might not be too bad, especially since there are practical gamut-mapping constraints.

6. Preferred color reproduction

Preferred color reproduction is reproduction in which the colors depart from equality of appearance to those in the original in order to give a more pleasing result. This might be applicable in situations such as consumer photography in which consumers prefer to have prints that reproduce colors closer to their memory colors for objects such as skin tones, vegetation, sky, bodies of water, and so on. However, as Hunt (1970) points out, "the concepts of spectral, colorimetric, exact, equivalent, and corresponding color reproduction provide a framework which is a necessary preliminary to any discussion of deliberate distortions of colour reproduction."

These objectives are a good summary of the problems encountered in color reproduction and how they can be addressed using concepts of basic and advanced colorimetry. It is interesting to note that these objectives were originally published long before issues in device-independent color imaging were commonly discussed (Hunt, 1970). A slight rearrangement and simplification of Hunt's objectives can be used to define five levels of color reproduction that provide a framework for modern color-imaging systems:

1. Color reproduction

   Color reproduction refers to simple availability devices capable of producing color graphics and images. There is usually great excitement surrounding the initial commercial availability of color devices of any given type. While this might not seem like much of an accomplishment, it is worth remembering that personal computers with reasonable color capabilities have been available for fewer than 10 years. The plethora of high-quality input and output devices is very recent. When these technologies are first introduced, users are excited simply by the fact that they now have color available where previously it was not. However, this "honeymoon period" quickly wears off and users begin to demand more from their color-imaging devices. They want devices that produce and reproduce colors with some semblance of control and accuracy. This pushes open-systems technology toward the next levels of color reproduction.

2. Pleasing color reproduction

   Pleasing color reproduction refers to efforts to adjust imaging devices and algorithms such that consumers find the resulting images acceptable. Such images might not be accurate reproductions, and they are probably not the preferred reproductions, but they look pleasing and are found acceptable to most consumers of the images. This level of reproduction can often be achieved through trial and error without requiring any of the concepts of device-independent color imaging. The approach to obtaining pleasing color reproduction in open systems would be similar to the approaches historically taken in closed imaging systems to achieve similar goals or, in some cases, preferred color reproduction. Pleasing color reproduction can be a reasonable final goal for a color-reproduction system in which observers have no knowledge of the original scene or image and, therefore, no expectations beyond obtaining a pleasing image.

3. Colorimetric color reproduction

Colorimetric color reproduction includes calibration and characterization of imaging devices. This means that for a given device signal, the colorimetric coordinates of the image element produced (or scanned) are known with a reasonable degree of accuracy and precision. Colorimetric color reproduction enables a user to put together a system in which an image is scanned and the data are converted to colorimetric coordinates (e.g., CIE XYZ). These coordinates are then transformed into appropriate RGB signals to display on a CRT or into CMYK signals for output to a printer.

Of course, it is not necessary for the image data to be actually transformed through the device-independent color space. Instead, the full transform from one device, through the device-independent space, to the second device can be constructed and implemented for enhanced computational efficiency and minimization of quantization errors. Such a system allows the CIE tristimulus values of the original image to be accurately reproduced on any given output device. This is similar to Hunt's definition of colorimetric color reproduction. To achieve colorimetric color reproduction, devices and techniques for the colorimetric characterization and calibration of input and output devices must be readily available. A variety of such techniques and devices is available commercially, but the degree to which colorimetric color reproduction can actually be achieved by typical users is dubious. Unfortunately, the state of the art for most users is just color reproduction; *colorimetric* color reproduction has yet to be reliably achieved.

Colorimetric color reproduction is useful only when the viewing conditions for the original and reproduced images are identical, since this is the only time that tristimulus matches represent appearance matches. When the viewing conditions differ, as they usually do, one must move from colorimetric color reproduction to the next level, color-appearance reproduction.

4. Color-appearance reproduction

Color-appearance reproduction requires a color appearance model, information about the viewing conditions of the original and reproduced images, and accurate colorimetric calibration and characterization of all the devices. In color-appearance reproduction, the tristimulus values of the original image are transformed to appearance correlates, such as lightness, chroma, and hue, using information about the viewing conditions, such as white point, luminance, surround, and so on. Information about the viewing conditions for the image to be reproduced is then used to transform these appearance correlates into the tristimulus values necessary to produce them on the output device. Color-appearance reproduction is necessary to account for the wide range of media and viewing conditions found in different imaging devices. This is similar to Hunt's equivalent color reproduction applied to lightness-chroma matches. Color-appearance reproduction has yet to become a commercial reality, and perhaps it cannot for typical users. However, even when reasonable color-appearance reproduction does become available, there will be cases in which users will desire reproductions that are not accurate appearance matches to the originals. Such cases enter the domain of color-preference reproduction.

5. Color-preference reproduction

Color-preference reproduction involves the purposeful manipulation of the colors in a reproduction such that the result, rather than an accurate appearance reproduction, is preferred by users. The objective is to produce the best possible reproduction for a given medium and subject. This is similar to Hunt's definition of preferred-color reproduction.

Note that to achieve each level of reproduction in open systems, one must first achieve the lower levels. To summarize, the five levels involve simply the reproduction of colors, the reproduction of pleasing colors, equality of tristimulus values, equality of appearance attributes, and the manipulation of appearance attributes to "improve" the result. In closed systems, technology need not progress through each of the five levels. This is because the path of image data is defined and controlled throughout the whole process. For example, in color photography, the film sensitivities, dyes, processing procedures, and printing techniques are all well defined. Thus it is possible to design a photographic negative film to produce pleasing or preferred color reproduction without anyone having the capability for colorimetric or color-appearance reproduction, since the processing and printing steps are well defined. A similar system exists in color television with standard camera sensitivities, signal processing, and output device setup. In open systems, an intractable number of combinations of input, processing, display, and output devices can be constructed and used together. The manufacturer of each subsystem cannot possibly anticipate all of the possible combinations of devices that might be used with its subsystem. Thus the only feasible solution is to have each device in the system develop through the five levels so that colorimetric, or color-appearance, data (or the information necessary for obtaining them) can be handed off from one device to the next in the process called device-independent color imaging.

17.3   General Solution

Figure 17-1 is a flow chart of the general process of device-independent color imaging. At the top of the chart is the original image as represented by some input device. (Note that this "input" could come from a display device such as a CRT.) First, the colorimetric characterization of the input device is used to transform the device coordinates (e.g., RGB) to colorimetric coordinates such as CIE XYZ or CIELAB, which are referred to as device-independent color spaces, since the colorimetric coordinates do not depend on any specific imaging device.

The second step is to apply a chromatic-adaptation and/or color appearance model to the colorimetric data, along with additional information on the viewing conditions of the original image, in order to transform the image data into dimensions that correlate with appearance such as lightness, hue, and chroma. These appearance coordinates, which have accounted for the influences of the particular device and the viewing conditions, are called *viewing-conditions-independent space*. At this point, the image is represented purely by its original appearance. Now is when it is most appropriate to perform manipulations on the image colors. These manipulations might include gamut mapping, preference editing, tone reproduction adjustments, spatial scaling operations, certain forms of error diffusion, and so on. The image is now in its final form with respect to the appearances that are to be reproduced. Next, the process must be reversed.

This highlights the utility of an analytically invertible color appearance model. The viewing conditions for the output image, along with the final image-appearance data, are used in an inverted color appearance model to transform from the viewing-conditions-independent space to a device-independent color space such as CIE XYZ tristimulus values. These values, together with the colorimetric characterization of the output device, are used to transform to the device coordinates (e.g., CMYK) necessary to produce the desired output image. The following sections provide some additional details on each step of this process.

Note that the literal implementation of the processes of device-independent color imaging as described here requires substantial computational resources. For example, to avoid severe quantization errors, image processing is usually performed on floating-point image data with floating-point computational precision when working within the intermediate color-appearance spaces. While this is acceptable for color-imaging research, it is not practical in most commercial color-imaging systems, particularly those that are limited to 24-bits-per-pixel color data. In such cases, the processes described here are used to construct the systems and algorithms, while implementation is left to multidimensional interpolation within 8-bits-per-channel lookup tables (LUTs). It is interesting to note that the computer graphics industry, as opposed to the color imaging/publishing industry, typically works with floating-point and higher-precision integer image data. Perhaps the confluence of the two fields will solve some historical computational issues and limitations.

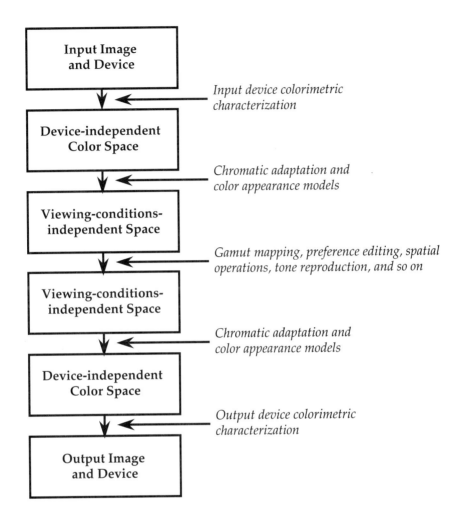

Figure 17-1.   A flow chart of the conceptual process of device-independent color imaging.

17.4   Device Calibration and Characterization

Device calibration is the setting of the imaging device to a known state. This might represent a certain white point, gain, and offset for a CRT or certain relationships between density and drive signal for a printer. Calibration ensures that the device is producing consistent results, both from day to day and from device to device. However, device calibration can be completed with absolutely no information about the relationship between device coordinates and the colorimetric coordinates of the input or output image. Colorimetric characterization of the device is required to obtain this information. *Characterization* is the creation of a relationship between device coordinates and a device-independent color space—the first step in Figure 17-1.

Device calibration is usually an issue for the manufacturer, rather than the user, and the techniques for doing it depend heavily on the technology. Thus calibration will not be discussed further except to stress its importance. If consistent results are necessary from day to day or from device to device, then careful and frequent device calibration is necessary. There are trade-offs that can be made between calibration and characterization. If careful calibration is not possible, then accuracy can be achieved through frequent characterization. If an extremely good calibration procedure is available, it might be possible to perform the colorimetric characterization just once, as long as the device is frequently calibrated.

Three Approaches to Device Characterization

There are three main approaches to device characterization:

1. Physical modeling
2. Empirical modeling
3. Exhaustive measurement

Of course, there are also procedures that combine aspects of one or more of these. In all cases, it is typical to use the characterization to build a three-dimensional LUT that is used in conjunction with an interpolation procedure to process the vast amounts of image data that are encountered.

PHYSICAL MODELING

Physical modeling of imaging devices involves building mathematical models that relate the colorimetric coordinates of the input or output image elements to the signals used to drive an output device or the signals originating from an input device. Such models can be derived for all types of imaging devices with varying degrees of difficulty. A physical model for a scanner would involve a step to first linearize the signals with respect to luminance, or perhaps absorbance, and then a second step to transform the signals to CIE tristimulus values. Depending on the scanner design, knowledge of the physical properties of the material being scanned might be required. This could be avoided if the scanner were designed as a colorimeter rather than with arbitrary *RGB* responsivities.

A physical model for a CRT display involves a nonlinear transform to convert drive voltages to the corresponding *RGB* phosphor luminances, followed by a linear transformation to CIE *XYZ* tristimulus values. A physical model for a hard-copy output device requires a transformation from drive signals to concentrations of dyes, pigments, or inks and then a color-mixing model to predict spectral reflectances or transmittances that can be used to calculate CIE *XYZ* tristimulus values.

The advantage of physical-device models is that they are robust, typically require few colorimetric measurements in order to characterize the device, and allow for easy recharacterization if some component of the imaging system is modified. The disadvantage is that the models are often quite complex to derive and can be complicated to implement. Physical models are often used for CRT-display characterization.

EMPIRICAL MODELING

Empirical modeling of imaging devices involves collecting a fairly large set of data and then statistically fitting a relationship between device coordinates and colorimetric coordinates. Such models are often implemented to transform directly to CIELAB coordinates to avoid quantization difficulties in CIE *XYZ* tristimulus values.

Empirical models are often high-order multidimensional polynomials or, alternatively, neural network models of significant complexity. They require fewer measurements than lookup table techniques do, but they need more than physical models. They also are often poorly behaved near the boundary of the device gamut and so produce very large systematic errors. Since empirical models have no relationship to the physics of the imaging devices, they must be recreated each time a change is made in any component of the system. Empirical models are often used for scanner characterization.

EXHAUSTIVE MEASUREMENT

The final class of characterization techniques involves exhaustive measurement of the output for a complete sampling of the device's gamut. (Signals for a large sampling of known input colors can be collected for scanner characterization.) Typically, something like a $9 \times 9 \times 9$ sampling of the device drive signals is output and colorimetrically measured. This results in a total of 729 measurements. Many more measurements might be used for devices with poor image-to-image or device-to-device repeatability. The array of colorimetric data must then be nonlinearly interpolated to populate a higher-density (e.g., $33 \times 33 \times 33$) LUT that can be used to process image data via multidimensional interpolation.

The disadvantages of such techniques include

- The large number of measurements that must be made
- Difficulties in interpolating the highly nonlinear data
- The need to redo the entire process if any aspect of the device changes
- The difficulty in creating the inverse solutions that are typically required

The advantage of exhaustive measurement techniques that makes them popular is that they require no knowledge of the device physics. Exhaustive measurement and LUT interpolation techniques are often used for printer characterization.

Types of Colorimetric Measurements

Different types of colorimetric measurements are required for the characterization of various imaging devices:

- *CRT-display characterization* requires spectroradiometric or colorimetric measurements of the phosphor chromaticities in order to derive the *RGB-to-XYZ* transformation and relative radiometric or photometric measurements to derive the nonlinear transfer functions for each channel. Berns (1996) reviews a practical procedure for the calibration and characterization of CRT displays. Berns et al. (1993a, b) provide further details on the measurement and characterization of CRT displays.

- *Printers and other output devices* require spectrophotometric measurements (spectral reflectance or transmittance) to characterize the device colorants or derive colorimetric coordinates for various illuminants or sources. Additional densitometric measurements might be of value to characterize tone-transfer functions. Issues in the colorimetric characterization of binary and multilevel display devices have been discussed by various authors, including Jarvis et al. (1976), Engeldrum (1986), Gentile et al. (1990a), Rolleston and Balasubramanian (1993), Berns (1993b), and Haneishi et al. (1996).

- *Scanners and digital cameras* require spectroradiometric evaluation of their channel spectral responsivities or empirical estimates of them. Spectroradiometric data on the illumination system are also required. Also, spectroradiometric linearity evaluation and characterization are required for the detector systems. Often, scanner data for well-characterized input targets are collected to derive relationships between scanner signals and colorimetric coordinates. The colorimetric calibration and characterization of input devices have been described by Hung (1991), Kang (1992), Engeldrum (1993), Rodriguez and Stockham (1993), and Berns and Shyu (1995).

Flare, Metamerism, and Fluorescence

Three additional issues regarding colorimetric measurements are often overlooked in device characterization, but require attention:

1. Flare
2. Metamerism
3. Fluorescence

FLARE

Typically, the spectrophotometric or colorimetric measurements made to characterize a device are performed with specialized instrumentation and specially prepared samples. Such measurements are not made in the actual viewing situation for the device. Any real viewing situation includes *flare*, an additive mixture of light with the image. The spectral energy distribution and level of the flare must be measured and added to any real colorimetric characterization of an imaging device. It can be treated as a simple addition of the tristimulus values of the flare to the tristimulus values of the image data. This addition might result in the need to recalculate the image white point and renormalize data appropriately.

In some cases, flare might be image-dependent and require a more sophisticated treatment. Alternatively, measurements of image color must be made *in situ,* using a telespectroradiometer that will include the flare of the viewing environment in the measurement.

METAMERISM

Metamerism causes difficulties in both input and output devices. For input devices, metamerism combined with noncolorimetric sensor responsivities can defeat all hope of obtaining reliable color reproduction. At the output end, devices must be characterized using spectral reflectance or transmittance functions integrated with the actual viewing spectral power distributions in order to derive colorimetric coordinates. This is necessary for reasonable accuracy even when using standardized viewing sources. For example, the colors observed under a fluorescent D50 simulator can differ dramatically from those calculated using CIE illuminant D50.

FLUORESCENCE

The colorimetry of fluorescent materials is a significant challenge because the energy emitted by the material is a function of the energy incident from the illuminating source. Since this is not the case for nonfluorescent materials, the light source used for spectrophotometric measurement has no impact on the colorimetric coordinates calculated for any particular illuminant. Fluorescent materials must be measured using illumination that closely simulates the illuminant to be used in colorimetric calculations in order to obtain reasonable accuracy. The best practical solution is to measure fluorescent materials in their final viewing conditions using a telespectroradiometer.

Fluorescence is an important issue in imaging applications because many substrates (i.e., most paper) and many inks and dyes are fluorescent. Grum and Bartleson (1980) provide an excellent overview of the colorimetry of fluorescent materials.

*   *   *

No matter what approach is taken to characterize an imaging device, the end result is typically used to construct a multidimensional LUT for practical implementations. This is because it is necessary to complete the many layers of nonlinear transformations and color-space conversions required with a computational precision that is significantly greater than the 8-bits-per-channel found in most imaging devices. Such computations take prohibitive amounts of time on typical desktop imaging systems. Thus multidimensional LUT interpolation is implemented for the end-to-end transform for convenience and efficiency. The construction of multidimensional LUTs and their use through interpolation have been described by Hung (1993), Kasson et al. (1993), and Kasson et al. (1995).

Multidimensional LUT interpolation is implemented in various ways, including proprietary software such as Adobe Photoshop and other "color-management" software that use these techniques for color-space transformations. Multidimensional LUTs are also implemented in the PostScript Level 2 (Adobe Systems Incorporated, 1990) page-description language in the form of color-rendering dictionaries. Another well-known open system that provides the framework for the implementation of multidimensional LUTs for device characterization is the ICC profile format (International Color Consortium, 1995) that serves as a cross-platform standard for a wide variety of system-level color-management systems.

17.5   The Need for Color Appearance Models

The process of device-independent color imaging described by Figure 17-1 illustrates the need for color appearance models. There are two main needs for these models: image editing and viewing-condition transformations. Image manipulations such as color-preference reproduction and gamut mapping are best performed in the perceptually significant dimensions (e.g., lightness, chroma, and hue) of a color appearance model. Clearly, the transformation of colorimetric coordinates from one set of viewing conditions (i.e., white point, luminance, surround, medium, etc.) to a second set of viewing conditions requires a color appearance model.

The only way to avoid the use of a color appearance model in device-independent color imaging is to specify a rather strong set of constraints. The original and the reproduction must be viewed in the same medium, under identical viewing conditions, with identical gamuts, and with the objective of colorimetric color reproduction. In such a constrained world, colorimetric and color-appearance reproduction are identical. Clearly, these constraints are far too severe for all but the most specialized applications. Thus the use of color appearance models in device-independent color imaging is unavoidable if high-quality, reliable results are to be obtained in open systems.

17.6   Definition of Viewing Conditions

One key unresolved issue in the implementation of color appearance models in device-independent color imaging is the definition and control of viewing conditions. Even a perfect color appearance model is of little use if the actual viewing conditions are not the same as those used in the model calculations. (The metamerism problems between CIE illuminants and their physical simulators is one straightforward example of this difficulty.)

Part of the difficulty in controlling the viewing conditions is the precise definition of the fields. Hunt (1991b) has done the most extensive job of defining the various components of the viewing field. However, even with Hunt's extended definitions, it is difficult to decide which portions of the field should be considered the proximal field, the background, and the surround when complex image displays in typical viewing conditions are viewed. For example, is the background of an image the area immediately adjacent to the image borders or should it be considered to be the areas adjacent to individual elements within the image? The latter definition might be more appropriate; however, it requires substantially more-complex image-wise computations that are often completely impractical. Regardless of how the particular aspects of the viewing conditions are defined, it is important that the treatment is consistent across all image transformations so as to avoid the introduction of bias due simply to the use of color appearance models. As a practical definition, the background for images should be defined as the area immediately around the image border, with the surround defined as the remainder of the viewing environment. This definition of background, however, is different from that used by Hunt as described in Chapter 7. The definition of proximal field is unnecessary in image reproduction, since the spatial relationships of the various image elements are constant in the original and the reproduction. The proximal field becomes important when it is desired to reproduce the color appearance of an image element in a completely different context (e.g., logo colors, trademark colors).

Even with strict definitions of the various components of the viewing field, it is of paramount importance that the viewing conditions be carefully controlled for successful device-independent color imaging. If users are unwilling to control the viewing conditions carefully, they should expect nothing less than unpredictable color reproduction. Viewing-condition parameters that must be carefully controlled for successful color-appearance reproduction include

- The spectral power distribution of the light source
- Luminance level
- Surround color and relative luminance
- Background color and relative luminance
- Image flare (if not already incorporated in the device characterization)

- Image size and viewing distance (i.e., solid angle)
- Viewing geometry

Also, observers must make critical judgments of the various images only after sufficient time has passed so as to allow full adaptation to the respective viewing conditions.

Braun et al. (1996) illustrated the importance of controlling the viewing conditions for cross-media image comparisons. They concluded that the best technique for critical judgments was *successive binocular viewing* in which the observer viewed first one image display with both eyes and then switched to the other display, allowing approximately one minute to adapt to the new viewing conditions. The arrangement was such that only one image display could be viewed at a time and the one-minute adaptation was required each time the observer changed from one display to the other. Unfortunately, the most common technique, *simultaneous binocular viewing*, produces unacceptable results. In simultaneous binocular viewing, the original and the reproduction (in a different medium and white point) are viewed simultaneously side by side. In such cases, the observer's state of chromatic adaptation cannot be reliably predicted, since it depends on the relative amount of time spent viewing each image. In general, the best results will be obtained if a single, intermediate adaptation point is assumed. However, the result of such a choice will be a reproduction that matches the original when both are viewed side by side but that looks quite strange when viewed by itself.

For example, if a CRT has a 9300K white point and a reproduced print is viewed under a D50 simulator, the required print to produce a simultaneously viewed match will have an overall blue cast. When this print is viewed in isolation, still under a D50 simulator, it will appear unacceptably bluish and be considered a poor match. Katoh (1995) has investigated the problems with simultaneous viewing of images in different media. However, if a successive viewing technique with sufficient adaptation time is used, an excellent neutrally balanced 9300K CRT image will be matched by a neutrally balanced print viewed under a D50 simulator. Thus both color-appearance matching and high individual-image quality can be obtained with appropriate viewing procedures.

Once the viewing conditions are appropriately defined and controlled, some computational advantage can be obtained through judicious precalculation procedures. Such procedures rely on parsing the implementation of the color appearance models into parts that need to be calculated only once for each viewing condition and those that require calculation for each image element. The most efficient implementation procedure is then to precalculate the model parameters that are viewing-condition-dependent and use this array of data for the individual appearance model calculations performed on each pixel or element of an LUT. For example, when RLAB is used for a change in white point and luminance with a constant surround, the change in viewing conditions can be precalculated down to a single 3×3 matrix transform that is applied to the CIE *XYZ* tristimulus values of the original in order to determine the tristimulus values of the reproduction. This is a significant computational simplification that makes it possible to allow users to interactively change the settings in a color appearance model such that they can choose the illuminant under which a given image will be viewed.

17.7   Viewing-conditions-independent Color Space

Device-independent color spaces are well understood as representations of color based on CIE colorimetry that are not specified in terms of any particular imaging device. Alternatively, a device-independent color space can be defined as a transform from CIE coordinates to those of some standardized device (e.g., Anderson et al., 1996). The introduction of color appearance models to the process, as illustrated in Figure 17-1, creates the additional concept of a viewing-conditions-independent color space. The viewing-conditions-independent coordinates extend CIE colorimetry to specify the color appearance of image elements at a level that does not rely on outside constraints. Such a representation encodes the perceptual correlates (e.g., lightness, chroma, and hue) of the image elements. This representation facilitates editorial adjustments to the image colors necessary for color-preference reproduction and gamut mapping.

It is also worth noting that the viewing conditions themselves might introduce "perceptual gamut limits." For example, the lightness, chroma, and hue of certain image elements viewed under a high luminance level cannot be reproduced in an image viewed at a low luminance level. In other words, certain color perceptions simply cannot be produced in certain viewing conditions.

The limitations of "perceptual gamut limits" do not in any way reduce the utility of color appearance models. In fact, such limits can be reliably defined only by using color appearance models. It is interesting to note that the concept of viewing-conditions-independent color space has a correlate in the field of cognitive science. Davidoff (1991) presents a model of object color representation that ultimately encodes color in terms of an *output lexicon*, the words we use to describe color appearance. Such a representation can be thought of as a high-level color appearance model in which colors are specified by name, as people specify them, rather than with the mathematically necessary reduction to scales of the five requisite color-appearance attributes.

17.8   Gamut Mapping

It would be misleading to suggest that all of the problems of device-independent color imaging would be solved by the use of a reliable, accurate color appearance model. Even if a perfect color appearance model were used, the critical question of color gamut mapping would remain. The development of robust algorithms for automated gamut specification and color mappings for various devices and intents remains as perhaps the most important unresolved issue in cross-media color reproduction (Fairchild, 1994a).

The gamut of a color-imaging device is defined as the range of colors that can be produced by the device as specified in some appropriate three-, or more, dimensional color space. (It is important to reiterate that color gamuts must be expressed in a three-dimensional color space, since two-dimensional representations such as those often plotted on chromaticity diagrams are misleading.) The most appropriate space for the specification of a device gamut is within the coordinates of a color appearance model, since the impact of viewing conditions on the perceived color gamut can be properly represented. For example, only a complete color appearance model will show that the color gamut of a printer shrinks to zero volume as the luminance decreases! Generally, a device gamut should be represented by the lightness, chroma, and hue dimensions within the chosen appearance model. However, in some cases, it might be more appropriate to express the gamut in terms of brightness, colorfulness, and hue. Examples of such cases include projection systems or other displays susceptible to ambient flare in which the absolute luminance level has a significant impact on the perceived color-image quality.

Figure 17-2 illustrates various views of three-dimensional models of two device gamuts. These gamuts are plotted in the CIELAB color space. The wireframe model represents the gamut of a typical monitor with SMPTE phosphors and a D50 white point. The solid model represents the color gamut of a typical dye-diffusion printer under CIE illuminant D50. Note that the gamut of the CRT display exceeds that of the printer for light colors, while the gamut of the printer exceeds that of the CRT for some darker colors. This three-dimensional gamut representation should clear up some misconceptions about color gamuts. For example, the figure illustrates the large extent to which color gamuts are not coincident and the fact that printer gamuts often extend outside the range of CRT gamuts. Typically, it is assumed that CRT gamuts are significantly larger than most printer gamuts. This is a result of examination of two-dimensional gamut boundaries in chromaticity diagrams while neglecting the third dimension of color space.

Gamut mapping is the process of adjusting the colors in an image such that it can be represented on a given device. For example, one might want to use a CRT display to reproduce a dark, saturated cyan that is present on a print. If the CRT cannot produce the desired color, the image element must be shifted to an appropriate color that is within the CRT gamut. The opposite problem might arise in which a device is capable of producing more-saturated colors than are present in the original image. If the full gamut of the output device is not utilized, users might be displeased with the results, since they know that the device is capable of producing a wider range of colors. Thus, image colors might also be adjusted to fill color gamuts as well. Therefore the problem of gamut mapping can be described as gamut *compression* in regions in

Figure 17-2.   Several views of three-dimensional representations of device gamuts in the CIELAB color space. The solid model is the gamut of a typical dye-diffusion thermal-transfer printer, and the wire-frame model is the gamut of a typical CRT display.

which the desired color falls outside the device gamut and gamut *expansion* in regions in which the gamut of image colors does not fully utilize the device gamut. Proper gamut expansion requires full knowledge of the source image's gamut and computationally expensive image-dependent processing. Thus it might not be fully implemented within practical systems for some time. This is somewhat counter to the common perception that gamut mapping is only a problem of gamut compression. Color adjustments in the opposite direction represent an equally important, and perhaps more challenging, problem. Clearly, a color appearance model is the best place in which to specify gamuts and perform mapping transformations, since the manipulations can be carried out on perceptually meaningful dimensions.

A variety of gamut-mapping techniques have been suggested, but a generalized, automated algorithm that can be used for a variety of applications has yet to be developed. Perhaps some lessons can be learned from the field of color photography (Evans et al., 1953; Hunt, 1995). There, optimum reproductions are thought to be those that preserve the hue of the original, that map lightness to preserve its relative reproduction and the mean level, and that map chroma such that the relationships between the relative chromas of various image elements are retained. Of course, such guidelines would be overruled by specific color preferences. Beginning with such approaches and considering other practical constraints, various researchers have discussed issues of color gamut mapping (Stone et al., 1988; Gentile et al., 1990b; Hoshino et al., 1993; Wolski et al., 1994; and Montag et al., 1996, 1997).

While a general solution to the gamut-mapping problem has not been derived, some fundamental concepts can be suggested. For pictorial images, a reasonable gamut-mapping solution can be obtained by first linearly scaling the lightnesses such that the white and black points match and the middle gray ($L^* = 50$) is kept constant. Next, hue is preserved as chroma is clipped to the gamut boundary for compression or linearly scaled for expansion. An alternative approach is to clip the out-of-gamut colors to the gamut boundary at a minimum distance in a uniform color space while eliminating the constant-hue constraint. Such approaches are likely to be too simplistic, and they do not produce the optimum results (e.g., Wolski et al., 1994; Montag et al., 1996, 1997).

For other image types, such as business graphics, other gamut-mapping strategies might be more appropriate. One such approach is to preserve the chroma of image elements while changing hue if necessary in order to retain the impact and intent of business-graphic images. These differences highlight the importance of understanding the intent for an image when making a reproduction. Depending on the intended application for a given image, the optimum gamut-mapping strategy will vary. The difference between pictorial images and business graphics is readily apparent. However, even among pictorial images, different gamut-mapping strategies might be more appropriate for various applications. For example, the best strategy for a scientific or medical image will differ from that for a fine-art reproduction, which will in turn differ from that for a consumer snapshot.

17.9   Color Preferences

Once the problems of color appearance and gamut mapping are solved, there will remain one last color operation—the mapping of colors to those that are preferred by observers for a given application. Thus accurate color reproduction might not be the ultimate goal, but rather a required step along the way. Color mapping for preference reproduction should be addressed simultaneously with gamut mapping, since the two processes deliberately force inaccurate color reproduction and will certainly impact each other. Like gamut mapping, color-preference mapping is intent-dependent, or application-dependent. In some applications, such as scientific and medical imaging, accurate color reproduction might be an objective that cannot be compromised. In pictorial imaging, preferred reproduction of certain object colors (e.g., sky, skin, foliage) might be biased toward the idealized memory color of these objects. In abstract images, such as business graphics, preferred color reproduction might depend more upon the device capabilities or intended message than on the colors of the original image.

An additional factor in color-preference reproduction is the cultural dependency of color preferences. It is well established in the color-reproduction industry that preferred color-reproduction systems sold into different cultures have different color capabilities and cannot be substituted between cultures without a loss in sales. While it seems certain that such cultural biases exist, they are not well documented (publicly) and their cause is not well understood. Many such effects have achieved the level of folklore and might exist only for historical reasons. For example, certain customers might have a strong preference for certain color-reproduction capabilities because they have grown accustomed to those properties and consider any change to be negative. Such biases are certainly cultural, but they are also learned responses. This illustrates the point that most, if not all, cultural biases in color-preference reproduction are learned in some way (this is the fundamental definition of culture). The topic of cultural biases is certainly an interesting one and worthy of additional research and exploration. It would be particularly interesting to see if such biases could be traced historically to see if they change with advances in communication and interchange of image information.

The concept of cultural dependency in color-preference reproduction sparks several interesting possibilities. However, there are also significant individual differences in color-preference reproduction as well. In fact, while there might be significant differences in color preference between cultures, it is almost certainly true that the range of color preferences of individuals within any given culture exceeds the differences between the mean levels. One needs only attempt to produce a single ideal image for two observers to understand the magnitude of such differences in preference.

17.10   Inverse Process

Thus far, the process of moving image data into the middle of the flow chart in Figure 17-1 has been described. Once all of the processing at this level is complete, the image data are conceptually in an abstract space that represents the appearances to be reproduced on the output image. At this point, the entire process must be reversed to move from the viewing-conditions-independent color space, to a traditional device-independent color space, to device coordinates, and ultimately to the reproduced image. This process highlights the importance of working in both the forward and reverse directions in order to successfully create reproductions. Clearly, the entire process is facilitated by the use of analytically invertible color appearance models and device characterizations. The main advantage is that such models allow the user to manipulate a setting on the imaging device or change the viewing conditions and still be able to recreate the process and produce an image in a reasonable amount of time. If the models must be iteratively inverted or recreated through exhaustive measurements, it might be completely impractical for a user to adjust any settings in order to obtain a desired result.

17.11   Example System

The previous discussions provide an overview of the process of device-independent color imaging. It is useful to examine an example of the results that can be obtained with such processes. The impact of various color appearance models in the chain was illustrated in Chapter 15 (Figures 15-4 through 15-6). Figure 17-3 illustrates the quality of color reproduction that can be obtained with high-quality device characterizations as described in the previous sections in comparison with the reproductions that would be obtained using the devices right out of the box with no additional calibration or characterization.

The example system consists of typical high-end image input, processing, display, and printing devices for the home-computer market. The images in the figure are synthesized representations of the colors that are actually obtained at various steps in the process. While the images are synthesized representations of the results, the colors are accurate representations of the results obtained in a real system. The original image is taken to be a photographic print of a Macbeth ColorChecker Chart (McCamy et al., 1976), as illustrated in the first row of the figure. The image is scanned using a 600-dpi flat-bed scanner with

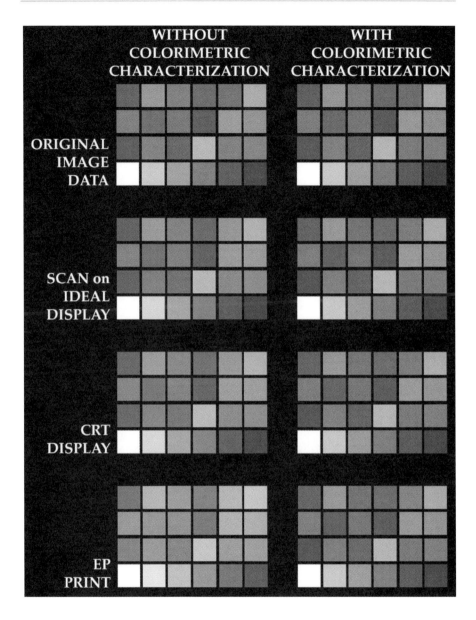

Figure 17-3. Examples of color-reproduction accuracy in a typical desktop imaging system without and with colorimetric calibration and characterization.

10-bits-per-channel quantization. The second row of the figure illustrates the accuracy of the scanned image as represented on a theoretically ideal display (i.e., the display introduces no additional error). The image on the left illustrates the result obtained with no colorimetric characterization (just gamma correction), while the image on the right illustrates the accuracy obtained with a characterization technique such as that outlined by Berns and Shyu (1995).

The next step involves display on a high-resolution CRT as illustrated in the third row of the figure. The left image illustrates the result of assuming the monitor and video driver are set up to the nominally defined system gamma (e.g., 1.8 for a Macintosh system, 1.3–1.5 for a Silicon Graphics system, 2.2–2.5 for a Windows system). It is assumed that any deviation from the nominal white point goes unnoticed due to chromatic adaptation. The image on the right illustrates the accuracy obtained with a specific characterization of the display systems using the techniques described by Berns (1996). Note that these images include both the scanner errors and the monitor errors as the full system is being constructed.

The final step is to print the image on a 600-dpi color laser printer. The image on the left illustrates the result obtained when the default printer driver and PostScript Printer Description (PPD) file is used. The image on the right illustrates the results obtained when a three-dimensional LUT is constructed using a measurement technique with a $9 \times 9 \times 9$ sampling of the gamut. These final images, too, illustrate errors that have been propagated through the entire imaging system. Clearly, careful colorimetric characterization of the three devices making up this system can result in significantly improved results. Unfortunately, the current state of technology does not allow typical users to achieve this level of colorimetric accuracy. However, the potential does exist. The ICC implementation described in the next section provides a framework. What remains to be implemented is the production of devices that can be accurately calibrated and characterized in the factory (and that remain stable) such that typical users will not have to be concerned about calibrating and characterizing the devices themselves.

17.12   ICC Implementation

The ICC (1996) has provided a framework for a more universal system to implement the process of device-independent color imaging, as illustrated in Figure 17-1, through their specification of the ICC profile format. The consortium consists of approximately 50 corporations and organizations involved in producing color-reproduction software, hardware, computer systems, and operating systems. The profile format is a specification of a data structure that can be used to describe device characterizations (both models and LUTs), viewing conditions, and rendering intent. Such profiles can be used in conjunction with various imaging devices or connected with images to facilitate communication of the colorimetric history of image data. The profile format provides a structure for communication of the required data such that the profiles can be easily interchanged among different computers, operating systems, and/or software applications. While the ICC profile provides the data necessary to implement device-independent color imaging, it is up to the software and hardware developers to build software to utilize this information to complete the system. Such software is referred to as a *color-management system* and is quickly becoming more and more integrated into operating systems. There is also a significant requirement for the development of profiles to accurately characterize various devices, appearance transformations, gamut-mapping transformations, and color-preference mappings. The quality of a system based on ICC profiles will depend on the capabilities of the color-management software and the quality of the profiles. With high-quality implementations, the ICC profile format specification provides the framework and potential for excellent results.

The construction and implementation of the ICC profile format and color-management systems and other compatible software are an evolving process. The current status and profile format documentation can be found at the ICC World Wide Web site, *http://www.color.org*. The ICC documents also contain information on other ongoing international standardization activities relevant to device-independent color-imaging applications.

Profile Connection Space

One important concept of the ICC specification is the profile connection space. This space is often misunderstood because its exact definition and implementation are still an issue of discussion and debate within ICC. For an up-to-date discussion of this topic, refer to the most recent ICC documentation.

Essentially, the profile connection space is defined by a particular set of viewing-condition parameters that are used to establish reference viewing conditions. The concept of the profile connection space is that a given input-device profile will provide the information necessary to transform device coordinates to a device-independent color specification (CIE *XYZ* or CIELAB) of the image data that represents the appearances of the original image data in the viewing conditions of the profile connection space (or from CIE specifications in profile connection space to device coordinates for an output-device profile). The technique for obtaining this transformation is not yet agreed upon. Here is the current definition of the ICC profile connection space reference viewing conditions:

> *Reference Reproduction Medium:* Idealized print with $D_{min}=0.0$
> *Reference Viewing Environment:* ANSI PH2.30 standard booth
> > *Surround:* Normal
> > *Illumination Color:* That of CIE illuminant D50
> > *Illuminance:* 2,200 ± 470 lux

Colorimetry: Ideal, flareless measurement
Observer: CIE 1931 Standard Colorimetric Observer (implied)
Measurement Geometry: Unspecified

An example of using ICC profiles with the profile connection space is a system with a digital camera calibrated for D65 illumination, a CRT display with a 9300-K white point, and a printer with output viewed under D50 illumination. An input profile for the digital camera would have to provide the information necessary to first transform from the camera device coordinates, say *RGB,* to CIE tristimulus values for illuminant D65 using typical device calibration and characterization techniques. The tristimulus values for illuminant D65 and the viewing conditions of image capture would then need to be transformed to corresponding tristimulus values for the profile connection space (illuminant D50, and so on) using some color appearance model. All of the information necessary to perform the transformation from device coordinates to corresponding tristimulus values in the profile connection space would have to be incorporated into the input-device profile.

The CRT display would have an output-device profile that would include the information necessary to transform from the profile connection space to corresponding tristimulus values for the CRT viewing conditions (9300-K white point, dim surround, and so on) and then through a device characterization to the *RGB* values required to display the desired colors. When implemented within a color-management system, the two device profiles would be concatenated such that the *RGB* data from the camera are transformed directly to the *RGB* data for the display without ever existing in any of the several intermediate spaces or, indeed, even in the profile connection space. Assuming the printer is set up and characterized for viewing conditions that match the profile connection space, the output-device profile for the printer needs to include only information for the transform from illuminant D50 CIE coordinates to the device coordinates.

An interesting "feature" of this process is that a device profile is required to provide the transformation from device coordinates to the profile connection space even for situations in which a color appearance model would normally not be required. For example, if the camera is set up for D65 illumination and the monitor is set up with a D65 white point, then it makes no sense to first transform through the appearance models to get to the D50 profile connection space and then come back out to a D65 display. Since profiles are concatenated by color-management systems, this is not a problem as long as a single color appearance model is agreed upon. The ICC is working toward this objective, but currently there is no single recommended color appearance model for the construction of ICC profiles. Thus an input-device-profile builder might implement the Hunt model to get into the profile connection space, while an output-device-profile builder might implement the RLAB model to come out of the profile connection space. Since there are significant differences between the various appearance models, processed images

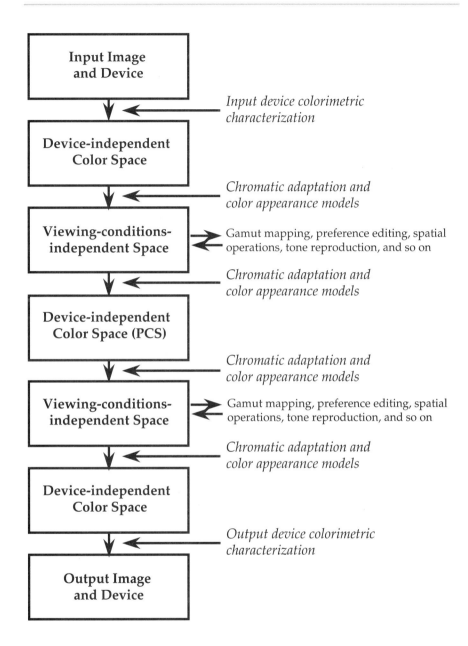

Figure 17-4. A revision of the flow chart in Figure 17-1 to accommodate the concept of a profile connection space as described by the ICC.

might change dramatically in situations for which a color appearance model was not even required. The latest ICC documentation should be examined to see how these and other issues have been addressed.

The concept of the profile connection space is completely compatible with the process described by Figure 17-1. The only addition required is a transformation out of the viewing-conditions-independent color space into a device-independent color space (CIE XYZ or CIELAB) for the viewing conditions of the profile connection space. At this point, the conceptual interchange of data from one device to the other occurs. Finally, a transformation from the device-independent color space for the profile connection space viewing conditions to the viewing-conditions-independent color space is made prior to adjustments for the viewing conditions of the output. This process is illustrated in Figure 17-4. The concept of the information processing is not changed. The profile connection space can be thought of as a virtual imaging device. This means that the original flow chart of Figure 17-1 is still used, but either the input device (for output situations) or the output device (for input situations) becomes the profile connection space "virtual device."

CHAPTER 18

The Future

18.1 Will There Be One Color Appearance Model?
18.2 Other Color Appearance Models
18.3 Ongoing Research to Test Models
18.4 Ongoing Model Development
18.5 What to Do Now

THE FORMULATION, TESTING, and application of color appearance models is an area of ongoing research. While the current status of this research makes it difficult to write a book on the topic, it is clear that enough of the fundamental background, concepts, and model formulation are fixed to warrant this more permanent and convenient record. Despite this, there is sure to be significant development in this field in the coming years. This chapter speculates on what might happen in the near future. For updates on the current status of various models and key references that appeared after publication of this book, refer to the associated World Wide Web site, *http://www.cis.rit.edu/people/faculty/ fairchild/* or *http://www.awl.com/cseng/*.

18.1 Will There Be One Color Appearance Model?

Will there ever be a single color appearance model that is universally accepted and used for all applications? Absolutely not! The problem is too complex to be solved for all imaginable applications by a single model. Even the specification of color differences continues to be performed with two color spaces, CIELAB and CIELUV. While CIELAB is clearly preferable for such applications, a single color-difference equation within the CIELAB space has yet to be widely accepted. There is no reason to expect the specification of a color appearance model to be any different.

The CIE, through TC1-34, is in the process of formulating a single color appearance model that will be designated as a CIE model and subjected to additional tests. It is hoped that such a model will promote uniformity of practice in industry and provide a significant step toward the establishment of a single, dominant technique for color-appearance specification. The status of the development of this model is described in Appendix A. Assuming the CIE activities are successful, the use of color appearance models should become simpler and more uniform. However, there is no question that, for specific applications, other models will continue to be used and developed. Perhaps the use of color appearance models will reach a status similar to today's specification of color-difference or color-order systems in which a small number of techniques are dominant, with a wider variety of techniques still being used in some specific applications.

18.2   Other Color Appearance Models

This book has concentrated on the Hunt, Nayatani et al., CIELAB, and RLAB models as the key available color appearance models. Also covered was the ATD model, with its unique historical perspective and formulation. LLAB was described as the most recently developed model. These models, along with the CIE model discussed in Appendix A, cover those that are likely to be considered for various applications in the foreseeable future. The probability of other color appearance models being published in the near future is low. The best possible outcome will be for all of those involved to contribute cooperatively to the development and comparative testing of the CIE model.

18.3   Ongoing Research to Test Models

Given the recent increase in interest in color appearance models and their application to practical problems, the amount of research dedicated to the evaluation of model performance has also grown. This research is being carried out by a variety of scientists in industry and academia through individual programs, product development, and the activities of CIE TC1-27 and TC1-34, as described in Chapter 15.

Of interest in the near future will be an evaluation of the success of color appearance models in real applications, such as device-independent color imaging through the ICC profile framework. The results of these "real-world" tests will set a practical standard for color appearance modeling. It is entirely possible that a very simple level of modeling, such as a von Kries transform, will suffice if the control and specification of viewing conditions and the accuracy of device characterizations are not first improved. These are necessary prerequisites to the application of color appearance models.

An analog for the future development of color appearance models can be found in the specification of the color-matching functions of the CIE 1931 Standard Colorimetric Observer. These functions were established nearly 70 years ago and have since been successfully used in industry. However, research on the measurement of more-accurate color-matching functions and the variability in those functions has continued. Such research has uncovered systematic errors in the CIE functions that are critical for some applications. In such applications, alternative color-matching functions are often used. However, none of the discrepancies found through this research to date has been significant enough to warrant the abandonment of the 1931 recommendation that is firmly, and successfully, entrenched in a variety of industries. Perhaps a similar path will be followed in the area of color appearance models. If the forthcoming CIE model finds wide acceptance and application, research on it and the other models presented in this book will continue. If an improvement that is truly significant for practical applications is found, it will be quickly adopted.

However, this might not happen until well after the other fundamental issues (characterization accuracy, viewing-condition control, and so on) are adequately addressed.

18.4   Ongoing Model Development

Development and formulation of color appearance models is continuing, although the rate of change seems to be diminishing. Perhaps this is an indication of a forthcoming relatively steady state in color appearance modeling. RLAB appears to have reached a stable level, with no forthcoming revisions expected. The Nayatani et al. model was recently revised, and ongoing work is aimed at incorporating empirical incomplete-adaptation factors into the model. The status of LLAB is uncertain. It is likely to remain somewhat stable in the form presented in Chapter 14. However, it was revised to its second version prior to the publication of the first version. Thus additional changes might be expected. The Hunt model is under revision; a new version might be published sometime in 1997. This new version will incorporate most of the features of the Hunt model, as described in Chapter 12, with the BFD chromatic-adaptation transform of LLAB. Hunt (1996), in collaboration with Luo, has developed the Bradford-Hunt 96 model, in simple and comprehensive forms, that is described in Appendix A. This model was submitted to TC1-34 for consideration as the starting point for the development of a CIE model. All of these models will continue to be subjected to an ever-widening array of tests.

An interesting direction that is likely to be pursued is the incorporation of spatial and temporal effects into color appearance models. Some of the issues to be addressed have been discussed by Wandell (1993), Poirson and Wandell (1993, 1996), and Bäuml and Wandell (1996). Another approach that is being investigated for the prediction of color appearance phenomena is the use of neural network models. Courtney et al. (1995a, b) presented interesting examples of such an approach.

18.5   What to Do Now

Given the current status of color-appearance specification, it is perfectly reasonable to ask, "What should I do now?" Some recommendations can be made based on the status of the models and tests described in this book. A point to remember is that a color appearance model should be used only when absolutely necessary. If the viewing conditions can be arranged to eliminate the need for an appearance transform, that is the best course of action. If a complete color appearance model is required, the Hunt model is probably the best choice. If that level of complexity is not required, RLAB might be a good choice. It is even possible that CIELAB might be adequate in some applications. The best recommendation is to work up this chain from the simplest solution to the higher levels of complexity until the problem is solved. It is also worth noting that models should not be "mixed and matched" due to the significant differences among them. A single model should be used throughout a given system or process. The following listing summarizes this recommendation in order of increasing complexity. Note that increasingly careful control of the viewing conditions is also required.

1. If possible, equate the viewing conditions such that simple tristimulus matches are also appearance matches.
2. If a white-point change is necessary, CIELAB can be used as a reasonable first-order approximation of an appearance model.
3. If CIELAB is found to be inadequate, it can be enhanced by using a von Kries chromatic-adaptation transform (on cone responses) to a reference viewing condition.
4. If a more flexible adaptation model is required (e.g., hard-copy to soft-copy changes) and/or there are surround changes, then RLAB (or ZLAB, described in Appendix A) can be used without too much added complexity.
5. If a full range of appearance phenomena and wide range of viewing conditions (e.g., very high or low luminances) must be addressed, then the Hunt model (or CIECAM97s, described in Appendix A) should be used. This model also should be used if brightness and colorfulness predictors are required (e.g., overall quality metrics for projected transparencies).

APPENDIX

The CIE Color Appearance Model (1997)

A.1   Historical Development, Objectives, and Approach
A.2   Bradford-Hunt 96S (Simple) Model
A.3   Bradford-Hunt 96C (Comprehensive) Model
A.4   The CIE TC1-34 Model, CIECAM97s
A.5   The ZLAB Color Appearance Model
A.6   Outlook

As STATED IN THE PREFACE to this book, there is a significant amount of interest in the establishment and use of a single, standardized color appearance model, but one has yet to be completely formulated. The industrial demand for such a model has led the CIE to step up its efforts to establish a model to be put into use, tested, and perhaps recommended as a standard in the coming years. This appendix is included to provide an update on the development of the CIE model as of the time this book went to press. Since the model is being developed by a committee, it is subject to change right up to the final moment it is published. To obtain a current update on the status of the CIE color appearance model, refer to the World Wide Web site associated with this book: *http://www.cis.rit.edu/people/faculty/fairchild/* or *http://www.awl.com/cseng/*.

A.1 Historical Development, Objectives, and Approach

In March 1996, the CIE held an expert symposium on *Colour Standards for Image Technology* in Vienna (CIE, 1996b). The symposium covered many aspects of image technology for which the CIE could provide guidance or standards to assist industry. But one of the most critical issues was the establishment of a color appearance model for general use. Industrial participants in the symposium recognized the need to apply a color appearance model. But they requested guidance from the CIE in establishing a single model that could be used throughout the industry to promote uniformity of practice and compatibility among various components in modern, open imaging systems.

The push toward a single model was highlighted and summarized in a presentation by Hunt at that symposium (CIE, 1996b) entitled, "The Function, Evolution, and Future of Colour Appearance Models." In that presentation, Hunt reviewed the current status and historical development of various models and presented 12 principles for consideration in establishing a single model. These principles are reproduced here verbatim (CIE, 1996b), as follows:

1. The model should be as comprehensive as possible, so that it can be used in a variety of applications; but at this stage, only static states of adaptation should be included, because of the great complexity of dynamic effects.
2. The model should cover a wide range of stimulus intensities, from very dark object colours to very bright self-luminous

colours. This means that the dynamic response function must have a maximum, and cannot be a simple logarithmic or power function.

3. The model should cover a wide range of adapting intensities, from very low scotopic levels, such as occur in starlight, to very high photopic levels, such as occur in sunlight. This means that rod vision should be included in the model; but because many applications will be such that rod vision is negligible, the model should be usable in a mode that does not include rod vision.

4. The model should cover a wide range of viewing conditions, including backgrounds of different luminance factors, and dark, dim, and average surrounds. It is necessary to cover the different surrounds because of their widespread use in projected and self-luminous displays.

5. For ease of use, the spectral sensitivities of the cones should be a linear transformation of the CIE $\bar{x}$, $\bar{y}$, $\bar{z}$ or $\bar{x}_{10}$, $\bar{y}_{10}$, $\bar{z}_{10}$ functions, and the $V'(\lambda)$ function should be used for the spectral sensitivity of the rods. Because scotopic photometric data is often unknown, methods of providing approximate scotopic values should be provided.

6. The model should be able to provide for any degree of adaptation between complete and none, for cognitive factors, and for the Helson-Judd effect, as options.

7. The model should give predictions of hue (both as hue-angle, and as hue-quadrature), brightness, lightness, saturation, chroma, and colourfulness.

8. The model should be capable of being operated in a reverse mode.

9. The model should be no more complicated than is necessary to meet the above requirements.

10. Any simplified version of the model, intended for particular applications, should give the same predictions as the complete model for some specified set of conditions.

11. The model should give predictions of colour appearance that are not appreciably worse than those given by the model that is best in each application.

12. A version of the model should be available for application to unrelated colours (those seen in dark surrounds in isolation from other colours).

The conclusion drawn at the symposium was that the CIE should immediately begin work on the formulation of such a model with the goal that it be completed prior to the International Colour Association (AIC) quadrennial meeting to be held in Kyoto in May 1997. The CIE decided that TC1-34 was the most appropriate committee to complete this work and expanded its terms of reference at the 1996 meeting of CIE Division 1 in Gothenburg to include:

> To recommend one colour appearance model. This model should give due consideration to the findings of other relevant Technical Committees.

TC1-34 immediately began work on the formulation of a CIE model (both simple and comprehensive versions). When complete, this model will then be subjected to further tests by the committee, as will the most recent versions of other published models. The committee then aims to prepare a final report on the performance of the CIE and other color appearance models during 1999.

TC1-34 members R. W. G. Hunt and M. R. Luo agreed to develop the first set of equations for consideration by the committee. The working philosophy of TC1-34 has been to essentially follow the 12 principles outlined by Hunt in the development of a comprehensive CIE model and a simplified version for practical applications. The general concept is to develop a comprehensive model (like the Hunt model) that can be applied to a wide range of color-appearance phenomena and a simplified version (like the RLAB model) that is sufficient for applications such as device-independent color imaging, with the additional constraint that the two versions of the model be compatible for some defined set of conditions.

During the preparation of these models, revised versions of the Hunt color appearance model were developed. These are referred to as the Bradford-Hunt 96S (simple) model and the Bradford-Hunt 96C (comprehensive) model. These models might be published by Hunt as the revised form of his model and have been kindly provided for inclusion in this book (Hunt, 1996). These models have not been approved by TC1-34 as the CIE model. However, they served as the starting point for the committee and provide good illustrations of how Hunt's 12 principles could be fulfilled. As expected, these models underwent some significant revision prior to consideration by the full committee. They are presented in the following sections so as to include the direction of revision for the Hunt model and to show the general form of the evolutionary process for the CIE model.

Hunt and Luo provided two revised models for TC1-34 consideration prior to the Kyoto meeting. M. Fairchild provided a third alternative and K. Richter provided a fourth. These four alternatives were considered at the May 1997 meeting of TC1-34 in Kyoto and an agreement was reached to adopt one of the Hunt and Luo alternatives as the simple form of the CIE 1997 Interim Color Appearance Model, tentatively designated CIECAM97s. This model is presented in Section A.4. It should be noted that the model has not yet undergone CIE ballot and publication and thus might be subject to minor changes prior to final approval. A comprehensive version of the model that extends upon CIECAM97s, to be designated CIECAM97c, will be formulated in the near future by TC1-34. A significantly simpler alternative with similar performance over a limited range of viewing conditions was prepared by

Fairchild, but it was not recommended by the committee, since it was not as extensible to a comprehensive form as the model selected to become CIECAM97s. This model has now been designated as the ZLAB color appearance model and is presented in section A.5, since it might prove useful in some simple image-reproduction applications.

A.2   Bradford-Hunt 96S (Simple) Model

This model is referred to as the *Bradford-Hunt model,* since it represents a combination of the Bradford chromatic-adaptation transform (see, e.g., Luo et al., 1996) and the Hunt (1994) model. The formulation of the Bradford-Hunt models includes features from the Hunt, LLAB, RLAB, and Nayatani et al. models, as was the aim of the CIE in developing a single model incorporating the best features of previously published models. It is useful to first describe the simple model and then describe the comprehensive model by just pointing out what features are added to the simple model.

Input Data

Following are the input data to the model:

- The luminance of the adapting field, L_A
- The tristimulus values of the sample in the source conditions, $X_s Y_s Z_s$
- The tristimulus values of the source white in the source conditions, $X_{ws} Y_{ws} Z_{ws}$
- The tristimulus values of the source background in the source conditions, $X_{bs} Y_{bs} Z_{bs}$
- The tristimulus values of the reference white in the reference conditions (usually taken to be the equal-energy illuminant), $X_{or} Y_{or} Z_{or}$

Chromatic Adaptation

The Bradford chromatic-adaptation transform is used to go from the source viewing conditions to the reference viewing conditions. First, all four sets of tristimulus values are normalized and transformed to "cone responses" using the Bradford transformation, as given in Equations A-1 and A-2:

$$\begin{bmatrix} R \\ G \\ B \end{bmatrix} = \mathbf{M}_{BFD} \begin{bmatrix} X/Y \\ Y/Y \\ Z/Y \end{bmatrix} \qquad \text{(A-1)}$$

$$\mathbf{M}_{BFD} = \begin{bmatrix} 0.8951 & 0.2664 & -0.1614 \\ -0.7502 & 1.7135 & 0.0367 \\ 0.0389 & -0.0685 & 1.0296 \end{bmatrix} \qquad \text{(A-2)}$$

The chromatic-adaptation transform is a modified von Kries transformation (performed on a type of chromaticity coordinates) with an exponential nonlinearity added to the short-wavelength sensitive channel, as given in Equations A-3 through A-6. In addition, the variable D is used to specify the degree of adaptation. D is set to 1.0 for complete adaptation or discounting-the-illuminant, to 0.0 for no adaptation, and to intermediate values for various degrees of incomplete chromatic adaptation.

$$R_r = \left[D\left(R_{or} / R_{ws} \right) + 1 - D \right] R_s \qquad \text{(A-3)}$$

$$G_r = \left[D\left(G_{or} / G_{ws} \right) + 1 - D \right] G_s \qquad \text{(A-4)}$$

$$B_r = \left[D\left(B_{or} / B_{ws}^p \right) + 1 - D \right] \left| B_s \right|^p \qquad \text{(A-5)}$$

$$p = \left(B_{ws} / B_{or} \right)^{0.0834} \qquad \text{(A-6)}$$

If B_s happens to be negative, then B_r is also set to be negative. Similar transformations are also made for the source white and the source background.

The next step is to calculate a luminance-level adaptation factor, as shown in Equations A-7 and A-8:

$$F_L = 0.2k^4 \left(5L_A \right) + 0.1 \left(1 - k^4 \right)^2 \left(5L_A \right)^{1/3} \qquad \text{(A-7)}$$

$$k = 1 / \left(5L_A + 1 \right) \qquad \text{(A-8)}$$

The reference tristimulus values for both the sample and the source white are then transformed from the Bradford "cone responses" to the Hunt-Pointer-Estevez cone responses, as shown in Equations A-9 and A-10, prior to the application of a nonlinear response compression. This is necessary because the Bradford responses are not physiologically plausible cone responsivities (they have negative values at many wavelengths).

$$\begin{bmatrix} R' \\ G' \\ B' \end{bmatrix} = \mathbf{M}_H \mathbf{M}_{BFD}^{-1} \begin{bmatrix} R_r Y_s F_L \\ G_r Y_s F_L \\ B_r Y_s F_L \end{bmatrix}$$

(A-9)

$$\mathbf{M}_H = \begin{bmatrix} 0.38971 & 0.68898 & -0.07868 \\ -0.22981 & 1.18340 & 0.04641 \\ 0.0 & 0.0 & 1.00 \end{bmatrix}$$

(A-10)

The relative luminance of the background is calculated as shown in Equation A-11 for achromatic backgrounds. If the background is not achromatic, Y_{br} and Y_{wr} must be explicitly calculated through the Bradford transform. The background relative luminance is used to calculate the induction factors given in Equations A-12 and A-13:

$$n = Y_{br} / Y_{wr} = Y_{bs} / Y_{ws}$$

(A-11)

$$N_{bb} = 0.725(1/n)^{0.2}$$

(A-12)

$$N_{cb} = 0.725(1/n)^{0.2}$$

(A-13)

The post-adaptation cone responses (for both the sample and the white) are then calculated using Equations A-14 through A-16:

$$R'_a = \frac{40(R'/100)^{0.73}}{\left[(R'/100)^{0.73} + 2\right]} + 1$$

(A-14)

$$G'_a = \frac{40(G'/100)^{0.73}}{\left[(G'/100)^{0.73} + 2\right]} + 1$$

(A-15)

$$B'_a = \frac{40(B'/100)^{0.73}}{\left[(B'/100)^{0.73} + 2\right]} + 1$$

(A-16)

Appearance Correlates

Preliminary red-green and yellow-blue opponent dimensions are calculated using Equations A-17 and A-18:

$$a' = R'_a - 12G'_a/11 + B'_a/11$$

(A-17)

$$b' = (1/9)(R'_a + G'_a - 2B'_a)$$

(A-18)

The hue angle, h, is then calculated from a' and b' using Equation A-19:

$$h = \tan^{-1}(b'/a')$$
(A-19)

Hue quadrature, H, and eccentricity factors, e, are calculated from the following unique hue data in the usual way (linear interpolation):

Red: $h = 20.14$, $e = 0.8$
Yellow: $h = 90.00$, $e = 0.7$
Green: $h = 164.25$, $e = 1.0$
Blue: $h = 237.53$, $e = 1.2$

The achromatic response is calculated as shown in Equation A-20 for both the sample and the white:

$$A = \left[2R'_a + G'_a + (1/20)B'_a\right]N_{bb}$$
(A-20)

Brightness, Q, for average, dim, and dark surrounds is calculated using Equations A-21 through A-23, respectively:

$$Q = 0.910A^{0.6}A_w^{0.5} - 0.167A_w$$
(A-21)

$$Q = 1.050A^{0.6}A_w^{0.5} - 0.112A_w$$
(A-22)

$$Q = 1.183A^{0.6}A_w^{0.5} - 0.081A_w$$
(A-23)

Lightness, J, is calculated from the brightnesses of the sample and white using Equation A-24:

$$J = 100(Q/Q_w)^{(1+n^{1/2})}$$
(A-24)

Final red-green and yellow-blue opponent responses are calculated using Equations A-25 and A-26 with $N_c = 1.0$ for average surrounds, 0.95 for dim surrounds, and 0.90 for dark surrounds:

$$a = 100a' e(10/13)N_c N_{cb}$$
(A-25)

$$b = 100b' e(10/13)N_c N_{cb}$$
(A-26)

Finally, saturation, s, chroma, C, and colorfulness, M, are calculated using Equations A-27 through A-29, respectively:

$$s = \left(\frac{50}{R'_a + G'_a + (21/20)B'_a}\right)\left(a^2 + b^2\right)^{1/2}$$
(A-27)

$$C = 2.44 s^{0.69} (Q/Q_w)^n (1.64 - 0.29^n)$$

(A-28)

$$M = CF_L^{0.15}$$

(A-29)

Inverse Model

The Bradford-Hunt 96S model can be analytically inverted. Beginning with lightness, J, chroma, C, and hue angle, h, the process is as follows:

1. From J obtain Q.
2. From Q obtain A.
3. From h obtain e.
4. Calculate s using C and Q.
5. Calculate a' and b' using s, h, and e.
6. Calculate R'_a, G'_a, and B'_a from A, a', and b'.
7. Calculate R', G', and B' from R'_a, G'_a, and B'_a.
8. Calculate $R_r Y_s F_L$, $G_r Y_s F_L$, and $B_r Y_s F_L$ from R', G', and B'.
9. Divide by F_L to obtain $R_r Y_s$, $G_r Y_s$, and $B_r Y_s$.
10. Calculate Y_r from $R_r Y_s$, $G_r Y_s$, and $B_r Y_s$.
11. Calculate $(Y_s/Y_r)R_s$, $(Y_s/Y_r)G_s$, and $(Y_s/Y_r)^{1/p}B_s$ from $(Y_s/Y_r)R_r$, $(Y_s/Y_r)G_r$, and $(Y_s/Y_r)B_r$.
12. Calculate Y_s' from $R_s Y_s$, $G_s Y_s$, and $(Y_s/Y_r)^{1/p}B_s Y_r$ (an approximation for the unavailable $B_s Y_s$).
13. Calculate X_s''/Y_r, Y_s''/Y_r, and Z_s''/Y_r from $(Y_s/Y_r)R_s$, $(Y_s/Y_r)G_s$, and $(Y_s/Y_r)^{1/p}B_s(Y_s'/Y_r)^{(1/p-1)}$.
14. Multiply by Y_r to obtain X_s'', Y_s'', and Z_s'', approximations to X_s, Y_s, and Z_s.

A.3   Bradford-Hunt 96C (Comprehensive) Model

The comprehensive model is based on the simple model, with the addition of the following:

- Cone bleach factors
- The Helson-Judd effect parameters
- The low-luminance tritanopia factor
- The rod contribution
- The Helmholtz-Kohlrausch effect prediction
- A whiteness-blackness option

Input Data

The input data are the same as for the Bradford-Hunt 96S model.

Chromatic Adaptation

The initial stages of the chromatic-adaptation model are identical to the Bradford-Hunt 96S model. The compressive nonlinearity functions of Equations A-14 through A-16 are extended to include cone bleach factors and the Helson-Judd effect, as shown in Equations A-30 through A-32. These additional factors are obtained exactly as described in the Hunt (1994) model given in Chapter 12.

$$R'_a = B_R\left[f_n(R'/100) + R'_D\right] + 1 \tag{A-30}$$

$$G'_a = B_G\left[f_n(G'/100) + G'_D\right] + 1 \tag{A-31}$$

$$B'_a = B_B\left[f_n(B'/100) + B'_D\right] + 1 \tag{A-32}$$

Appearance Correlates

The low-level tritanopia factor, F_t, calculated using Equation A-33, is applied to the yellow-blue opponent response:

$$F_t = L_A / (L_A + 0.1) \tag{A-33}$$

For the rod contribution to be included, Q_s is calculated using Equations A-21 through A-23 with scotopic achromatic responses for the sample and the white calculated as in the Hunt (1994) model given in Chapter 12. Total brightness, Q_T, is then calculated as the sum of the photopic and scotopic brightnesses, as given in Equation A-34:

$$Q_T = Q + Q_S \tag{A-34}$$

Lightness predictors, J_T and J_S, are calculated in the same way as in the simple model (Equation A-24). The scotopic lightness or brightness must be retained to facilitate inversion of the model. Since the rod response is included, model inversion becomes a task of solving for four unknowns, and therefore, four appearance correlates must be retained.

The Helmholtz-Kohlrausch effect can be predicted using Equation A-35 for lightness or Equations A-36 and A-37 for brightness (where $z = 1+n^{1/2}$):

$$J_{HK} = J + \left(|100 - J|\right)(C/300)\left(\sin|(h - 90)/2|\right)$$

(A-35)

$$Q_{HK} = \left[Q^z + \left(|Q_W^z - Q^z|\right)(C/300)\left(\sin|(h - 90)/2|\right)\right]^{1/z}$$

(A-36)

$$Q_{THK} = Q_{HK} + Q_S$$

(A-37)

The optional whiteness-blackness correlate is calculated using Equation A-38, where Q_b is the brightness of the background:

$$Q_{WB} = 20\left(Q^{0.7} - Q_b^{0.7}\right)$$

(A-38)

Inverse Model

The Bradford-Hunt 96C model is analytically invertible in a fashion similar to the Bradford-Hunt 96S model, as long as the scotopic lightness, J_s, or brightness, Q_s, is retained.

A.4   The CIE TC1-34 Model, CIECAM97s

Some slight, but important, revisions were made to the Bradford-Hunt 96S model to derive the model agreed upon by TC1-34 to become the CIECAM97s model (i.e., the simple version of the CIE 1997 Interim Color Appearance Model). These include a reformulation of the surround compensation to use power functions in order to avoid predictions of corresponding colors with negative CIE tristimulus values and a clear definition of the adaptation level factor, D.

It is important to note that the formulation of CIECAM97s builds upon the work of many researchers in the field of color appearance. This was a key issue in TC1-34's establishment of this model as the best of what is currently available. Various aspects of the model can be traced to the work of (in alphabetical order) Bartleson, Breneman, Estevez, Fairchild, Hunt, Lam, Luo, Nayatani, Rigg, Seim, and Valberg, among others. The comprehensive model, CIECAM97c, will be derived from the simple model, CIECAM97s, by adding features in a process similar to that used to derive the Bradford-Hunt 96C model from the 96S version.

Input Data

Following are the input data to the model:

- The luminance of the adapting field (normally taken to be 20% of the luminance of white in the adapting field), L_A
- The tristimulus values of the sample in the source conditions, XYZ
- The tristimulus values of the source white in the source conditions, $X_w Y_w Z_w$
- The relative luminance of the source background in the source conditions, Y_b

Also, the constants c, for the impact of surround; N_c, a chromatic-induction factor; F_{LL}, a lightness contrast factor; and F, a factor for degree of adaptation, must be selected according to the guidelines in the following chart.

Viewing Condition	c	N_c	F_{LL}	F
Average surround, samples subtending > 4°	0.69	1.0	0.0	1.0
Average surround	0.69	1.0	1.0	1.0
Dim surround	0.59	1.1	1.0	0.9
Dark surround	0.525	0.8	1.0	0.9
Cut-sheet transparencies	0.41	0.8	1.0	0.9

Chromatic Adaptation

An initial chromatic-adaptation transform is used to go from the source viewing conditions to the equal-energy-illuminant reference viewing conditions (although tristimulus values need never be expressed in the reference conditions). First, tristimulus values for both the sample and white are normalized and transformed to spectrally sharpened cone responses using the transformation given in Equations A-39 and A-40:

$$\begin{bmatrix} R \\ G \\ B \end{bmatrix} = \mathbf{M}_B \begin{bmatrix} X/Y \\ Y/Y \\ Z/Y \end{bmatrix} \tag{A-39}$$

$$\mathbf{M}_B = \begin{bmatrix} 0.8951 & 0.2664 & -0.1614 \\ -0.7502 & 1.7135 & 0.0367 \\ 0.0389 & -0.0685 & 1.0296 \end{bmatrix} \tag{A-40}$$

Then the chromatic-adaptation transform is a modified von Kries transformation (performed on a type of chromaticity coordinates) with an exponential nonlinearity added to the short-wavelength sensitive channel, as given in Equations A-3 through A-6. In addition, the variable D is used to specify the degree of adaptation. D is set to 1.0 for complete adaptation or discounting-the-illuminant and to 0.0 for no adaptation. It takes on intermediate values for various degrees of incomplete chromatic adaptation. Equation A-45 allows calculation of D for various luminance levels and surround conditions:

$$R_c = \left[D(1.0 / R_w) + 1 - D \right] R$$

(A-41)

$$G_c = \left[D(1.0 / G_w) + 1 - D \right] G$$

(A-42)

$$B_c = \left[D(1.0 / B_w^p) + 1 - D \right] B \Big|^p$$

(A-43)

$$p = (B_w / 1.0)^{0.0834}$$

(A-44)

$$D = F - F / \left[1 + 2 \left(L_A^{1/4} \right) + \left(L_A^2 / 300 \right) \right]$$

(A-45)

If B happens to be negative, then B_c is also set to be negative. Similar transformations are also made for the source white, since they are required in later calculations. Various factors must be calculated prior to further calculations, as shown in Equations A-46 and A-50. These include a background induction factor, n, the background and chromatic brightness induction factors, N_{bb} and N_{cb}, and the base exponential nonlinearity, z.

$$k = 1 / (5L_A + 1)$$

(A-46)

$$F_L = 0.2k^4 (5L_A) + 0.1(1 - k^4)^2 (5L_A)^{1/3}$$

(A-47)

$$n = Y_b / Y_w$$

(A-48)

$$N_{bb} = N_{cb} = 0.725(1/n)^{0.2}$$

(A-49)

$$z = 1 + F_{LL} n^{1/2}$$

(A-50)

The post-adaptation signals for both the sample and the source white are then transformed from the sharpened cone responses to the Hunt-Pointer-Estevez cone responses, as shown in Equations A-51 and A-52, prior to application of a nonlinear response compression:

$$\begin{bmatrix} R' \\ G' \\ B' \end{bmatrix} = \mathbf{M}_H \mathbf{M}_B^{-1} \begin{bmatrix} R_c Y \\ G_c Y \\ B_c Y \end{bmatrix}$$

(A-51)

$$\mathbf{M}_H = \begin{bmatrix} 0.38971 & 0.68898 & -0.07868 \\ -0.22981 & 1.18340 & 0.04641 \\ 0.0 & 0.0 & 1.00 \end{bmatrix}$$

(A-52)

The post-adaptation cone responses (for both the sample and the white) are then calculated using Equations A-53 through A-55:

$$R'_a = \frac{40(F_L R'/100)^{0.73}}{\left[(F_L R'/100)^{0.73} + 2\right]} + 1 \tag{A-53}$$

$$G'_a = \frac{40(F_L G'/100)^{0.73}}{\left[(F_L G'/100)^{0.73} + 2\right]} + 1 \tag{A-54}$$

$$B'_a = \frac{40(F_L B'/100)^{0.73}}{\left[(F_L B'/100)^{0.73} + 2\right]} + 1 \tag{A-55}$$

Appearance Correlates

Preliminary red-green and yellow-blue opponent dimensions are calculated using Equations A-56 and A-57:

$$a = R'_a - 12G'_a/11 + B'_a/11 \tag{A-56}$$

$$b = (1/9)(R'_a + G'_a - 2B'_a) \tag{A-57}$$

The hue angle, h, is then calculated from a and b using Equation A-58:

$$h = \tan^{-1}(b/a) \tag{A-58}$$

Hue quadrature, H, and eccentricity factors, e, are calculated from the following unique hue data in the usual way (linear interpolation):

Red: $h = 20.14$, $e = 0.8$, $H = 0$ or 400
Yellow: $h = 90.00$, $e = 0.7$, $H = 100$
Green: $h = 164.25$, $e = 1.0$, $H = 200$
Blue: $h = 237.53$, $e = 1.2$. $H = 300$

Equations A-59 and A-60 illustrate the calculation of e and H for arbitrary hue angles, where the quantities subscripted 1 and 2 refer to the unique hues with hue angles just below and just above the hue angle of interest:

$$e = e_1 + (e_2 - e_1)(h - h_1)/(h_2 - h_1) \tag{A-59}$$

$$H = H_1 + \frac{100(h - h_1)/e_1}{(h - h_1)/e_1 + (h_2 - h)/e_2}$$

(A-60)

The achromatic response is calculated as shown in Equation A-61 for both the sample and the white:

$$A = \left[2R'_a + G'_a + (1/20)B'_a - 2.05\right]N_{bb}$$

(A-61)

Lightness, J, is calculated from the achromatic signals of the sample and white using Equation A-62:

$$J = 100(A/A_w)^{cz}$$

(A-62)

Brightness, Q, is calculated from lightness and the achromatic for the white using Equation A-63:

$$Q = (1.24/c)(J/100)^{0.67}(A_w + 3)^{0.9}$$

(A-63)

Finally, saturation, s, chroma, C, and colorfulness, M, are calculated using Equations A-64 through A-66, respectively:

$$s = \frac{50\left(a^2 + b^2\right)^{1/2} 100e(10/13)N_c N_{cb}}{R'_a + G'_a (21/20)B'_a}$$

(A-64)

$$C = 2.44s^{0.69}(J/100)^{0.67n}(1.64 - 0.29^n)$$

(A-65)

$$M = CF_L^{0.15}$$

(A-66)

Inverse Model

The CIECAM97s model can be analytically inverted. Beginning with lightness, J, chroma, C, and hue angle, h, the process is as follows:

1. From J obtain A.
2. From h obtain e.
3. Calculate s using C and J.
4. Calculate a and b using s, h, and e.
5. Calculate R'_a, G'_a, and B'_a from A, a, and b.
6. Calculate R', G', and B' from R'_a, G'_a, and B'_a.
7. Calculate $R_c Y$, $G_c Y$, and $B_c Y$ from R', G', and B'.
8. Calculate Y from $R_c Y$, $G_c Y$, and $B_c Y$ using M_B^{-1}.
9. Calculate R_c, G_c, and B_c from $R_c Y$, $G_c Y$, and $B_c Y$ and Y.

10. Calculate R, G, and B from R_c, G_c, and B_c.
11. Calculate X, Y, and Z from R, G, B, and Y.

A.5   The ZLAB Color Appearance Model

A simple model was derived from the various models submitted to TC1-34 for consideration by the committee. Ultimately, the committee determined that a more extensible model was required for recommendation as CIECAM97s. Thus the simpler model was abandoned by the committee and has been renotated by the author as the ZLAB color appearance model. It is derived from the CIECAM97s, LLAB, and RLAB models with significant input from the work of Luo and Hunt submitted to TC1-34.

ZLAB is designed to perform nearly as well as the CIECAM97s model for a limited set of viewing conditions. These limitations include a restriction to intermediate values of adapting luminance, since the hyperbolic nonlinearity has been replaced with a square-root function that describes the hyperbolic function well for intermediate luminance levels. Also, ZLAB is limited to medium gray backgrounds in order to further simplify computation.

Lastly, ZLAB is limited to the prediction of the relative appearance attributes of lightness, chroma, saturation, and hue. It cannot be used to predict colorfulness or brightness. This is due to the removal of most of the luminance dependencies, thereby resulting in significantly simplified equations.

ZLAB performs identically to CIECAM97s for most corresponding-colors calculations, since it utilizes the same chromatic-adaptation transform. It also performs nearly as well for the prediction of appearance-scaling data. It is expected that ZLAB might find useful application in image-reproduction applications, where gamut mapping and lack of viewing-conditions control are major limiting factors.

Input Data

Following are the input data to the model:

- The luminance of the adapting field, L_A (taken to be 0.2 times the luminance of a reference white)
- The tristimulus values of the sample in the source conditions, XYZ
- The tristimulus values of the source white in the source conditions, $X_w Y_w Z_w$

Chromatic Adaptation

As with CIECAM97s, the Bradford chromatic-adaptation transform is used to go from the source viewing conditions to corresponding colors under the reference (equal-energy illuminant) viewing conditions. First, all three sets of tristimulus values are normalized and transformed to sharpened cone responses by using the Bradford transformation, as given in Equations A-67 and A-68:

$$\begin{bmatrix} R \\ G \\ B \end{bmatrix} = M \begin{bmatrix} X/Y \\ Y/Y \\ Z/Y \end{bmatrix} \tag{A-67}$$

$$M = \begin{bmatrix} 0.8951 & 0.2664 & -0.1614 \\ -0.7502 & 1.7135 & 0.0367 \\ 0.0389 & -0.0685 & 1.0296 \end{bmatrix} \tag{A-68}$$

The chromatic-adaptation transform is a modified von Kries transformation (performed on a type of chromaticity coordinates) with an exponential nonlinearity added to the short-wavelength sensitive channel, as given in Equations A-69 through A-72. In addition, the variable D is used to specify the degree of adaptation. D is set to 1.0 for complete adaptation or discounting-the-illuminant, to 0.0 for no adaptation, and to intermediate values for various degrees of incomplete chromatic adaptation. The D variable could be left as an empirical parameter or calculated using Equation A-73, as in CIECAM97s, with $F = 1.0$ for average surrounds and $F = 0.9$ for dim or dark surrounds. If Equation A-73 is used, it is the only place absolute luminance is required in ZLAB.

$$R_c = \left[D(1.0/R_w) + 1 - D \right] R \tag{A-69}$$

$$G_c = \left[D(1.0/G_w) + 1 - D \right] G \tag{A-70}$$

$$B_c = \left[D\left(1.0/B_w^p\right) + 1 - D \right] |B|^p \tag{A-71}$$

$$p = (B_w/1.0)^{0.0834} \tag{A-72}$$

$$D = F - F / \left[1 + 2\left(L_A^{1/4}\right) + \left(L_A^2/300\right) \right] \tag{A-73}$$

If B happens to be negative, then B_c is also set to be negative. R_c, G_c, and B_c represent the corresponding colors of the test stimulus under the reference condition (i.e., illuminant E).

The final step in the adaptation transform is to convert from the sharpened cone responses back to CIE XYZ tristimulus values for the reference condition, as illustrated in Equation A-74:

$$\begin{bmatrix} X_c \\ Y_c \\ Z_c \end{bmatrix} = \mathbf{M}^{-1} \begin{bmatrix} R_c Y \\ G_c Y \\ B_c Y \end{bmatrix}$$

(A-74)

Appearance Correlates

Opponent responses are calculated using modified CIELAB-type equations with the power-function nonlinearity defined by the surround relative luminances. These were derived from a simplification of the CIECAM97s model by recalling that the hyperbolic nonlinear function in CIECAM97s can be approximated by a square-root function for intermediate luminances. Thus the opponent responses reduce to the forms given in Equations A-75 and A-76:

$$A = 500\left[(X_c/100)^{1/2\sigma} - (Y_c/100)^{1/2\sigma}\right]$$

(A-75)

$$B = 200\left[(Y_c/100)^{1/2\sigma} - (Z_c/100)^{1/2\sigma}\right]$$

(A-76)

The exponents are directly related to those used in CIECAM97s, as illustrated in the following table. The values of $1/\sigma$ (called c) in CIECAM97s are modified to $1/2\sigma$ in ZLAB in order to incorporate the square-root approximation to the hyperbolic nonlinearity of CIECAM97s.

		Surround	
	Average	Dim	Dark
$1/\sigma$	0.69	0.59	0.525
$1/2\sigma$	0.345	0.295	0.2625

Hue angle is calculated in the typical manner, as illustrated in Equation A-77:

$$h^z = \tan^{-1}\left(\frac{B}{A}\right)$$

(A-77)

Hue composition is also determined in the usual way via linear interpolation between the defined angles for the unique hues. These are $h_r=25°$, $h_y=93°$, $h_g=165°$, and $h_b=254°$.

ZLAB is specified only for a background of medium (20%) luminance factor. Thus the z parameter in the CIECAM97s model takes on a constant value of 1.45 and lightness, L^z, is expressed as shown in Equation A-78:

$$L^z = 100(Y_c/100)^{1.45/2\sigma}$$

(A-78)

Chroma, C^z, is given by Equation A-79 as originally defined in LLAB to predict magnitude-estimation data well. Saturation, s^z, is simply the ratio of chroma to lightness, as illustrated in Equation A-80:

$$C^z = 25\log_e\left[1+0.05(A^2+B^2)^{1/2}\right]$$

(A-79)

$$s^z = C^z/L^z$$

(A-80)

If rectangular coordinates are required for color-space representations, they can easily be obtained from C^z and h^z using Equations A-81 and A-82:

$$a^z = C^z\cos(h^z)$$

(A-81)

$$b^z = C^z\sin(h^z)$$

(A-82)

Inverse Model

ZLAB is extremely simple to operate in the inverse direction. Starting with lightness, chroma, and hue angle, follow these steps:

1. Calculate $(A^2+B^2)^{1/2}$ from C^z.
2. Calculate A and B from $(A^2+B^2)^{1/2}$ and h^z.
3. Calculate X_c, Y_c, and Z_c from L^z, A, and B.
4. Calculate R_c, G_c, and B_c from X_c, Y_c, and Z_c.
5. Calculate R, G, and B from R_c, G_c, and B_c.
6. Calculate X, Y, and Z from R, G, and B.

A.6   Outlook

It is a truly unprecedented event that CIE TC1-34 was able to agree upon the derivation of the CIECAM97s model within its goal of 1 year. At the time this book went to press, the CIECAM97s model had not been formally reported and voted upon. The CIE approval procedures are not expected to introduce any significant changes to the model. However, it would be prudent to verify that the previous equations indeed made it unchanged into the final CIE publication of the model. (It is particularly feasible that some of the notation might have changed.)

The work of CIE TC1-34 is ongoing toward the goal of deriving a comprehensive version of the model and continuing to test these and other models. It is important to note that CIECAM97s and CIECAM97c are considered interim models. They will be revised if needed as more data and theoretical understanding become available. Industrial response to the CIECAM97s model has already been strong, and the model is expected to be quickly brought to bear on commercial applications, particularly in the imaging industry. Refer to updates on the World Wide Web at *http://www.awl.com/cseng/* or *http://www.cis.rit.edu/people/faculty/fairchild/* for the current status of the model.

References

I. Abramov, J. Gordon, and H. Chan, "Color appearance across the retina: Effects of a white surround," *J. Opt. Soc. Am. A* **9**, 195–202 (1992).

E. H. Adelson, "Perceptual organization and the perception of brightness," *Science* **2**, 2042–2044 (1993).

Adobe Systems Incorporated, *PostScript® Language Reference Manual*, 2nd Ed., Addison-Wesley, Reading, Mass. (1990).

J. Albers, *Interaction of Color*, Yale University Press, New Haven, Conn. (1963).

P. J. Alessi, "CIE guidelines for coordinated research on evaluation of colour appearance models for reflection print and self-luminous display comparisons," *Color Res. Appl.* **19**, 48–58 (1994).

R. L. Alfvin and M. D. Fairchild, "Observer variability in metameric color matches using color reproduction media," *Color Res. Appl.* **21**, 174–188 (1997).

D. H. Alman, R. S. Berns, G. D. Snyder, and W. A. Larson, "Performance testing of color-difference metrics using a color tolerance dataset," *Color Res. Appl.* **14**, 139–151 (1989).

M. Anderson, S. Chandrasekar, R. Motta, and M. Stokes, "Proposal for a standard color space for the Internet—sRGB," *IS&T/SID 4th Color Imaging Conference*, Scottsdale, Ariz., 238–246 (1996).

ANSI PH2.30-1989, *American National Standard for Graphic Arts and Photography—Color Prints, Transparencies, and Photomechanical Reproductions—Viewing Conditions*, American National Standards Institute, New York, N.Y. (1989).

L. E. Arend and A. Reeves, "Simultaneous color constancy," *J. Opt. Soc. Am. A* **3**, 1743–1751 (1986).

L. E. Arend and R. Goldstein, "Simultaneous constancy, lightness and brightness," *J. Opt. Soc. Am. A* **4**, 2281–2285 (1987).

L. E. Arend and R. Goldstein, "Lightness and brightness over spatial illumination gradients," *J. Opt. Soc. Am. A* **7**, 1929–1936 (1990).

L. E. Arend, A. Reeves, J. Schirillo, and R. Goldstein, "Simultaneous color constancy: Papers with diverse Munsell values," *J. Opt. Soc. Am. A* **8**, 661–672 (1991).

L. E. Arend, "How much does illuminant color affect unattributed colors?" *J. Opt. Soc. Am. A* **10**, 2134–2147 (1993).

ASTM, *Standard Terminology of Appearance*, E284–95a (1995).

ASTM, *Standard Guide for Designing and Conducting Visual Experiments*, E1808–96 (1996).

M. Ayama, T. Nakatsue, and P. K. Kaiser, "Constant hue loci of unique and binary balanced hues at 10, 100, and 1000 td," *J. Opt. Soc. Am. A* **4**, 1136–1144 (1987).

H. B. Barlow and J. D. Mollon, *The Senses*, Cambridge University Press, Cambridge, Mass. (1982).

C. J. Bartleson, "Memory colors of familiar objects," *J. Opt. Soc. Am.* **50**, 73–77 (1960).

C. J. Bartleson and E. J. Breneman, "Brightness perception in complex fields," *J. Opt. Soc. Am.* **57**, 953–957 (1967).

C. J. Bartleson, "Optimum Image Tone Reproduction," *J. SMPTE* **84**, 613–618 (1975).

C. J. Bartleson, "Brown," *Color Res. Appl.* **1**, 181–191 (1976).

C. J. Bartleson, "A review of chromatic adaptation," *AIC Proceedings, Color 77*, 63–96 (1978).

C. J. Bartleson and F. Grum, *Optical Radiation Measurements, Vol 5: Visual Measurements*, Academic Press, Orlando, Fla. (1984).

K.-H. Bäuml and B. A. Wandell, "Color appearance of mixture gratings," *Vision Res.* **36**, 2894–2864 (1996).

A. Berger–Schunn, *Practical Color Measurement*, John Wiley & Sons, New York, N.Y. (1994).

R. S. Berns, "The mathematical development of CIE TC1–29 proposed color difference equation: CIELCH," *AIC Proceedings, Color 93*, C19–1 (1993a).

R. S. Berns, "Spectral modeling of a dye diffusion thermal transfer printer," *J. Electronic Imaging* **2**, 359–370 (1993b).

R. S. Berns, R. J. Motta, and M. E. Gorzynski, "CRT colorimetry, part I: Theory and practice," *Color Res. Appl.* **18**, 299–314 (1993).

R. S. Berns, M. E. Gorzynski, and R. J. Motta, "CRT colorimetry, part II: Metrology," *Color Res. Appl.* **18**, 315–325 (1993).

R. S. Berns and M. J. Shyu, "Colorimetric characterization of a desktop drum scanner using a spectral model," *J. Electronic Imaging* **4**, 360–372 (1995).

R. S. Berns, "Methods for characterizing CRT displays," *Displays* **16**, 173–182 (1996).

K. T. Blackwell and G. Buchsbaum, "The effect of spatial and chromatic parameters on chromatic induction," *Color Res. Appl.* **13**, 166–173 (1988a).

K. T. Blackwell and G. Buchsbaum, "Quantitative studies in color constancy," *J. Opt. Soc. Am. A* **5**, 1772–1780 (1988b).

R. M. Boynton, *Human Color Vision*, Optical Society of America, Washington, D.C. (1979).

R. M. Boynton, "History and current status of a physiologically-based system of photometry and colorimetry," *J. Opt. Soc. Am. A* **13**, 1609–1621 (1996).

D. H. Brainard and B. A. Wandell, "Analysis of the retinex theory of color vision," *J. Opt. Soc. Am. A* **3**, 1651–1661 (1986).

D. H. Brainard and B. A. Wandell, "Asymmetric color matching: How color appearance depends on the illuminant," *J. Opt. Soc. Am. A* **9**, 1433–1448 (1992).

K. M. Braun and M. D. Fairchild, "Evaluation of five color-appearance transforms across changes in viewing conditions and media," *IS&T/SID 3rd Color Imaging Conference*, Scottsdale, Ariz., 93–96 (1995).

K. M. Braun, M. D. Fairchild, and P. J. Alessi, "Viewing environments for cross-media image comparisons," *Color Res. Appl.* **21**, 6–17 (1996).

K. M. Braun and M. D. Fairchild, "Testing five color appearance models for changes in viewing conditions," *Color Res. Appl.* **21**, 165–173 (1997a).

K. M. Braun and M. D. Fairchild, "Psychophysical generation of matching images for cross-media color reproduction," *J. Soc. Info. Disp.* **5**, submitted (1997b).

E. J. Breneman, "Corresponding chromaticities for different states of adaptation to complex visual fields," *J. Opt. Soc. Am. A* **4**, 1115–1129 (1987).

P. Bressan, "Revisitation of the luminance conditions for the occurrence of the achromatic neon color spreading illusion," *Perception & Psychophysics* **54**, 55–64 (1993).

M. H. Brill and G. West, "Chromatic adaptation and color constancy: A possible dichotomy," *Color Res. Appl.* **11**, 196–227 (1986).

M. E. Chevreul, *The Principles of Harmony and Contrast of Colors* (1839). Reprinted, Van Nostrand Reinhold, New York, N.Y., 1967.

E.-J. Chichilnisky and B. A. Wandell, "Photoreceptor sensitivity changes explain color appearance shifts induced by large uniform backgrounds in dichoptic matching," *Vision Res.* **35**, 239–254 (1995).

CIE, *Colorimetry*, CIE Publ. No. 15.2, Vienna (1986).

CIE, *International Lighting Vocabulary*, CIE Publ. No. 17.4, Vienna (1987).

CIE, *Special Metamerism Index: Change in Observer*, CIE Publ. No. 80, Vienna (1989).

CIE, *CIE 1988 2° Spectral Luminous Efficiency Function for Scotopic Vision*, CIE Publ. No. 86, Vienna (1990).

CIE, *A Method of Predicting Corresponding Colours under Different Chromatic and Illuminance Adaptations*, CIE Tech. Rep. 109, Vienna (1994).

CIE, *Method of Measuring and Specifying Colour Rendering Properties of Light Sources*, CIE Publ. No. 13.3, Vienna (1995a).

CIE, *Industrial Colour–Difference Evaluation*, CIE Tech. Rep. 116, Vienna (1995b).

CIE, *Report to CIE Division 1 from TC1–31 Colour Notations and Colour-Order Systems* (1996a).

CIE, *CIE Expert Symposium '96 Color Standards for Image Technology*, CIE Pub. x010, Vienna (1996b).

F. J. J. Clarke, R. McDonald, and B. Rigg, "Modification to the JPC 79 colour-difference formula," *J. Soc. Dyers Colourists* **100**, 128–132 (1984).

F. W. Cornelissen and E. Brenner, "On the role and nature of adaptation in chromatic induction," *Channels in the Visual Nervous System: Neurophysiology, Psychophysics and Models*, B. Blum, Ed., Freund Publishing, London, 109–123 (1991).

F. W. Cornelissen and E. Brenner, "Simultaneous colour constancy revisted: An analysis of viewing strategies," *Vision Res.* **35**, 2431–2448 (1995).

S. M. Courtney, L. H. Finkel, and G. Buchsbaum, "A multistage neural network for color constancy and color induction," *IEEE Trans. on Neural Networks* **6**, 972–985 (1995a).

S. M. Courtney, L. H. Finkel, and G. Buchsbaum, "Network simulations of retinal and cortical contributions to color constancy," *Vision Res.* **35**, 413–434 (1995b).

B. J. Craven and D. H. Foster, "An operational approach to color constancy," *Vision Res.* **32**, 1359–1366 (1992).

J. Davidoff, *Cognition Through Color*, MIT Press, Cambridge, Mass. (1991).

G. Derefeldt, "Colour appearance systems," Chapter 13 in *The Perception of Colour*, P. Gouras, Ed., CRC Press, Boca Raton, La., 218–261 (1991).

R. L. DeValois, C. J. Smith, S. T. Kitai, and S. J. Karoly, "Responses of single cells in different layers of the primate lateral geniculate nucleus to monochromatic light," *Science* **127**, 238–239 (1958).

M. S. Drew and G. D. Finlayson, "Device-independent color via spectral sharpening," *IS&T/SID 2nd Color Imaging Conference*, Scottsdale, Ariz., 121–126 (1994).

M. D'Zmura and P. Lennie, "Mechanisms of color constancy," *J. Opt. Soc. Am. A* **3**, 1662–1672 (1986).

P. G. Engeldrum, "Four color reproduction theory for dot formed imaging systems," *J. Imag. Tech.* **12**, 126–131 (1986).

P. G. Engeldrum, "Color scanner colorimetric design requirements," *Proc. SPIE* **1909**, 75–83 (1993).

P. G. Engeldrum, "A framework for image quality models," *J. Imag. Sci. Tech.* **39**, 312–318 (1995).

R. M. Evans, "Visual processes and color photography," *J. Opt. Soc. Am.* **33**, 579–614 (1943).

R. M. Evans, *An Introduction to Color*, John Wiley & Sons, New York, N.Y. (1948).

R. M. Evans, W. T. Hanson, and W. L. Brewer, *Principles of Color Photography*, John Wiley & Sons, New York, N.Y. (1953).

M. D. Fairchild, *Chromatic Adaptation and Color Appearance*, Ph.D. Dissertation, University of Rochester, N.Y. (1990).

M. D. Fairchild, "A model of incomplete chromatic adaptation," *Proceedings of the 22nd Session of the CIE*, Melbourne, 33–34 (1991a).

M. D. Fairchild, "Formulation and testing of an incomplete-chromatic-adaptation model," *Color Res. Appl.* **16**, 243–250 (1991b).

M. D. Fairchild and E. Pirrotta, "Predicting the lightness of chromatic object colors using CIELAB," *Color Res. Appl.* **16**, 385–393 (1991).

M. D. Fairchild, "Chromatic adaptation and color constancy," in *Advances in Color Vision Technical Digest*, Vol. 4 of the OSA Technical Digest Series (Optical Society of America, Washington, D.C.), 112–114 (1992a).

M. D. Fairchild, "Chromatic adaptation to image displays," *TAGA* **2**, 803–824 (1992b).

M. D. Fairchild and P. Lennie, "Chromatic adaptation to natural and artificial illuminants," *Vision Res.* **32**, 2077–2085 (1992).

M. D. Fairchild, "Chromatic adaptation in hard-copy/soft-copy comparisons," *Color Hard Copy and Graphic Arts II, Proc. SPIE* **1912**, 47–61 (1993a).

M. D. Fairchild, "Color Forum: The CIE 1931 Standard Colorimetric Observer: Mandatory retirement at age 65?" *Color Res. Appl.* **18**, 129–134 (1993b).

M. D. Fairchild and R. S. Berns, "Image color appearance specification through extension of CIELAB," *Color Res. Appl.* **18**, 178–190 (1993).

M. D. Fairchild, E. Pirrotta, and T. G. Kim, "Successive-*Ganzfeld* haploscopic viewing technique for color-appearance research," *Color Res. Appl.* **19**, 214–221 (1994).

M. D. Fairchild, "Some hidden requirements for device-independent color imaging," *SID International Symposium,* San Jose, Calif., 865–868 (1994a).

M. D. Fairchild, "Visual evaluation and evolution of the RLAB color space," *IS&T/SID 2nd Color Imaging Conference,* Scottsdale, Ariz., 9–13 (1994b).

M. D. Fairchild, "Testing colour-appearance models: Guidelines for coordinated research," *Color Res. Appl.* **20**, 262–267 (1995a).

M. D. Fairchild, "Considering the surround in device-independent color imaging," *Color Res. Appl.* **20**, 352–363 (1995b).

M. D. Fairchild and R. L. Alfvin, "Precision of color matches and accuracy of color matching functions in cross-media color reproduction," *IS&T/SID 3rd Color Imaging Conference*, Scottsdale, Ariz., 18–21 (1995).

M. D. Fairchild and L. Reniff, "Time-course of chromatic adaptation for color-appearance judgements," *J. Opt. Soc. Am. A* **12**, 824–833 (1995).

M. D. Fairchild, R. S. Berns, and A. A. Lester, "Accurate color reproduction of CRT-displayed images as projected 35mm slides," *J. Elec. Imaging* **5**, 87–96 (1996).

M. D. Fairchild, "Refinement of the RLAB color space," *Color Res. Appl.* **21**, 338–346 (1996).

M. D. Fairchild and L. Reniff, "A pictorial review of color appearance models," *IS&T/SID 4th Color Imaging Conference*, Scottsdale, Ariz., 97–100 (1996).

H. S. Fairman, "Metameric correction using parameric decomposition," *Color Res. Appl.* **12**, 261–265 (1987).

G. Fechner, *Elements of Psychophysics Vol. I* (trans. by H. E. Adler), Holt, Rinehart and Winston, New York, N.Y. (1966).

G. D. Finlayson, M. S. Drew, and B. V. Funt, "Spectral sharpening: Sensor transformations for improved color constancy," *J. Opt. Soc. Am. A* **11**, 1553–1563 (1994a).

G. D. Finlayson, M. S. Drew, and B.V. Funt, "Color constancy: Generalized diagonal transforms suffice," *J. Opt. Soc. Am. A* **11**, 3011–3019 (1994b).

D. J. Finney, *Probit Analysis,* 3rd Ed., Cambridge University Press, Cambridge, England (1971).

J. D. Foley, A. van Dam, S. K. Feiner, and J. F. Hughes, *Computer Graphics: Principles and Practice*, 2nd Ed., Addison-Wesley, Reading, Mass. (1990).

D. H. Foster and S. M. C. Nascimento, "Relational colour constancy from invariant cone-excitation ratios," *Proc. R. Soc. Lond. B* **257**, 115–121 (1994).

K. Fuld, J. S. Werner, and B. R. Wooten, "The possible elemental nature of brown," *Vision Res.* **23**, 631–637 (1983).

R. S. Gentile, E. Walowit, and J. P. Allebach, "Quantization multilevel halftoning of color images for near-original image quality," *J. Opt. Soc. Am. A* **7**, 1019–1026 (1990a).

R. S. Gentile, E. Walowit, and J. P. Allebach, "A comparison of techniques for color gamut mismatch compensation," *J. Imaging Tech.* **16**, 176–181 (1990b).

G. A. Gescheider, *Psychophysics: Method, Theory, and Application*, 2nd. Ed., Lawrence Erlbaum Associates, Hillsdale, N.J. (1985).

A. L. Gilchrist, "When does perceived lightness depend on perceived spatial arrangement," *Perception & Psychophysics* **28**, 527–538 (1980).

E. Giorgianni and T. Madden, *Digital Color Management: Encoding Solutions*, Addison-Wesley, Reading, Mass. (1997).

E. M. Granger, "ATD, appearance equivalence, and desktop publishing," *SPIE Vol. 2170*, 163–168 (1994).

E. M. Granger, "Gamut mapping for hard copy using the ATD color apace," *SPIE Vol. 2414*, 27–35 (1995).

F. Grum and C. J. Bartleson, *Optical Radiation Measurements Vol. 2, Color Measurement*, Academic Press, New York, N.Y. (1980).

J. Guild, "The colorimetric properties of the spectrum," *Phil. Trans. Roy. Soc. A* **230**, 149–187 (1931).

S. L. Guth, "Model for color vision and light adaptation," *J. Opt. Soc. Am. A* **8**, 976–993 (1991).

S. L. Guth, "ATD model for color vision I: Background," *SPIE Vol. 2170*, 149–152 (1994a).

S. L. Guth, "ATD model for color vision II: Applications," *SPIE Vol. 2170*, 153–168 (1994b).

S. L. Guth, "Further applications of the ATD model for color vision," *SPIE Vol. 2414*, 12–26 (1995).

H. Haneishi, T. Suzuki, N. Shimoyama, and Y. Miyake, "Color digital halftoning taking colorimetric color reproduction into account," *J. Elec. Imaging* **5**, 97–106 (1996).

A. Hard and L. Sivik, "NCS-Natural Color System: A Swedish standard for color notation," *Color Res. Appl.* **6**, 129–138 (1981).

M. M. Hayhoe, N. I. Benimoff, and D. C. Hood, "The time-course of multiplicative and subtractive adaptation processes," *Vision Res.* **27**, 1981–1996 (1987).

M. M. Hayhoe and M. V. Smith, "The role of spatial filtering in sensitivity regulation," *Vision Res.* **29**, 457–469 (1989).

H. v. Helmholtz, *Handbuch der physiologischen Optick*, 1st Ed. Voss, Hamburg (1866).

H. Helson, "Fundamental problems in color vision. I. The principle governing changes in hue, saturation, and lightness of non-selective samples in chromatic illumination," *J. Exp. Psych.* **23**, 439–477 (1938).

H. Helson, D. B. Judd, and M. H. Warren, "Object color changes from daylight to incandescent filament illumination," *Illum. Eng.* **47**, 221–233 (1952).

E. Hering, *Outlines of a theory of the light sense*, Harvard Univ. Press, Cambridge, Mass. (1920) (trans. by L. M. Hurvich and D. Jameson, 1964).

T. Hoshino and R. S. Berns, "Color gamut mapping techniques for color hard copy images," *Proc. SPIE* **1909**, 152–165 (1993).

P.-C. Hung, "Colorimetric calibration for scanners and media," *Proc. SPIE* **1448**, 164–174 (1991).

P.-C. Hung, "Colorimetric calibration in electronic imaging devices using a look-up-table model and interpolations," *J. Electronic Imaging* **2**, 53–61 (1993).

P.-C. Hung and R. S. Berns, "Determination of constant hue loci for a CRT gamut and their predictions using color appearance spaces," *Color Res. Appl.* **20**, 285–295 (1995).

D. M. Hunt, S. D. Kanwaljit, J. K. Bowmaker, and J. D. Mollon, "The chemistry of John Dalton's color blindness," *Science* **267**, 984–988 (1995).

R. W. G. Hunt, "The effects of daylight and tungsten light-adaptation on color perception," *J. Opt. Soc. Am.* **40**, 362–371 (1950).

R. W. G. Hunt, "Light and dark adaptation and the perception of color," *J. Opt. Soc. Am.* **42**, 190–199 (1952).

R. W. G. Hunt, "Objectives in Colour Reproduction," *J. Phot. Sci.* **18**, 205–215 (1970).

R. W. G. Hunt, I. T. Pitt, and L. M. Winter, "The preferred reproduction of blue sky, green grass and caucasian skin in colour photography," *J. Phot. Sci.* **22**, 144–150 (1974).

R. W. G. Hunt and L.M. Winter, "Colour adaptation in picture-viewing situations," *J. Phot. Sci.* **23**, 112–115 (1975).

R. W. G. Hunt, "Sky–blue pink," *Color Res. Appl.* **1**, 11–16 (1976).

R. W. G. Hunt, "The specification of colour appearance. I. Concepts and terms," *Color Res. Appl.* **2**, 55–68 (1977).

R. W. G. Hunt, "Colour terminology," *Color Res. Appl.* **3**, 79–87 (1978).

R. W. G. Hunt, "A model of colour vision for predicting colour appearance," *Color Res. Appl.* **7**, 95–112 (1982).

R. W. G. Hunt and M. R. Pointer, "A colour-appearance transform for the CIE 1931 Standard Colorimetric Observer," *Color Res. Appl.* **10**, 165–179 (1985).

R. W. G. Hunt, "A model of colour vision for predicting colour appearance in various viewing conditions," *Color Res. Appl.* **12**, 297–314 (1987).

R. W. G. Hunt, "Hue shifts in unrelated and related colours," *Color Res. Appl.* **14**, 235–239 (1989).

R. W. G. Hunt, *Measuring Colour*, 2nd Ed., Ellis Horwood, New York, N.Y. (1991a).

R. W. G. Hunt, "Revised colour-appearance model for related and unrelated colours," *Color Res. Appl.* **16**, 146–165 (1991b).

R. W. G. Hunt, "Standard sources to represent daylight," *Color Res. Appl.* **17**, 293–294 (1992).

R. W. G. Hunt, "An improved predictor of colourfulness in a model of colour vision," *Color Res. Appl.* **19**, 23–26 (1994).

R. W. G. Hunt and M. R. Luo, "Evaluation of a model of colour vision by magnitude scalings: Discussion of collected results," *Color Res. Appl.* **19**, 27–33 (1994).

R. W. G. Hunt, *The Reproduction of Colour*, 5th Ed., Fountain Press, England (1995).

R. W. G. Hunt, Personal Communication, October 14 (1996).

R. S. Hunter and R. W. Harold, *The Measurement of Appearance*, 2nd Ed., John Wiley & Sons, New York, N.Y. (1987).

L. M. Hurvich and D. Jameson, "A psychophysical study of white. III. Adaptation as a variant," *J. Opt. Soc. Am.* **41**, 787–801 (1951).

L. M. Hurvich, *Color Vision*, Sinauer Associates, Sunderland, Mass. (1981).

T. Indow, "Multidimensional studies of Munsell color solid," *Psych. Rev.* **95**, 456–470 (1988).

International Color Consortium, *ICC Profile Format Specification*, Version 3.3 (1996) (http://www.color.org).

D. Jameson and L. M. Hurvich, "Some quantitative aspects of an opponent-colors theory: I. Chromatic responses and spectral saturation," *J. Opt. Soc. Am.* **45**, 546–552 (1955).

D. Jameson and L. M. Hurvich, "Essay concerning color constancy," *Ann. Rev. Psychol.* **40**, 1–22 (1989).

J. F. Jarvis, C. N. Judice, and W. H. Ninke, "A survey of techniques for the display of continuous tone images on bilevel displays," *Comp. Graphics Image Proc.* **5**, 13–40 (1976).

E. W. Jin and S. K. Shevell, "Color memory and color constancy," *J. Opt. Soc. Am. A* **13**, 1981–1991 (1996).

D. J. Jobson, Z. Rahman, and G. A. Woodell, "A multi-scale retinex for bridging the gap between color images and the human observation of scenes," *IEEE Trans. Im. Proc.* **6**, 965–976 (1997).

D. B. Judd, "Hue, saturation, and lightness of surface colors with chromatic illumination," *J. Opt. Soc. Am.* **30**, 2–32 (1940).

D. B. Judd, "Appraisal of Land's work on two-primary color projections," *J. Opt. Soc. Am.* **50**, 254–268 (1960).

P. K. Kaiser and R. M. Boynton, *Human Color Vision,* 2nd Ed., Optical Society of America, Washington, D.C. (1996).

H. Kang, "Color scanner calibration," *J. Imaging Sci. Tech.* **36**, 162–170 (1992).

H. Kang, *Color Technology for Electronic Imaging Devices,* SPIE, Bellingham, Wash. (1997).

J. M. Kasson, W. Plouffe, and S. I. Nin, "A tetrahedral interpolation technique for color space conversion," *Proc. SPIE* **1909**, 127–138 (1993).

J. M. Kasson, S. I. Nin, W. Plouffe, and J. L. Hafner, "Performing color space conversions with three-dimensional linear interpolation," *J. Elec. Imaging* **4**, 226–250 (1995).

N. Katoh, "Practical method for appearance match between soft copy and hard copy," in *Device-Independent Color Imaging, SPIE Vol. 2170,* 170–181 (1994).

N. Katoh, "Appearance match between soft copy and hard copy under mixed chromatic adaptation," *IS&T/SID 3rd Color Imaging Conference,* Scottsdale, Ariz., 22–25 (1995).

D. Katz, *The World of Colour,* Trubner & Co., London (1935).

T. G. Kim, R. S. Berns, and M. D. Fairchild, "Comparing appearance models using pictorial images," *IS&T/SID 1st Color Imaging Conference,* Scottsdale, Ariz. 72–77 (1993).

J. J. Koenderink and W. A. Richards, "Why is snow so bright?" *J. Opt. Soc. Am. A* **9**, 643–648 (1992).

J. M. Kraft and J. S. Werner, "Spectral efficiency across the life span: Flicker photometry and brightness matching," *J. Opt. Soc. Am. A* **11**, 1213–1221 (1994).

J. B. Kruskal and M. Wish, *Multidimensional Scaling,* Sage Publications, Thousand Oaks, Calif. (1978).

W.-G. Kuo, M. R. Luo, and H. E. Bez, "Various chromatic-adaptation transformations tested using new colour appearance data in textiles," *Color Res. Appl.* **20**, 313–327 (1995).

I. Kuriki and K. Uchikawa, "Limitations of surface-color and apparent-color constancy," *J. Opt. Soc. Am. A* **13**, 1622–1636 (1996).

E. H. Land, "Color vision and the natural image part II," *Proc. Nat. Acad. Sci.* **45**, 636–644 (1959).

E. H. Land and J. J. McCann, "Lightness and the retinex theory," *J. Opt. Soc. Am.* **61**, 1–11 (1971).

E. H. Land, "The retinex theory of color vision," *Scientific American* **237**, 108–128 (1977).

E. H. Land, "Recent advances in retinex theory," *Vision Res.* **26**, 7–21 (1986).

P. Lennie and M. D'Zmura, "Mechanisms of color vision," *CRC Critical Reviews in Neurobiology* **3**, 333–400 (1988).

Y. Liu, J. Shigley, E. Fritsch, and S. Hemphill, "Abnormal hue-angle change of the gemstone tanzanite between CIE illuminants D65 and A in CIELAB color space," *Color Res. Appl.* **20**, 245–250 (1995).

M.-C. Lo, M. R. Luo, and P. A. Rhodes, "Evaluating colour models' performance between monitor and print images," *Color Res. Appl.* **21**, 277–291 (1996).

A. Logvinenko and G. Menshikova, "Trade-off between achromatic colour and perceived illumination as revealed by the use of pseudoscopic inversion of apparent depth," *Perception* **23**, 1007–1023 (1994).

M. R. Luo, A. A. Clarke, P. A. Rhodes, A. Schappo, S. A. R. Scrivner, and C. J. Tait, "Quantifying colour appearance. Part I. LUTCHI colour appearance data," *Color Res. Appl.* **16**, 166–180 (1991a).

M. R. Luo, A. A. Clarke, P. A. Rhodes, A. Schappo, S. A. R. Scrivner, and C. J. Tait, "Quantifying colour appearance. Part II. Testing colour models performance using LUTCHI color appearance data," *Color Res. Appl.* **16**, 181–197 (1991b).

M. R. Luo, X. W. Gao, P. A. Rhodes, H. J. Xin, A. A. Clarke, and S. A. R. Scrivner, "Quantifying colour appearance. Part III. Supplementary LUTCHI color appearance data," *Color Res. Appl.* **18**, 98–113 (1993a).

M. R. Luo, X. W. Gao, P. A. Rhodes, H. J. Xin, A. A. Clarke, and S. A. R. Scrivner, "Quantifying colour appearance. Part IV. Transmissive media," *Color Res. Appl.* **18**, 191–209 (1993b).

M. R. Luo, X. W. Gao, and S. A. R. Scrivner, "Quantifying colour appearance. Part V. Simultaneous contrast," *Color Res. Appl.* **20**, 18–28 (1995).

M. R. Luo and J. Morovic, "Two unsolved issues in colour management—colour appearance and gamut mapping," *5th International Conference on High Technology*, Chiba, Japan, 136–147 (1996).

M. R. Luo, M.-C. Lo, and W.-G. Kuo, "The LLAB(l:c) colour model," *Color Res. Appl.* **21**, 412–429 (1996).

D. L. MacAdam, "Chromatic adaptation," *J. Opt. Soc. Am.* **46**, 500–513 (1956).

D. L. MacAdam, "A nonlinear hypothesis for chromatic adaptation," *Vis. Res.* **1**, 9–41 (1961).

D. L. MacAdam, "Uniform color scales," *J. Opt. Soc. Am.* **64**, 1691–1702 (1974).

D. L. MacAdam, "Colorimetric data for samples of the OSA uniform color scales," *J. Opt. Soc. Am.* **68**, 121–130 (1978).

D. L. MacAdam, Ed., *Selected Papers on Colorimetry—Fundamentals,* SPIE Milestone Series, Vol. MS 77, SPIE, Bellingham, Wash. (1993).

L. T. Maloney and B. A. Wandell, "Color constancy: A method for recovering surface spectral reflectance," *J. Opt. Soc. Am.* A **3**, 29–33 (1986).

R. Mausfeld and R. Niederée, "An inquiry into relational concepts of colour, based on incremental principles of colour coding for minimal relational stimuli," *Perception* **22**, 427–462 (1993).

B. Maximus, A. De Metere, and J. P. Poels, "Influence of thickness variations in LCDs on color uniformity," *SID 94 Digest,* 341–344 (1994).

J. C. Maxwell, "On the theory of three primary colors," *Proc. Roy. Inst.* **3**, 370–375 (1858–62).

C. S. McCamy, H. Marcus, and J. G. Davidson, "A color rendition chart," *J. App. Phot. Eng.* **11**, 95–99 (1976).

J. McCann, "Color sensations in complex images," *IS&T/SID 1st Color Imaging Conference,* Scottsdale, Ariz., 16–23 (1993).

E. D. Montag and M. D. Fairchild, "Simulated color gamut mapping using simple rendered images," *Proc. SPIE* **2658**, 316–325 (1996).

E. D. Montag and M. D. Fairchild, "Evaluation of gamut mapping techniques using simple rendered images and artificial gamut boundaries," *IEEE Transactions on Image Processing* **6**, 977–989 (1997).

L. Mori and T. Fuchida, "Subjective evaluation of uniform color spaces used for color-rendering specification," *Color Res. Appl.* **7**, 285–293 (1982).

L. Mori, H. Sobagaki, H. Komatsubara, and K. Ikeda, "Field trials on CIE chromatic adaptation formula," *Proceedings of the CIE 22nd Session,* Melbourne, Australia, 55–58 (1991).

Y. Nayatani, K. Takahama, and H. Sobagaki, "Estimation of adaptation effects by use of a theoretical nonlinear model," *Proceedings of the 19th CIE Session,* Kyoto, Japan (1979), CIE Publ. No. 5, 490–494 (1980).

Y. Nayatani, K. Takahama, and H. Sobagaki, "Formulation of a nonlinear model of chromatic adaptation," *Color Res. Appl.* **6**, 161–171 (1981).

Y. Nayatani, K. Takahama, H. Sobagaki, and J. Hirono, "On exponents of a nonlinear model of chromatic adaptation," *Color Res. Appl.* **7**, 34–45 (1982).

Y. Nayatani, K. Takahama, and H. Sobagaki, "Prediction of color appearance under various adapting conditions," *Color Res. Appl.* **11**, 62–71 (1986).

Y. Nayatani, K. Hashimoto, K. Takahama, and H. Sobagaki, "A nonlinear color-appearance model using Estévez-Hunt-Pointer primaries," *Color Res. Appl.* **12**, 231–242 (1987).

Y. Nayatani, K. Takahama, and H. Sobagaki, "Field trials on color appearance of chromatic colors under various light sources," *Color Res. Appl.* **13**, 307–317 (1988).

Y. Nayatani, K. Takahama, H. Sobagaki, and K. Hashimoto, "Color-appearance model and chromatic adaptation transform," *Color Res. Appl.* **15**, 210–221 (1990).

Y. Nayatani, T. Mori, K. Hashimoto, K. Takahama, and H. Sobagaki, "Comparison of color-appearance models," *Color Res. Appl.* **15**, 272–284 (1990).

Y. Nayatani, Y. Gomi, M. Kamei, H. Sobagaki, and K. Hashimoto, "Perceived lightness of chromatic object colors including highly saturated colors," *Color Res. Appl.* **17**, 127–141 (1992).

Y. Nayatani, "Revision of chroma and hue scales of a nonlinear color-appearance model," *Color Res. Appl.* **20**, 143–155 (1995).

Y. Nayatani, H. Sobagaki, K. Hashimoto, and T. Yano, "Lightness dependency of chroma scales of a nonlinear color-appearance model and its latest formulation," *Color Res. Appl.* **20**, 156–167 (1995).

Y. Nayatani, "A simple estimation method for effective adaptation coefficient," *Color Res. Appl.* **22**, submitted (1997).

S. M. Newhall, "Preliminary report of the O. S. A. subcommittee on the spacing of the Munsell colors," *J. Opt. Soc. Am.* **30**, 617–645 (1940).

D. Nickerson, "History of the Munsell Color System and its scientific application," *J. Opt. Soc. Am.* **30**, 575–586 (1940).

D. Nickerson, "History of the Munsell Color System, Company, and Foundation, I.," *Color Res. Appl.* **1**, 7–10 (1976a).

D. Nickerson, "History of the Munsell Color System and its scientific application," *Color Res. Appl.* **1**, 69–77 (1976b).

D. Nickerson, "History of the Munsell Color System," *Color Res. Appl.* **1**, 121–130 (1976c).

T. H. Nilsson and T. M. Nelson, "Delayed monochromatic hue matches indicate characteristics of visual memory," *J. Exp. Psych.: Human Perception and Performance* **7**, 141–150 (1981).

I. Nimeroff, J. R. Rosenblatt, and M. C. Dannemiller, "Variability of spectral tristimulus values," *J. Res. NBS* **65**, 475–483 (1961).

OSA, "Psychological concepts: Perceptual and affective aspects of color," Chapter 5 in *The Science of Color,* Optical Society of America, Washington, D.C., 145–171 (1963).

H. Pauli, "Proposed extension of the CIE recommendation on 'Uniform color spaces, color difference equations, and metric color terms,'" *J. Opt. Soc. Am.* **36**, 866–867 (1976).

E. Pirrotta and M. D. Fairchild, "Directly testing chromatic-adaptation models using object colors," *Proceedings of the 23rd Session of the CIE,* New Delhi, Vol. 1, 77–78 (1995).

A. B. Poirson and B. A. Wandell, "Appearance of colored patterns: Pattern-color separability," *J. Opt. Soc. Am. A* **10**, 2458–2470 (1993).

A. B. Poirson and B. A. Wandell, "Pattern-color separable pathways predict sensitivity to simple colored patterns," *Vision Res.* **36**, 515–526 (1996).

J. Pokorny, V. C. Smith, and M. Lutze, "Aging of the human lens," *Appl. Opt.* **26**, 1437–1440 (1987).

D. M. Purdy, "Spectral hue as a function of intensity," *Am. J. Psych.* **43**, 541–559 (1931).

K. Richter, "Cube-root color spaces and chromatic adaptation," *Color Res. Appl.* **5**, 7–11 (1980).

K. Richter, *Farbempfindungsmerkmal Elementarbuntton und Buntheitsabstände als Funktion von Farbart und Leuchtdichte von In– und Umfeld,* Bundesanstalt für Materialprüfung (BAM) Forschungsbericht 115, Berlin (1985).

M. Richter and K. Witt, "The story of the DIN color system," *Color Res. Appl.* **11,** 138–145 (1986).

A. R. Robertson, "A new determination of lines of constant hue," *AIC Color 69,* Stockholm, 395–402 (1970).

A. R. Robertson, "The CIE 1976 color-difference formulae," *Color Res. Appl.* **2,** 7–11 (1977).

A. R. Robertson, "Historical development of CIE recommended color difference equations," *Color Res. Appl.* **15,** 167–170 (1990).

A. R. Robertson, Figure 6-2 presented at the 1996 ISCC Annual Meeting, Orlando, Fla. (1996).

M. A. Rodriguez and T. G. Stockham, "Producing colorimetric data from densitometric scans," *Proc. SPIE* **1913,** 413–418 (1993).

R. Rolleston and R. Balasubramanian, "Accuracy of various types of Neugebauer Model," *Proceedings IS&T/SID Color Imaging Conference,* Scottsdale, Ariz., 32–37 (1993).

B. E. Schefrin and J. S. Werner, "Age-related changes in the color appearance of broadband surfaces," *Color Res. Appl.* **18,** 380–389 (1993).

J. Schirillo, A. Reeves, and L. Arend, "Perceived lightness, but not brightness, of achromatic surfaces depends on perceived depth information," *Perception & Psychophysics* **48,** 82–90 (1990).

J. Schirillo and S. K. Shevell, "Lightness and brightness judgments of coplanar retinally noncontiguous surfaces," *J. Opt. Soc. Am. A* **10,** 2442–2452 (1993).

J. Schirillo and L. Arend, "Illumination changes at a depth edge can reduce lightness constancy," *Perception & Psychophysics* **57,** 225–230 (1995).

J. Schirillo and S. K. Shevell, "Brightness contrast from inhomogeneous surrounds," *Vision Res.* **36,** 1783–1796 (1996).

T. Seim and A. Valberg, "Towards a uniform colorspace: A better formula to describe the Munsell and OSA Color Scales," *Color Res. Appl.* **11,** 11–24 (1986).

C. C. Semmelroth, "Prediction of lightness and brightness on different backgrounds," *J. Opt. Soc. Am.,* **60,** 1685–1689 (1970).

G. Sharma and H. J. Trussel, "Digital color imaging," *IEEE Trans. Im. Proc.* **6,** 901–932 (1997).

S. K. Shevell, "The dual role of chromatic backgrounds in color perception," *Vision Res.* **18,** 1649–1661 (1978).

S. K. Shevell, "Color and brightness: Contrast and context," *IS&T/SID 1st Color Imaging Conference,* Scottsdale, Ariz., 11–15 (1993).

J. M. Speigle and D. H. Brainard, "Is color constancy task independent?" *IS&T/SID 4th Color Imaging Conference,* Scottsdale, Ariz., 167–172 (1996).

L. Spillman and J. S. Werner, *Visual Perception: The Neurophysiological Foundations,* Academic Press, San Diego, Calif. (1990).

R. Stanziola, "The Colorcurve System®," *Color Res. Appl.* **17,** 263–272 (1992).

S. S. Stevens, "To honor Fechner and repeal his law," *Science* **133,** 80–86 (1961).

J. C. Stevens and S. S. Stevens, "Brightness functions: Effects of adaptation," *J. Opt. Soc. Am.* **53**, 375–385 (1963).

W. S. Stiles and J. M. Burch, "N. P. L. colour-matching investigation: Final report (1958)," *Optica Acta* **6**, 1–26 (1959).

M. Stokes, M. Fairchild, and R. S. Berns, "Precision requirements for digital color reproduction," *ACM Trans. Graphics* **11**, 406–422 (1992).

M. C. Stone, W. B. Cowan, and J. C. Beatty, "Color gamut mapping and the printing of digital images," *ACM Trans. Graphics* **7**, 249–292 (1988).

G. Svaetichin, "Spectral response curves from single cones," *Acta Physiologica Scandinavica* **39** (Suppl. 134), 17–46 (1956).

K. Takahama, H. Sobagaki, and Y. Nayatani, "Analysis of chromatic adaptation effect by a linkage model," *J. Opt. Soc. Am.* **67**, 651–656 (1977).

K. Takahama, H. Sobagaki, and Y. Nayatani, "Formulation of a nonlinear model of chromatic adaptation for a light-gray background," *Color Res. Appl.* **9**, 106–115 (1984).

R. Taya, W. H. Ehrenstein, and C. R. Cavonius, "Varying the strength of the Munker-White effect by stereoscopic viewing," *Perception* **24**, 685–694 (1995).

H. Terstiege, "Chromatic adaptation: A state-of-the-art report," *J. Col. & Appear.* **1**, 19–23 (1972).

L. L. Thurstone, "A law of comparative judgment," *Psych. Review* **34**, 273–286 (1927).

L. L. Thurstone, *The Measurement of Values*, University of Chicago Press, Chicago, Ill. (1959).

W. S. Torgerson, "A law of categorical judgment," in *Consumer Behavior*, L. H. Clark, Ed., New York University Press, New York, N.Y., 92–93 (1954).

W. S. Torgerson, *Theory and Methods of Scaling*, John Wiley & Sons, New York, N.Y. (1958).

A. Valberg and B. Lange-Malecki, "'Colour constancy' in Mondrian patterns: A partial cancellation of physical chromaticity shifts by simultaneous contrast," *Vision Res.* **30**, 371–380 (1990).

J. von Kries, "Chromatic adaptation," *Festschrift der Albrecht–Ludwig–Universität* (Fribourg) (1902) [Translation: D. L. MacAdam, *Sources of Color Science*, MIT Press, Cambridge, Mass. (1970).]

J. Walraven, "Discounting the background—the missing link in the explanation of chromatic induction," *Vision Res.* **16**, 289–295 (1976).

B. A. Wandell, "Color appearance: The effects of illumination and spatial pattern," *Proc. Natl. Acad. Sci. USA* **90**, 9778–9784 (1993).

B. A. Wandell, *Foundations of Vision*, Sinauer, Sunderland, Mass. (1995).

M. A. Webster and J. D. Mollon, "The influence of contrast adaptation on color appearance," *Vision Res.* **34**, 1993–2020 (1994).

J. S. Werner and B. E. Schefrin, "Loci of achromatic points throughout the life span," *J. Opt. Soc. Am. A* **10**, 1509–1516 (1993).

D. R. Williams, N. Sekiguchi, W. Haake, D. Brainard, and O. Packer, "The cost of trichromacy for spatial vision," in *From Pigments to Perception*, A. Valberg and B. B. Lee, Eds., Plenum Press, New York, N.Y., 11–22 (1991).

M. Wolski, J. P. Allebach, and C. A. Bouman, "Gamut mapping: Squeezing the most out of your color system," *IS&T/SID 2nd Color Imaging Conference,* Scottsdale, Ariz., 89–92 (1994).

W. D. Wright, "A re-determination of the trichromatic coefficients of the spectral colours," *Trans. Opt. Soc.* **30,** 141–161 (1928–29).

W. D. Wright, "Why and how chromatic adaptation has been studied," *Color Res. Appl.* **6,** 147–152 (1981a).

W. D. Wright, "50 years of the 1931 CIE standard observer for colorimetry," *AIC Color 81,* Paper A3 (1981b).

G. Wyszecki, "Current developments in colorimetry," *AIC Color 73,* 21–51 (1973).

G. Wyszecki and W. S. Stiles, *Color Science: Concepts and Methods, Quantitative Data and Formulae,* John Wiley & Sons, New York, N.Y. (1982).

G. Wyszecki, "Color appearance," Chapter 9 in *Handbook of Perception and Human Performance,* John Wiley & Sons, New York, N.Y. (1986).

X. Zhang and B. A. Wandell, "A spatial extension of CIELAB for digital color image reproduction," *SID 96 Digest* (1996).

Index

Numbers followed by the letter f indicate figures; numbers followed by the letter t indicate tables.

Abney effect, 140–141, 141f, 142f
Absorptance, 73, 74f
Achromatic color, defined, 101
Adaptation
  chromatic. *See* Chromatic adaptation
  and chromatic induction, 135, 136f
  dark, 23, 25, 25f, 176–177
  light, 26–27, 176
  mechanisms of, 25–26
  motion, 186
  spatial frequency, 185–186, 185f
Additive mixtures, 113
Additivity, Grassmann's law of, 82
Adjustment, method of, 51
Advanced colorimetry, 63–64
Afterimages, 179–180, 179f
Albers, Josef, 135
Amacrine cells, of retina, 8
Aqueous humor, 5
Art, color-order systems in, 126
ASTM Standard Guide for Designing and Conducting Visual Experiments, 43
Asymmetric matching, 54, 190
ATD color appearance model
  adaptation model for, 287–289
  brightness equation in, 290
  chroma equation in, 290
  data for, 287
  example of, 321f, 322
  hue equation in, 290
  limitations of, 291–292
  objectives and approach of, 286–287
  opponent-color dimensions in, 289
  perceptual correlates in, 290
  predictions of, 290–291, 291t
  saturation equation in, 290
  testing of, 306t
Axons, defined, 7

Background, defined, 164–165
Bartleson-Breneman equations, 149–150, 150f
Basic colorimetry, x, 63
Bezold-Brücke hue shift, 139–140, 139f
Bidirectional reflectance distribution functions (BRDF), 74–75
Bidirectional transmittance distribution functions, 74–75
Black-body radiator, 68
Blackness, NCS value, 121
Blind spot, 8, 15–16
Bradford-Hunt 96C color appearance model
  appearance correlates in, 383–384
  chromatic adaptation in, 383
  data for, 383
  invertibility of, 384
Bradford-Hunt 96s color appearance model, 371
  appearance correlates in, 380–382
  chromatic adaptation in, 378–380
  data for, 378
  invertibility of, 382
Brain
  visual area 1 of, 19–21
  and visual signal processing, 16, 20f
Brightness, 102f
  chromaticity and, 141–143
  in color appearance, 107–109
  defined, 101
  luminance and, 141–143

Cameras, digital, 350
Category scaling, 55
Characterization
  approaches to, 348–351
  defined, 348
  goals of, 352
Chroma, 104
  in color appearance, 107–109
  defined, 103
  equation for, 106
  Munsell, 117, 118f
Chromatic adaptation, 26–28, 28f, 29f
  cognitive mechanisms of, 186,
    187–188
  defined, 175
  example of, 177–180, 178f, 179f, 307f
  high-level, 184
  mechanics of, 177
  models of, 183–184
  photoreceptors in, 180–182
  physiology of, 180–186
  receptor gain control in, 182
  sensory mechanisms of, 186, 187
  subtractive mechanisms of, 182–183
  time course of, 189
  treatises on, 175–176
Chromatic-adaptation models, 191–
    192
  applications of, 194, 194f
  concerns of, 199–201
  equations for, 193
  Fairchild's model, 211–214
  MacAdam's model, 205–206
  Nayatani's model, 205–208
  retinex theory, 204–205
  treatises on, 199
  von Kries model, 201–204
Chromatic color, defined, 101
Chromatic contrast, 31–32
  contrast sensitivity functions for, 31f,
    33f
Chromatic induction, 135, 136f
Chromaticity
  and brightness, 141–143
  diagrams of, 90–91
Chromaticness, NCS value, 121
CIE illuminants, 68–69
  characteristics of, 69f, 70f, 70–73t,
    166–167, 166t, 167f
CIE 1931 Standard Colorimetric
    Observer, 70t–73t, 370
CIE 1964 Supplementary Standard
    Colorimetric Observer, 90

CIE 1976 Uniform Chromaticity Scales,
    91
CIECAM97s color appearance model,
    372
  appearance correlates in, 387–388
  chromatic adaptation in, 385–387
  data for, 384–385
  future of, 393
  historical background for, 375–378
  invertibility of, 388–389
CIELAB color space, vi, 92, 93f
  color differences in, 94–95
  coordinates of, 219–225
  described, 219
  example of, 321f, 322
  lightness scale of, 224, 226f
  limitations of, 221, 224–225, 225f,
    228–229
  mapping of, 220, 221f, 222f, 223f
  recommended usage of, 372
  testing of, 305, 306, 306t, 309, 310,
    311, 313, 314, 316
  uniformity of, 221, 224f, 269, 270f
  utility of, 369
  wrong von Kries transforms in,
    225–228
CIELAB ΔE^*_{ab}, 94
CIELUV color space, vi, 93, 229–230,
    230f
  example of, 321f, 322
  testing of, 309, 313, 317
CIE94 color-difference model, 94–95
CMC color-difference equation, 94
CMYK, device coordinates, 340, 341
Color(s)
  adaptation to, 27–28, 28f, 29f
  characteristics of, 73–77
  constancy of, 29, 156–157, 195
  corresponding, 189–192, 192f,
    308–310
  defined, 64–65, 199
  memory, 28, 30, 154
  related, 105, 172
  triangle of, 64, 65f
  tristimulus values and, 90–91
  unrelated, 105, 172
Color appearance
  cognitive aspects of, 156
  spatial influences on, 152, 153f, 154
Color appearance models
  applications of, v, 328–329, 331
  ATD, 286–292
  Bradford-Hunt models, 378–384

CIECAM97s, 384–389
CIELAB, 219–228
CIELUV, 229–230
color attributes and, 107–109
construction of, 218–219
corresponding-colors data and,
    308–310
current cutting edge, 371–372
defined, ix, 217–218
developments for the future, 371,
    392–393
and device-independent color imaging,
    352
direct-model testing of, 312–317
Hunt's 245–265
LLAB, 292–299
magnitude-estimation experiments
    and, 310–312
Nayatani's, 231–244
pictorial review of, 320–324, 321f,
    323f, 324f
psychophysics and, 60
qualitative testing of, 304–306
research on, 370–371
RLAB, 267–282
testing of, 128, 217, 303–324
utility of, 114
variety of, 369–370
visual system and, 38–39
ZLAB, 384–392
Color-appearance phenomena, 133–134
Color-appearance reproduction, 344–
    345
Color constancy, 28–29, 156–157
    computational, 195
Color-difference specifications, 94–95
Color differences measurement, 329–330
    application of color appearance models
    to, 331
    future directions for, 331
    recommendations for, 330
    techniques of, 330
Color gamut, 340
Color management, 363
    based on ICC Profile Format, vi
Color matching, 77–78
Color-matching functions, 82–83
    average, 84–85, 86t–88t, 88
Color-naming systems, 114
    types of, 129–130
Color-order systems, 113–126
    applications of, 126
    in art and design, 126

in communication, 127
in education, 127
in evaluation of color appearance mod-
    els, 128
and imaging systems, 128
limitations of, 128–129
in visual experiments, 126
Color-preference reproduction, 345
Color preferences, 359–360
Color rendering, 327
    application of color appearance models
    to, 328–329
    CIE technical committee on, 319, 329
    future directions for, 329
    indices for, 328t
    recommendations for, 328
    techniques of, 328
Color reproduction
    defined, 343
    enhancing accuracy of, 360, 361f, 362
    levels of, 341–345
Color spaces(s)
    CIE, vi, 91–94
    color differences in, 94–95
    viewing-conditions-independent, 355
Color temperature, 68
Color vision
    deficiencies of, 34–38
    historic theories of, 21–23
    spatial properties of, 30–32, 31f, 32f
    temporal properties of, 32–33
Color vision deficiencies, 34–36
    gender and, 36–37, 37f
    screening for, 37–38
Colorcurve system, 122–123
Colorfulness, 104
    in color appearance, 107–109
    defined, 103
    varying directly with luminance,
    144–145, 144f
Colorimetric color reproduction, 341,
    344
Colorimetry
    absolute, 85
    advanced, 63–64
    basic, x, 63
    defined, 63
    future directions in, 334–335
    normalized, 88
    origin of, v–vi
    relative, 85, 88
Colour Standards for Image Technology,
    375

Commission Internationale de l'Eclairage (CIE), v
 activities of, 318–319
 See also CIE entries
Communication, color-order systems in, 127
Computational color constancy, 195
Cone monochromatism, 35
 incidence of, 37t
Cone responses
 transformation of tristimulus values to, 199, 200f
Cones, 6, 8, 10f
 distinguished from rod, 10–11
 function of, 16–17
 relative abundance of, 14f, 38
 role in light and dark adaptation, 180–182
 spectral responsivities of, 77–78
 types of, 11–13, 38
Constant stimuli, method of, 52
Contrast
 varying directly with luminance, 145–146, 146f
 varying directly with surround, 149–150, 150f
Contrast sensitivity functions (CSFs), 30
 and eye movements, 34
 spatial, 31–32, 31f, 32f
 temporal, 32–33
Cool white fluorescent lamp, and color rendering, 328t
Cornea
 described, 4
 function of, 3
Correlated color temperature (CCT), 68
Corresponding color reproduction, 342
Corresponding colors, 189–192, 192f
 color appearance models and, 308–310
Cowan equation, 142
Crispening, 137, 137f
CRT display, 188–189
 characterization of, 350
Cubo-octahedron, 123

Dalton, John, 37
Dark adaptation, 23, 25–26, 176–177
 curve of, 25f
Daylight fluorescent lamp, and color rendering, 328t
Definitions, importance about being enormously careful about, 99–100

Design, color-order systems in, 126
Deuteranomaly, 35
 incidence of, 37t
Deuteranopia, 35, 36f
 incidence of, 37t
Deutsches Institut für Normung. *See* DIN
Device calibration, 347–348
Device characterization
 approaches to, 348–351
 defined, 348
 goals of, 352
Device-independent color imaging
 color appearance models and, 352
 defined, 339
 device calibration and characterization in, 347–352
 example system for, 360, 362
 gamut mapping and, 355–359
 general solution to, 346
 as goal, 339–340
 ICC implementation of, 362–366
 inverse process, 360
 process of, 347f
 viewing conditions and, 353–355
Diffuse/normal viewing geometry, 75
Digital cameras, characterization of, 350
DIN color system, 125
Discounting the illuminant, 29, 150–151

Education, color-order systems in, 127
Empirical modeling, 349
Equivalent color reproduction, 342
Exact color reproduction, 342
Eye
 anatomy of, 4f, 4–8
 optics of, 3

Fairchild, M., 377
Fairchild's chromatic-adaptation model, 211
 equations of, 211–213
 predictions by, 213–214, 214f
Farnsworth-Munsell 100-hue test, 37, 38
Fechner's law, 45–45
Filling-in, 15
Film mode of appearance, 172
Flare, 351
Fluorescence, 76–77
 in output devices, 351
Fluorescent lamp, and color rendering, 328t

Forced-choice method, 52–53
45/normal viewing geometry, 75–76
Fovea, 7
Frequency-of-seeing curve, 52

Gamut, 357f
    color, 340
    defined, 356
    limits to, 355
Gamut mapping, 357f
    defined, 356
    techniques for, 358–359
Ganglion cells, 18–19, 18f
Gender, and color-vision deficiencies,
        36–37, 37t
Graphical rating, 55
Grassmann's laws, 81, 82
Guth's chromatic-adaptation model, 208
    equations of, 209
    predictions by, 209–210, 210f

Haploscopic matching, 54, 190
Hard-copy output, 188
Helmholtz-Kohlrausch effect, 141–143,
        143f
Helson-Judd effect, 146–148, 147f
Hering opponent-colors theory, 21–22,
        22f
High pressure sodium lamp, and color
        rendering, 328t
Horizontal cells, of retina, 8
Hue
    in color appearance, 107–109
    defined, 101
    Munsell, 116–117, 118f
    NCS value, 119, 121
    of nonselective samples, 146–148
    varying directly with colorimetric pu-
        rity, 140, 141f
    varying directly with luminance,
        139–140
Human visual response, quantification
        of, 77–78
Hunt, Pointer, Estevez transformation,
        250
Hunt, R. W. G., 247, 377
Hunt color appearance model
    adaptation model for, 249–255
    advantages of, 371, 372
    brightness equations in, 258–259
    chroma equation in, 260
    colorfulness equation in, 260–261
    data for, 248–249

    example of, 321, 321f, 322
    hue equations in, 255–256, 256t
    inversion of, 261–263
    lightness equations in, 260
    limitations of, 264–265
    objectives and approach of, 247–248
    opponent-color dimensions in, 255
    predictions of, 263, 264t
    revision of, 371
    saturation equations in, 257
    testing of, 305–306, 306t, 309,
        311–312, 313, 314, 316
Hunt effect, 144–145, 144f

ICC Profile Format, vi
Illuminant A, 68
Illuminant C, 68
Illuminant D50, 68
    spectral power of, 69f
Illuminant D65, 68
    and color rendering, 328t
    spectral power of, 69f
Illuminant F2, 69
    spectral power of, 70f
Illuminant F8, 69
    spectral power of, 70f
Illuminant F11, 69
    spectral power of, 70f
Illuminant metamerism, 332
Illuminant mode of appearance, 171
Illuminants
    CIE, 68–69, 69f, 70f, 70–73t
    discounting, 29, 150–151
Illumination, standard, 75–76
Illumination mode of appearance, 171
Imaging systems, color-order systems in,
        128
Indices of metamerism, 332
    application of color appearance models
        to, 333
    future directions for, 333
    recommendations for, 332
    techniques of, 333
Interaction of Color, 134
International Colour Association (AIC)
    1997 meeting, 376
International Lighting Vocabulary,
        99–100
Interval scale, defined, 49
Invertibility, 360
Iris, described, 6
Irradiance, 66
Ishihara's Tests for Colour-Blindness, 38

Just-noticeable difference (JND), 46
  experiments testing, 50

L cones, 11, 13
  relative abundance of, 13f, 39
  spectral responsivities of, 12f
LABHNU2 model, example of, 321f,
    322
Land two-color projection, 154
Lateral geniculate nucleus (LGN), 16, 19
Lens
  described, 5
  function of, 3
Light adaptation, 26, 176
  curve of, 27f
  role of rods and cones in, 180–182
Light source
  defined, 65
  types of, 66
Lightness, 102f
  in color appearance, 107–109
  defined, 102
  equation for, 106–107
Limits, method of, 51
Linkage chromatic-adaptation model,
    212
LLAB color appearance model
  adaptation model for, 294–295
  chroma equation in, 296
  color differences in, 297
  colorfulness equation in, 296
  current status of, 371
  data for, 293, 293t
  example of, 321f, 322
  hue equations in, 297
  lightness equation in, 295
  limitations of, 299
  objectives and approach of, 292–293
  opponent-color dimensions in,
    295–296
  perceptual correlates in, 296
  predictions of, 297–298, 298t
  saturation equation in, 296–297
  testing of, 306t
Lookup tables (LUTs), 346
  multidimensional, 352
Lovibond Tintometer, 113
Luminance, 31–32
  and brightness, 141–143
  contrast sensitivity functions for, 31f,
    33f
  varying directly with colorfulness,
    144–145, 144f

varying directly with contrast, 145–146
varying directly with hue, 139–140
Luo, M. R., 377
LUTCHI study, 310–312

M cones, 11, 13
  relative abundance of, 13f, 39
  spectral responsivities of, 12f
MacAdam's chromatic-adaptation
    model, 205–206
Macbeth Color Checker Chart, 128
Macula, 7
Magnitude estimation, 56, 191
Matching techniques, 53–54, 190
Maxwell's color process, 154, 155f
McCollough effect, 184
Measurement
  colorimetric, 350–351
  exhaustive, 349–350
Memory color, 28, 30, 154
Memory matching, 54, 190–191
Mercury lamp, and color rendering, 328t
Mesopic vision, 11
Metal halide lamp, and color rendering,
    328t
Metamerism, 78
  indices of, 332
  in input and output devices, 351
Metameric, defined, 53
Method of adjustment, 51
Method of constant stimuli, 52
Method of limits, 51
Modeling
  empirical, 349
  physical, 348–349
Modulation transfer functions, 30
Monochromatism, 35
  incidence of, 37t
Mori experiment, 148
Motion adaptation, 186
Multidimensional scaling (MDS), 56–58,
    57t, 58t
Munsell Book of Color, 115–117, 118f,
    119f
Munsell Value Scale, 115–116

Natural Color System (NCS), 114, 120f,
    126
  chromaticness in, 121
  hue in, 119
Nayatani's chromatic-adaptation model
  as basis for color appearance model,
    235–237

equations of, 206
forebears of, 205–206
predictions by, 207–208, 207f
Nayatani's color appearance model
   adaptation model for, 235–237
   brightness equations in, 238–239
   chroma equations in, 241
   colorfulness equations in, 241–242
   data for, 234–235
   example of, 321f, 322
   hue equations in, 240
   inversion of, 242
   lightness equations in, 239
   limitations of, 244
   objectives and approach of, 233–234
   opponent-color dimensions in,
      237–238
   predictions of, 242–243, 243t
   revision of, 371
   saturation equations in, 240–241
   testing of, 304, 306t, 309, 311
Nominal scale, defined, 48
Normal/diffuse viewing geometry, 75
Normal/45 viewing geometry, 75–76
Normalization constant, in colorimetry,
   85, 88
Normalized colorimetry, 88

Objects, recognition of, 30
Oblique effect, 33–34
Observer metamerism, 332
Off-center ganglion cells, 18–19, 18f
On-center ganglion cells, 18–19, 18f
One-dimensional scaling, 54–56
Opponent-colors theory
   Hering's, 21–22, 22f
   modern, 22–23, 24f
Opsin, 17
Optic nerve, 7–8
Optical illusions, 151–152, 152f, 153f
Optical Society of America Uniform
      Color Scales (OSA UCS), 123–124,
      124f
Ordinal scale, defined, 48–49
Ostwald system, 113, 125
Output lexicon, defined, 355

Paired comparison experiment, 55–56
Pantone Process Color Formula Guide,
   114, 129
Partition scaling, 56
Pass-fail method, 52
Perceptual threshold, 45

Perfect reflecting diffuser (PRD), 76
Photometry, 78–79, 81
Photopic luminous efficiency function,
   80t–81t
Photopic vision, 11
Photoreceptors, 6
   energy responses of, 17f
   gain control of, 182.
   See also Cones; Rods
Physical modeling, 348–349
Pigmented epithelium, 6–7
Planckian radiator, 68
Pleasing color reproduction, 343
PostScript Process Color Guide, 130
Power laws, 47
Preferred color reproduction, 342–343
Presbyopia, 5
Printers, characterization of, 350
Probit analysis, 53
Profile connection space, 363–364, 366
Proportionality, Grassmann's law of, 82
Protanomaly, 35
   incidence of, 37t
Protanopia, 35, 36f
   incidence of, 37t
Proximal field, defined, 164
Pseudoisochromatic plates, 37, 38
Psychometric function, 52
Psychophysics
   defined, 43–44
   experimental concepts of, 48–58
   experimental design in, 58–60
   history of, 45–48
   relation to color appearance modeling,
      60
Pupil, in adaptation to light and dark,
   180
Purkinje shift, 11, 81

Radiance, 66
   equation of, 67
Radiator, black-body, 68
Rank order experiment, 55
Ratio estimation, 56
Ratio scale, defined, 49
Receptive fields, 17–19
Receptor gain control, 182
Reflectance, 73, 74f
   distribution functions of, 74
Related color, 105–106
   defined, 105, 172
Relative spectral power distribution, 67
Response compression, 183, 183f

Retina
  described, 6–7
  light perception in, 8–16
  light processing in, 16–17
  structure of, 9f
Retinal, 17
Retinex theory, 204–205
RGB, device coordinates, 340
Rhodopsin, 11, 16
Richter, K., 377
RLAB color appearance model
  adaptation model for, 272–274
  chroma equation in, 277
  data for, 271
  example of, 321, 321f, 322
  hue equation in, 276–277, 277t, 278f
  inversion of, 279–280
  lightness equation in, 274, 276
  limitations of, 281–282
  objectives and approach of, 269, 271
  opponent-color dimensions in,
      274–274
  predictions of, 280–281
  saturation equations in, 277–278
  testing of, 306, 306t, 309, 313, 314,
      316
  usefulness of, 372
Rod monochromatism, 35
  incidence of, 37t
Rods, 6, 8, 10f, 14–15
  distinguished from cones, 10–11
  function of, 14–16
  luminous efficiency function for, 79, 81
  relative abundance of, 14f
  role in light and dark adaptation,
      180–182

S cones, 11, 13
  relative abundance of, 13f, 39
  spectral responsivities of, 12f
Saturation, 104–105
  defined, 104
  equation for, 106, 107
Scales
  types of, 48–49
  uses of, 49–50
Scaling
  multidimensional, 56–58
  one-dimensional, 54–56
Scanners, characterization of, 350
Scotopic luminous efficiency function,
      80t–81t
Scotopic vision, 11

Simultaneous binocular viewing, 354
Simultaneous contrast, 134–136
  complexity of, 136f
  example of, 134f
  and shape, 152, 152f
  vs. spreading, 138, 138f
Sodium lamp, and color rendering,
      328t
Soft-copy display, 188–189
Spatial CSFs, 30–32, 31f, 32f
Spatial frequency adaptation, 185–186,
      185f
Spatial perception, and chromatic percep-
      tion, 152, 153f, 154
Specification of Colour Appearance for
      Reflective Media and Self-Luminous
      Display Comparisons (CIE technical
      committee), 319
Spectral color reproduction, 341
Spectral luminous efficiency, 78, 79, 81
Spectral power distribution, 67
Spectrophotometry, 73
  CIE standards for, 75–76
Spectroradiometry, 66
Spreading, 138–139
  vs. simultaneous contrast, 138f
Stage theory, 23
Staircase procedures, 53
Stevens effect, 145–146, 146f
Stevens power law, 46–48
Stimulus
  characteristics of, 163–164
  defined, 162
Subtractive mixtures, 113
Successive binocular viewing, 354
Surface mode of appearance, 171
Surround
  contrast and, 149–150, 150f
  defined, 165

Tapetum, 7
TC1–27, Specification of Colour
      Appearance for Reflective Media
      and Self-Luminous Display
      Comparisons, 319, 370
TC1-33, Colour Rendering, 319
TC1-34, Testing Colour-Appearance
      Models, 318–319, 369, 370
  model preparation by, 377, 384–389
Temporal CSFs, 32–33
10° observer, 90
Testing Color Appearance Models,
      217

Testing Colour-Appearance Models (CIE technical committee), 318–319, 369, 370
model preparation by, 377, 384–389
Threshold experiments, 50
types of, 51–53
Thurstone's law, 56
Tintometer, 113
Transmittance, 73, 74f
distribution functions of, 74
Tri-band fluorescent lamp, and color rendering, 328t
Trichromatic theory, 21
Trichromia, 35
incidence of, 37t
Tristimulus values, 81–82
extension of, 82–83
measurement of, 166–168
transformation to cone responses, 199, 200f
XYZ, vi, 84, 89
Tritanomaly, 35
incidence of, 37t
Tritanopia, 34, 35, 36f
incidence of, 37t
Troland, defined, 287
Trumatch Colorfinder, 129–130
Tungsten halogen lamp, and color rendering, 328t
Two-color projections, 154, 155f
2° color-matching functions, 89–90
Unrelated color, 105–106
defined, 105, 172

Viewing
anomalies of, 168–170, 169f
modes of, 168, 170–172, 170f
Viewing conditions, definition of, 353–355

Viewing-conditions-independent space, 346, 355
Viewing field
colorimetric specifications of, 165–168
components of, 162–165, 163f
configuration of, 161–162
Viewing geometries, 75–76
Vision
experiments on, 44, 126
human, 77–81
Visual area 1, 19–21
Visual experiments, 126
types of, 44
Visual threshold, 45
Vitreous humor, 5
Volume mode of appearance, 171–172
von Kries chromatic-adaptation model
equations of, 201–203
history of, 201
predictions by, 203f, 204
testing of, 313, 314, 316

Ware equation, 142
Weber, E. H., 45
Weber's law, 45
Whiteness, NCS value, 121
Wrong von Kries transforms, 225–228

Xenon lamp, and color rendering, 328t
XYZ tristimulus values, vi, 84, 89

Yes-no method, 52

ZLAB color appearance model, 372
appearance correlates in, 391–392
chromatic adaptation in, 390–391
data for, 389
invertibility of, 392
similarities to CIECAM97s, 389